ALASKA'S CHANGING ARCTIC

LONG-TERM ECOLOGICAL RESEARCH NETWORK SERIES
LTER Publications Committee

Grassland Dynamics: Long-Term Ecological Research in Tallgrass Prairie
Editors: Alan K. Knapp, John M. Briggs, David C. Hartnett, and Scott L. Collins

Standard Soil Methods for Long-Term Ecological Research
Editors: G. Philip Robertson, David C. Coleman, Caroline S. Bledsoe, and Phillip Sollins

Structure and Function of an Alpine Ecosystem: Niwot Ridge, Colorado
Editors: William D. Bowman, and Timothy R. Seastedt

Climate Variability and Ecosystem Response at Long-Term Ecological Sites
Editors: David Greenland, Douglas G. Goodin, and Raymond C. Smith

Biodiversity in Drylands: Toward a Unified Framework
Editors: Moshe Shachak, James R. Gosz, Steward T.A. Pickett, and Avi Perevolotsky

Long-Term Dynamics of Lakes in the Landscape: Long-Term Ecological Research on North Temperate Lakes
Editors: John J. Magnuson, Timothy K. Kratz, and Barbara J. Benson

Alaska's Changing Boreal Forest
Editors: F. Stuart Chapin, III, Mark W. Oswood, Keith Van Cleve, Leslie A. Viereck, and David L. Verbyla

Structure and Function of a Chihuahuan Desert Ecosystem: The Jornada Basin Long-Term Ecological Research Site
Editors: Kris M. Havstad, Laura F. Huenneke, and William H. Schlesinger

Principles and Standards for Measuring Net Primary Production in Long-Term Ecological Studies
Editors: Timothy J. Fahey, and Alan K. Knapp

Agrarian Landscapes in Transition: Comparisons of Long-Term Ecological and Cultural Change
Editors: Charles L. Redman and David R. Foster

Ecology of the Shortgrass Steppe: A Long-Term Perspective
Editors: William K. Lauenroth, and Ingrid C. Burke

A Caribbean Forest Tapestry: The Multidimensional Nature of Disturbance and Response
Editors: Nicholas Brokaw, Todd A. Crowl, Ariel E. Lugo, William H. McDowell, Frederick N. Scatena, Robert B. Waide, and Michael R. Willig

Alaska's Changing Arctic: Ecological Consequences for Tundra, Streams, and Lakes
Editors: John Hobbie, and George Kling

Alaska's Changing Arctic

Ecological Consequences for Tundra, Streams, and Lakes

Edited by

JOHN E. HOBBIE
AND
GEORGE W. KLING

LTER

OXFORD
UNIVERSITY PRESS

OXFORD
UNIVERSITY PRESS

Oxford University Press is a department of the University of Oxford.
It furthers the University's objective of excellence in research, scholarship,
and education by publishing worldwide.

Oxford New York
Auckland Cape Town Dar es Salaam Hong Kong Karachi
Kuala Lumpur Madrid Melbourne Mexico City Nairobi
New Delhi Shanghai Taipei Toronto

With offices in
Argentina Austria Brazil Chile Czech Republic France Greece
Guatemala Hungary Italy Japan Poland Portugal Singapore
South Korea Switzerland Thailand Turkey Ukraine Vietnam

Published in the United States of America by
Oxford University Press
198 Madison Avenue, New York, NY 10016

Library of Congress Cataloging-in-Publication Data
Alaska's changing arctic : ecological consequences for tundra, streams, and lakes / edited by
John E. Hobbie and George W. Kling.
pages cm
ISBN 978–0–19–986040–1 (alk. paper)
1. Tundra ecology—Alaska. 2. Stream ecology—Alaska. 3. Lake ecology—Alaska.
4. Alaska—Environmental conditions. I. Hobbie, John E.
QH541.5.T8A53 2014
577.5'8609798—dc23
2013031620

9 8 7 6 5 4 3 2 1
Printed in the United States of America
on acid-free paper

Contents

Preface vii
Contributors xi

1 *Introduction* 1
John E. Hobbie

2 Climate and Hydrometeorology of the Toolik Lake Region and the Kuparuk River Basin: Past, Present, and Future 21
Jessica E. Cherry, Stephen J. Déry, Yiwei Cheng, Marc Stieglitz, Amy S. Jacobs, and Feifei Pan

3 Glacial History and Long-Term Ecology in the Toolik Lake Region 61
Donald A. Walker, Thomas D. Hamilton, Hilmar A. Maier, Corinne A. Munger, and Martha K. Raynolds

4 Late-Quaternary Environmental and Ecological History of the Arctic Foothills, Northern Alaska 81
W. Wyatt Oswald, Linda B. Brubaker, Feng Sheng Hu, and George W. Kling

5 Terrestrial Ecosystems at Toolik Lake, Alaska 90
Gaius R. Shaver, James A. Laundre, M. Syndonia Bret-Harte, F. Stuart Chapin, III, Joel A. Mercado-Díaz, Anne E. Giblin, Laura Gough , William A. Gould, Sarah E. Hobbie, George W. Kling, Michelle C. Mack, John C. Moore, Knute J. Nadelhoffer, Edward B. Rastetter, and Joshua P. Schimel

6 Land–Water Interactions 143
George W. Kling, Heather E. Adams, Neil D. Bettez, William B. Bowden, Byron C. Crump, Anne E. Giblin, Kristin E. Judd, Katy Keller, George W. Kipphut, Edward R. Rastetter, Gaius R. Shaver, and Marc Stieglitz

7 Ecology of Streams of the Toolik Region 173
William B. Bowden, Bruce J. Peterson, Linda A. Deegan, Alex D. Huryn, Jonathan P. Benstead, Heidi Golden, Michael Kendrick, Stephanie M. Parker, Elissa Schuett, Joseph J. Vallino, and John E. Hobbie

8 The Response of Lakes Near the Arctic LTER to Environmental Change 238
Chris Luecke, Anne E. Giblin, Neil D. Bettez, Greta A. Burkart, Byron C. Crump, Mary Anne Evans, Gretchen Gettel, Sally MacIntyre, W. John O'Brien, Parke A. Rublee, and George W. Kling

9 Mercury in the Alaskan Arctic 287
William F. Fitzgerald, Chad R. Hammerschmidt, Daniel R. Engstrom, Prentiss H. Balcom, Carl H. Lamborg, and Chun-Mao Tseng

10 Ecological Consequences of Present and Future Changes in Arctic Alaska 303
John E. Hobbie, and George W. Kling

Index 325

Preface

This book in the Long Term Ecology Research (LTER) Synthesis series, reports results from ecological studies at a site in northern Alaska, the region around Toolik Lake. When the study began in the mid-1970s, ecological research in northern Alaska had been restricted by the difficulty of access in a region with no roads. Accordingly, research was concentrated on the coastal ocean, shallow ponds and lakes, and the wet coastal tundra near the Barrow research laboratory, where there was an airport. In addition to research at nearby sites, the Office of Naval Research funded Barrow laboratory supported a few temporary field camps in the mountains where small planes could land on lakes and snowfields. This era of limited access suddenly changed in 1975 when the construction of the oil pipeline and the adjacent Dalton Highway gave scientists easy access to a transect of the coastal plain, foothills, and mountains of the Brooks Range. A foothills site with tussock tundra, the deep Toolik Lake, and the Kuparuk River was chosen for detailed investigation.

The Toolik research began in 1975, the year the National Science Foundation (NSF)'s International Biological Programme study (IBP) ended its intensive research at Barrow. By coincidence, this was the same year as the opening of the Dalton Highway alongside the pipeline. This gravel road allowed sampling the organisms along the road in the coastal plain, the foothills, and the Brooks Range mountains. The road also allowed a base camp to be established at Toolik Lake so that observations could be made over time, meals and sanitation could be provided, and a small laboratory set up. During the first two summers, power and an emergency phone was provided by a nearby pipeline construction camp. Today Toolik is a modern year-round research facility operated by the University of Alaska with dormitories, laboratories, electricity, and fiber-optic cable internet access.

The goal of the early NSF-funded research was to document the region's organisms and ecological processes as examples of ecology of the Low Arctic. Organisms and processes are identical or very similar throughout northern Alaska, but only the wet sedge and coastal freshwaters had been studied in detail (e.g., in the IBP at Barrow). At the Toolik site to the south in the warmer foothills, tundra, deep lakes, and rivers could be investigated and the ecology compared with the well-known processes in the cooler coastal regions. After the initial documentation phase, the terrestrial and aquatic research asked questions about nutrient controls of primary productivity, temperature effects, and top-down and bottom-up control of populations. Both the aquatic and terrestrial researchers made use of large-scale experimental manipulations, including nutrient additions to streams, lakes, and tundra; the heating of tundra plots; and the manipulation of food webs and plant communities. Mathematical models help to understand the integration of these processes and help to project the results to large areas of the Arctic and into the next century.

In 1987, the research became the Arctic LTER project, a part of the NSF Long Term Ecological Research program (LTER) that now includes 25 sites. The Arctic LTER project is responsible for the collection of environmental data and the measurement of samples of various types from the tundra and aquatic systems as well as for the archiving of environmental data in computer files accessible to all. In addition, the LTER maintains several large experiments on tundra and streams and facilitates research projects at Toolik by a large number of investigators who cooperate closely and use the LTER data and experiments. Goals of the Arctic LTER include continued long-term measurements of biota and of ecological processes. In recent years the goals have included the ecological effects of environmental changes, especially those linked to the long-term changes of the Arctic such as a possible increase in frequency of wildfire and to effects of warming permafrost. Another goal is to examine the movements of organisms and material across different parts of the landscape. New questions include the fate of carbon produced in photosynthesis each year or sequestered in permafrost thousands of years ago; the ecology of Toolik in the next century; and the changes in the extent of thermokarst, the physical change in the land surface caused by melting of ice inclusions in permafrost.

How typical of the Arctic is the Toolik Lake region? The short answer is that no single site can be completely typical of a vast circumarctic region where climate variability south to north produces annual average temperatures ranging from 0°C to –18°C or both lakes with four months of open water and lakes with a year-round ice cover. The same latitudinal gradient leads to plant cover ranging from an unbroken blanket of tussock tundra to a polar desert of scattered individual plants. Despite this variability, the types of tundra, ecological processes, and relationships investigated in detail at Toolik occur arctic-wide and are very relevant to understanding the ecology of all arctic ecosystems. The Toolik region lies in the warmest of the arctic vegetation zones, where productivity and diversity is higher than in the other zones. On the aquatic side, freshwaters across the Arctic all have low primary productivity, but also the same species of algae and zooplankton tend to make up lake food webs, while fish, such as arctic char and grayling, are also similar across the entire Arctic.

The research at Toolik Lake began as a follow-up to the IBP at Barrow, Alaska. The director of the IBP project, Dr. Jerry Brown, played a major role in contacting the pipeline and oil companies so that researchers could take advantage of the newly accessible pipeline road. Drs. George Llano and Polly Penhale at the NSF's Office of Polar Programs recognized the value of intensive investigations of streams and lakes and encouraged coordinated research at Toolik. Terrestrial research in the early days of the research was funded by NSF's Division of Ecological Programs. The U.S. Department of Energy supported a large ecological project at nearby Imnavait Creek.

We thank the Long-Term Ecological Research program of the National Science Foundation for providing major funding for this research, with John Hobbie and Gus Shaver as principle investigators. The University of Alaska's Toolik Field Station has provided field support with funding from the NSF's Office of Polar Programs. We acknowledge the contributions of hundreds of scientists from the United States, Canada, and Europe who have made use of LTER data and experiments in their own projects at Toolik and greatly expanded the ecological understanding incorporated into this book. We also acknowledge our late Toolik colleagues and friends Dr. Robie Vestal and Dr. W. John O'Brien; we remember them for their lasting contributions to the project and for a sense of humor and humanity that they gave to us all.

Contributors

Heather E. Adams, School of Natural Resources and Environment, University of Michigan, Ann Arbor, MI 48109

Prentiss H. Balcom, Department of Marine Sciences, University of Connecticut, Groton, CT 06340

Jonathan P. Benstead, Department of Biology, University of Alabama, Tuscaloosa, AL 35404

Neil D. Bettez, Cary Institute of Ecosystem Studies, Millbrook, NY 12545

William B. Bowden, The Rubenstein School of Environment and Natural Resources, University of Vermont, Burlington, VT 05405

M. Syndonia Bret-Harte, Institute of Arctic Biology, University of Alaska, Fairbanks, AK 99775

Linda B. Brubaker, School of Environmental and Forest Sciences, University of Washington, Seattle, WA 98195

Greta A. Burkart, Arctic National Wildlife Refuge, U.S. Fish and Wildlife Service, Fairbanks, AK 99701

F. Stuart Chapin, III, Institute of Arctic Biology, University of Alaska, Fairbanks, AK 99775

Yiwei Cheng, School of Civil and Environmental Engineering, Georgia Institute of Technology, Atlanta, GA 30332

Jessica E. Cherry, International Arctic Research Center and Arctic Region Supercomputing Center, University of Alaska, Fairbanks, AK 99775

Byron C. Crump, Center for Environmental Science, University of Maryland, Cambridge, MD 21613

Linda A. Deegan, The Ecosystems Center, Marine Biological Laboratory, Woods Hole, MA 02543

Stephen J. Déry, Environmental Science and Engineering Program, University of Northern British Columbia, Prince George, BC, V2N 4Z9, Canada

Daniel R. Engstrom, St. Croix Watershed Research Station, Science Museum of Minnesota, Marine on St. Croix, MN 55047

Mary Anne Evans, School of Natural Resources and Environment, University of Michigan, Ann Arbor, MI 48109

William F. Fitzgerald, Department of Marine Sciences, University of Connecticut, Groton, CT 06340

Gretchen Gettel, Institute for Water Education, UNESCO-IHE, Delft, The Netherlands

Anne E. Giblin, The Ecosystem Center, Marine Biological Laboratory, Woods Hole, MA 02543

Heidi Golden, Department of Ecology and Evolutionary Biology, University of Connecticut, Storrs, CT 06269

Laura Gough, Department of Biology, University of Texas, Arlington, TX 76019

William A. Gould, USDA Forest Service International Institute of Tropical Forestry, 1201 Calle Ceiba, Río Piedras, PR 00926

Thomas D. Hamilton, U.S. Geological Survey (Emeritus), 4200 University Drive, Anchorage, AK 99508

Chad R. Hammerschmidt, Department of Earth and Environmental Sciences, Wright State University, Dayton, OH 45435

John E. Hobbie, The Ecosystems Center, Marine Biological Laboratory, Woods Hole, MA 02543

Sarah E. Hobbie, Department of Ecology, Evolution, and Behavior, University of Minnesota, Saint Paul, MN 55108

Feng Sheng Hu, Department of Plant Biology and Department of Geology, University of Illinois, Urbana, IL 61801

Alex D. Huryn, Department of Biology, University of Alabama, Tuscaloosa, AL 35404

Amy S. Jacobs, International Arctic Research Center and Arctic Region Supercomputing Center, University of Alaska, Fairbanks, AK 99775

Kristin E. Judd, Department of Biology, Eastern Michigan University, Ypsilanti, MI 48197

Katy Keller, School of Natural Resources and Environment, University of Michigan, Ann Arbor, MI 48109

Michael Kendrick, Department of Biology, University of Alabama, Tuscaloosa, AL 35404

George W. Kipphut, Department of Geosciences, Murray State University, Murray, KY 42071

George W. Kling, Department of Ecology and Evolutionary Biology, University of Michigan, Ann Arbor, MI 48109

Carl H. Lamborg, Department of Marine Chemistry and Geochemistry, Woods Hole Oceanographic Institution, Woods Hole, MA 02543

James A. Laundre, The Ecosystems Center, Marine Biological Laboratory, Woods Hole, MA 02543

Chris Luecke, Department of Watershed Sciences, Utah State University, Logan, UT 84322

Sally MacIntyre, Department of Ecology, Evolution and Marine Biology, University of California, Santa Barbara, CA 93106

Michelle C. Mack, Department of Biology, University of Florida, Gainesville, FL 32611

Hilmar A. Maier, Alaska Geobotany Center, Institute of Arctic Biology, University of Alaska, Fairbanks, AK 99775

Joel A. Mercado-Díaz, USDA Forest Service International Institute of Tropical Forestry, 1201 Calle Ceiba, Río Piedras, PR 00926

John C. Moore, Natural Resource Ecology Laboratory, Colorado State University, Fort Collins, CO 80523

Corinne A. Munger, Alaska Geobotany Center, Institute of Arctic Biology, University of Alaska, Fairbanks, AK 99775

Knute J. Nadelhoffer, Department of Ecology and Evolutionary Biology, University of Michigan, Ann Arbor, MI 48109

W. John O'Brien (deceased), Department of Biology, University of North Carolina, Greensboro, NC 27402

W. Wyatt Oswald, Department of Communication Sciences and Disorders, Emerson College, Boston, MA 02116

Feifei Pan, Department of Geography, University of North Texas, Denton, TX 76203

Stephanie M. Parker, Department of Biology, University of Alabama, Tuscaloosa, AL 35404

Bruce J. Peterson, The Ecosystems Center, Marine Biological Laboratory, Woods Hole, MA 02543

Edward B. Rastetter, The Ecosystems Center, Marine Biological Laboratory, Woods Hole, MA 02543

Martha K. Raynolds, Alaska Geobotany Center, Institute of Arctic Biology, University of Alaska, Fairbanks, AK 99775

Parke A. Rublee, Department of Biology, University of North Carolina, Greensboro, NC 27402

Joshua P. Schimel, Department of Ecology, Evolution and Marine Biology, University of California, Santa Barbara, CA 93106

Gaius R. Shaver, The Ecosystems Center, Marine Biological Laboratory, Woods Hole, MA 02543

Elissa Schuett, The Rubenstein School of Environment and Natural Resources, University of Vermont, Burlington, VT 05405

Marc Stieglitz, School of Earth and Atmospheric Sciences and School of Civil and Environmental Engineering, Georgia Institute of Technology, Atlanta, GA 30332

Chun-Mao Tseng, Institute of Oceanography, National Taiwan University, Taipei 106, Taiwan

Joseph J. Vallino, The Ecosystems Center, Marine Biological Laboratory, Woods Hole, MA 02543

Donald A. Walker, Alaska Geobotany Center, Institute of Arctic Biology, University of Alaska, Fairbanks, AK 99775

ALASKA'S CHANGING ARCTIC

1

Introduction

John E. Hobbie

Description of the Arctic LTER Site and Project

Toolik, the field site of the Arctic Long Term Ecological Research (LTER) project, lies 170 km south of Prudhoe Bay in the foothills of Alaska's North Slope near the Toolik Field Station (TFS) of the University of Alaska, Fairbanks (UAF). The project goal is to describe the communities of organisms and their ecology, to measure changes that are occurring, and to predict the ecology of this region in the next century. Research at the Toolik Lake site began in the summer of 1975 when the completion of the gravel road alongside the Trans-Alaska Pipeline, now called the Dalton Highway, opened the roadless North Slope for research. This book synthesizes the research results from this site since 1975, as supported by various government agencies but mainly by the U.S. National Science Foundation (NSF), and focuses on the Arctic LTER project started in 1987.

Toolik Lake is located in the foothills of the North Slope (Figure 1.1), the arctic region of northern Alaska, at 68°38' N and 149°43' W and an altitude of 719 m. The LTER research site is defined as the catchment south of the junction of two rivers: the Toolik Lake outlet and the headwaters of the Kuparuk River (Figure 1.2). The view looking south from above Toolik Lake shows the Toolik Field Station, the inlet stream, numerous lakes in the inlet basin, and the Brooks Range (Figure 1.3).

The North Slope (about the size of Nebraska) encompasses the 207,000 km^2 of Alaska that drain to the Arctic Ocean. There are three physiographic divisions: coastal plain (71,000 km^2), foothills (62,400 km^2), and mountains (73,600 km^2). The North Slope is underlain by hundreds of meters of permafrost, which is defined as earth material (soil, rock, ice, and organic matter) that remains at or below 0°C for at least two consecutive years (van Everdingen 1998). Permafrost is a product of the low annual temperature of –8°C at Toolik; it limits the plants and roots to the upper layer of soil, called the "active layer," which thaws during the three months of summer. Permafrost also holds the ~300 mm of annual precipitation at the surface so the tundra is moist, rivers and streams plentiful, and wetlands, ponds, and lakes abundant.

The largest disturbance affecting this site occurred when a series of glaciers advanced northward from the valleys of the Brooks Range during the last ice age. Today, the age of the land surface across the research site is either >300,000 years, ~60,000 years, or ~10,000 years (chapter 3). The age affects the chemistry and

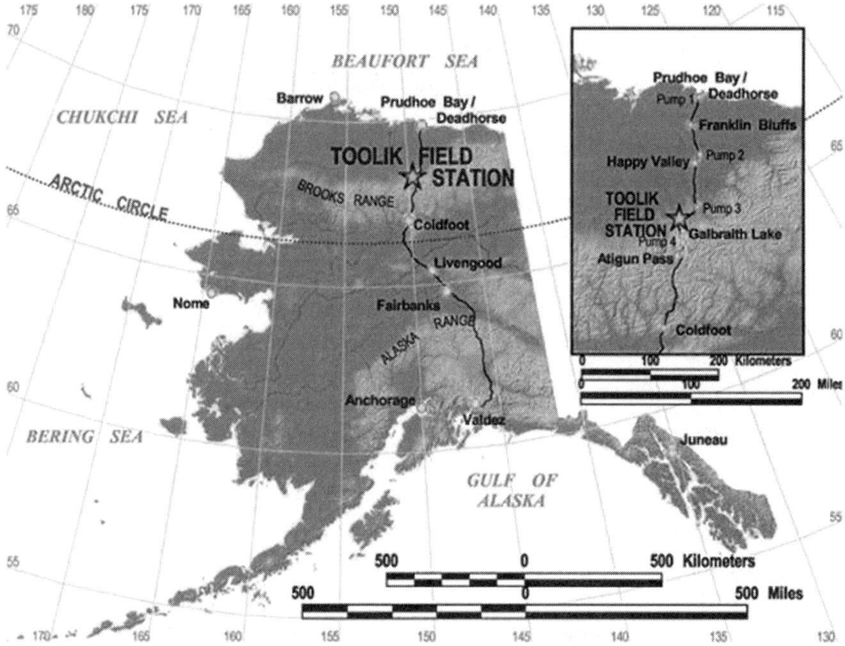

Figure 1.1 Location of the Arctic LTER field site.

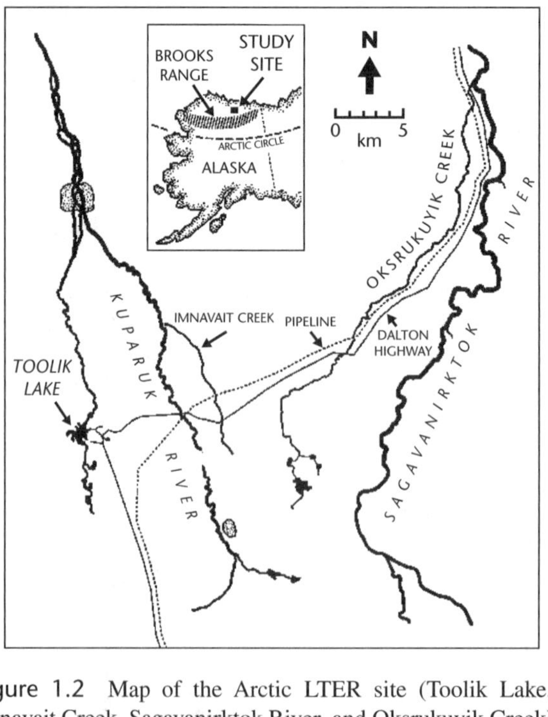

The Arctic LTER site is the upper Kuparuk River drainage basin. This includes the entire drainage basin of the river draining Toolik Lake. Measurements have also been made on Oksrukuyik Creek and along the Sag River

Figure 1.2 Map of the Arctic LTER site (Toolik Lake, watersheds, Kuparuk River, Imnavait Creek, Sagavanirktok River, and Oksrukuyik Creek).

Figure 1.3 Toolik Lake, the Toolik Field Station (TFS), inlet stream, inlet lakes ("I-series"), and the Brooks Range.

topography with the result that there are characteristic drainage patterns, landforms, and even vegetation communities on each of these glacial tills.

The dominant plants of the foothills are the sedges and grasses of tussock tundra, a vegetation type that covers some 80% of arctic Alaska. Low shrubs, including birches and willows, grow between the tussocks and abundantly along the streams. Nutrients, particularly nitrogen (N), are in short supply and primary productivity low. Low concentrations of N and phosphorus (P) are also found in streams and lakes; both are oligotrophic. During the nine months of winter the lakes are covered with ~1.5 m of ice and the streams are frozen solid.

Advantages of the Site for LTER Research

The LTER program includes more than 20 sites located in most of the major biomes of the United States, surrounding oceans, and of the polar regions. In the north, there is the boreal Bonanza Creek site near Fairbanks, Alaska, and the Arctic LTER site at Toolik Lake. For comparative studies of ecological questions in the LTER network, the Toolik site is towards the cold end of a temperature gradient and also is a remote, high latitude site where natural communities of fish and wildlife continue to exist and there is little airborne pollution. There is virtually no nitrogen deposition from acid rain, and the western Arctic location of the site minimizes the cross-pole deposition of pollutants (Ford et al. 1995). This contrasts with parts of the Canadian Arctic where deposition of pollutants from Europe can be high.

The site has become the location for a large number of non-LTER research projects, many of which cooperate with the LTER. It is quite probably the best known arctic research site because of logistics that include an excellent living and research facility, helicopter and truck transportation, the availability of the LTER database containing all the data since 1975, and availability of climate information from the LTER and the TFS. The LTER data set provides researchers with climate data, soil chemistry information, vegetation distribution information, the results of experiments, and other ecological information from many projects. In addition, scientists make use of the long-term terrestrial experiments (some running for more than 20 years) of warming the tundra, fertilizing the tundra and streams, reducing the available solar radiation to plants, and excluding grazing animals.

History of Research at the Toolik LTER Site

Research on the Coastal Plain: Naval Arctic Research Laboratory

Environmental research on the North Slope of Alaska, summarized by Hobbie (1997) and Shaver (1996), began during the 1882 First International Polar Year when a research station was set up at Barrow. Over the next 60 years, expeditions collected plants and animals and surveyed the geological features. "Big science" in the north began in 1947 when the Arctic Research Laboratory, the predecessor of the Naval Arctic Research Laboratory (NARL), was established at the edge of the coastal plain at Barrow. The Office of Naval Research (ONR) at that time operated as a source of federal money for basic research at university laboratories (summarized from England 1982). Over the next three decades, NARL-ONR-supported research covered every aspect of the arctic environment including soils, vegetation, permafrost, beach formation, microbiology, taxonomy, marine biology, limnology, and atmospheric concentrations of carbon dioxide. The enlightened attitude of ONR was stated by Britton (1973:11) that to the Navy "fell a first responsibility, and what might be considered a moral commitment, to learn both how to use the environment and how to protect it." Not only was research carried out near Barrow, but also float planes were used to set up field camps in the mountains on deep lakes in the Brooks Range (Livingstone et al. 1958; Hobbie 1964).

"Big international science" reached the North Slope in the three years (1971–1973) of the NSF-funded International Biological Programme (IBP) at Barrow. The overall themes were (1) to develop a predictive understanding of the arctic ecosystem, (2) to obtain a database for modeling and comparison, and (3) to use environmental knowledge for problems of degradation, maintenance, and restoration of ecosystems. All of the major ecosystem components such as primary producers, decomposers, herbivores, predators, climate and microclimate, and soils, were studied on the same site. For the aquatic study, investigators focused on three small ponds (results synthesized in Hobbie 1980). In the terrestrial study, the focus site was the wet sedge tundra (reported in Brown et al. 1980). One of the ways the entire project was integrated was the whole-system goal of understanding carbon, nitrogen, and phosphorus cycling. The scale of support and facilities enabled scientists to go far beyond descriptive ecology and investigate the processes and controls of carbon and nutrient fluxes. And the scale of support was immense; there were hundreds of scientists at Barrow during the IBP, 32 authors for the terrestrial book, and 19 for the aquatic study.

In many ways, some of the IBP goals were ahead of the available technology. Computer programs for a database had to be written *de novo*, took years to complete, and never worked well. Ecological modeling of large systems was just beginning and computers were small; model results were inconclusive, although the aquatic program pioneered the development of stochastic versions of deterministic models (e.g., Tiwari et al. 1978). However, the processes necessary for the modeling were studied in well-documented detail, many for the first time in the Arctic. The relationships and parameters from the IBP studies are still used in ecological modeling today (e.g., the effects of temperature and moisture on decomposition determined by Flanagan and Veum 1974). The integration of research caused by the modeling efforts, by the concentration of research onto a single site, and by the crush of scientists and students into the close living and laboratory quarters at NARL produced close interactions and sharing of information.

Research in the Foothills: Toolik Lake, R4D, LTER

At the end of the IBP project in 1974, the aquatic scientists decided to look for a new site on the North Slope where they could investigate deeper lakes in the foothills, an area never before studied in arctic Alaska. At the same time the opening of the Dalton Highway in September 1974, the Prudhoe to Fairbanks supply route for the oil fields, suddenly created access to a magnificent environmental transect across the heart of northern Alaska. In June 1975 an ecological team surveyed the habitats along the road and chose Toolik Lake because it was suitably deep, contained fish, and contrasted with the shallow coastal ponds studied under IBP. Later the other advantages of the site became obvious: small lakes and headwater streams were nearby, and a complex glacial history provided a variety of soils and vegetations.

In 1975, research supported by the NSF's Division of Polar Programs began at Toolik Lake at a small tent camp next to the lake on a former airstrip near the 400-person Toolik Lake Camp pipeline construction facility. The construction camp, built in 1970, used the lake for drinking water and the camp wastewater (from tertiary treatment) drained downstream of the lake. One likely impact of the construction camp was recreational fishing for lake trout. An impact of the construction of the Dalton Highway was the mining of gravel deposits from periglacial kames in the catchment of Toolik Lake. When this occurred the underlying permafrost thawed and exposed previously frozen soil to weathering. One stream affected by gravel removal still showed elevated concentrations of alkalinity and phosphate 30 years later (see chapters 6 and 10; Hobbie et al. 1999).

The aquatic NSF research initially focused on the determination of the biomass and cycles of the organisms in the lake as well as the rates of primary productivity and chemical cycles of the nutrients. Some years later, the NSF research emphasized manipulations of the predators or fertilization of whole lakes to allow for comparisons of biotic controls of trophic structure by top-down and bottom-up mechanisms. The lake work was augmented by studies of the nearby Kuparuk River, begun in the late 1970s. Here, too, initial studies of the biology of organisms were followed by manipulations of nutrients and fish. O'Brien et al. (1997) and Hershey et al. (1997) synthesized the results of the observational studies of lakes and streams but not the results of the experiments, which are now described in chapters 7 and 8.

By 1979 terrestrial ecologists were using the Toolik Field Station as a site for research on nutritional controls over plant growth. Experiments on effects of increased temperature and nutrients and decreased light levels on tundra plants began in 1982. Additional baseline information for the LTER terrestrial studies came from a large multi-investigator study sponsored by the U.S. Department of Energy (DOE) in the late 1980s. The aim of this "R4D" study of response, resistance, resilience, and recovery was to predict the effects of disturbance. Results were synthesized in a special journal issue (Oechel 1989) and a book (Reynolds and Tenhunen 1996).

The Arctic LTER project was funded in 1987 to carry out a long-term study of streams, lakes, and tundra. The goal of the project is to understand all of the arctic ecosystems around Toolik Lake, their structure, function, and interactions, in order to predict the effects of changing landscapes and climate. A major approach of the Arctic LTER group continues to be the analysis of results of long-term, whole-ecosystem experiments, along with monitoring of key ecological variables over long time periods (chapters 5, 7, and 8). A growing emphasis in the LTER research is the linkages between terrestrial and aquatic ecosystems, which requires a scaling-up of knowledge from individual plots, lakes, and streams to hill slopes, larger basins, and heterogeneous landscapes (chapter 6). Shaver (1996) has written a detailed history of the early terrestrial research at Toolik Lake.

The research reported in this volume was possible only because of the continued support of the NSF, especially of the Office of Polar Programs, for projects and for the construction and support of the Toolik Field Station operated by the Institute of Arctic Biology of the University of Alaska Fairbanks. LTER funds from NSF's Division of Environmental Biology also have provided vital support for administration, for long-term observations, and for synthesis. The long-term help of the administration and logistics staff of UAF's Institute of Arctic Biology is gratefully acknowledged.

The Natural Setting

The Kuparuk Basin

The LTER site lies in the Kuparuk River basin, a 9000 km^2 catchment stretching from the edge of the Brooks Range on the south to the Arctic Ocean on the north (Figure 1.4). The elevation varies from sea level at the north to 720 m at Toolik Lake to several thousand meters at the southern crest of the basin in the Brooks

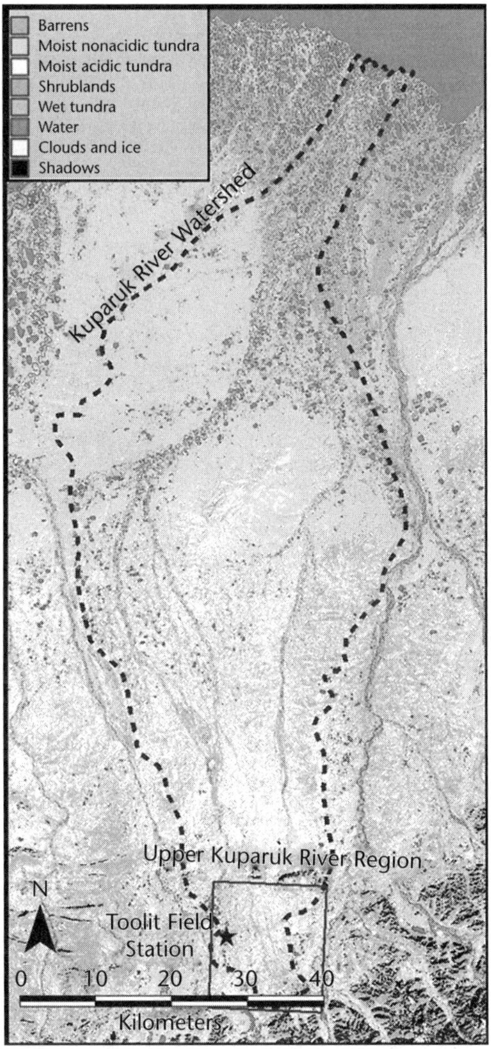

Figure 1.4 The Kuparuk River watershed and the distribution of vegetation. Based on Muller et al. (1998). [REFER COLOR MAP]

Range. The combination of distance from the ice-covered Arctic Ocean in the north and the high elevations of the Brooks Range in the south results in a summer air temperature range of 5°C; it is coldest closest to the ocean and in the mountains at the south end of the basin (Figure 1.5) and warmest mid-basin (see chapter 2). Irradiance is closely correlated with temperatures. Zhang et al. (1996) pointed out, however, that in the winter a temperature inversion is often set up with a maximum height of around 500 m. The result is that the coastal region, lying beneath the inversion, is often colder than the southern half of the drainage basin where the elevation is above 500 m.

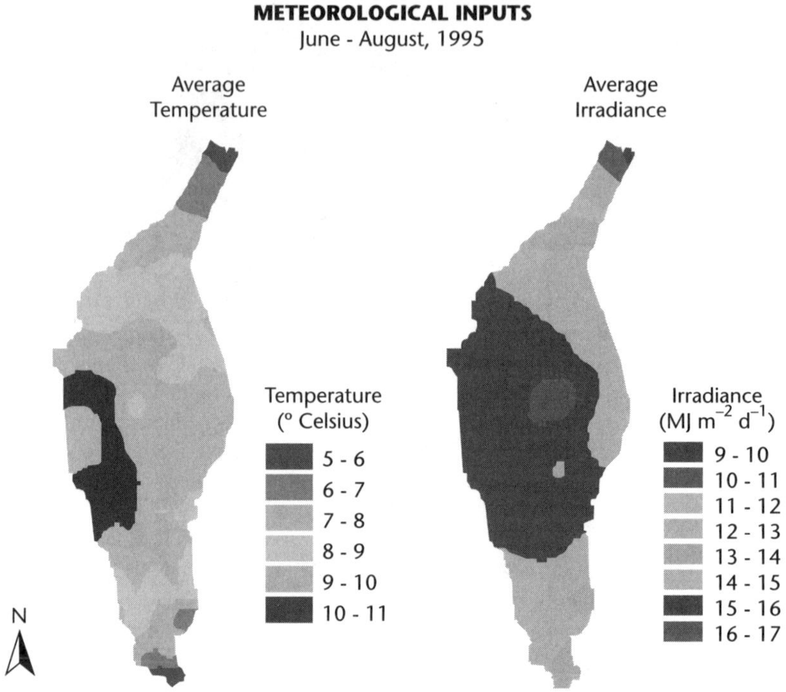

Figure 1.5 Summer (June, July, August) average temperatures and irradiance in the Kuparuk River basin (data from Williams et al. 2001).

Tundra vegetation covers the Kuparuk River basin, but there are differences in tundra communities and in the amounts of plant material. Walker's vegetation maps (www.ArcticAtlas.org) show that wet tundra (e.g., sedges) dominates the northern part or coastal plain part of the basin (Figure 1.4). There are areas of moist non-acidic tundra in the northern half of the basin as well as small areas near Toolik Lake in the south. Much of the basin is moist acidic tundra dominated by cotton grass (*Eriophorum vaginatum),* also called "tussock" tundra. Large shrubs (e.g., willows), found along the rivers south of the coastal plain, reach their maximum development in the central part of the watershed. The satellite view shows that the maximum "greenness" (normalized difference vegetation index, or NDVI) occurs in the mid-part of the basin. The NDVI can be converted to leaf area index (LAI), which is the area of leaf per area of land surface; the LAI is also highest in the middle parts of the river basin and, reasonably enough, so is the primary productivity as modeled by Williams et al. (2001) at the 1 km^2 scale.

The Kuparuk River flows northward from the edge of the Brooks Range to enter the Arctic Ocean at Prudhoe Bay. Flow begins in late May or early June with the snowmelt peak. Typically there are a few increases in flow during the summer caused by rainstorms and then a cessation of flow in late September. The chemistry of the river is described in Kling et al. (1992) and McNamara et al. (2008).

Shallow lakes and ponds formed by thermokarst activity, which is the collapse and settling of soil caused by the melting of buried ice, are abundant on the coastal plain of the Kuparuk basin even though the region was never glaciated. Ponds with a maximum depth of 50 cm often lie in patterned ground created by ice wedges formed in the permafrost. These ponds, which freeze solid every winter, were extensively studied at Barrow during the IBP (see Hobbie 1980). Lakes with a maximum depth of around 2.5 m form on the coastal plain when the ice within the soil melts. The biology and chemistry of some of these ponds and lakes is reported in Kling et al. (1992).

The LTER Site

The site encompasses the upper headwaters of the Kuparuk River including the drainage basin of Toolik Lake and the Kuparuk River basin above the confluence of the outlet stream from Toolik Lake (Figure 1.2). In practical terms, the research is concentrated in the drainage basins of Toolik Lake, Imnavait Creek, and of the Kuparuk River above the Dalton Highway bridge. The Sagavanirktok River and its tributary, Oksrukuyik Creek, are also sites of research but are not officially part of the LTER site. The U.S. Bureau of Land Management (BLM) has designated nearly the entire LTER research site a Research Natural Area.

Natural History

The climate at Toolik Lake is characterized as low arctic (Hobbie et al. 2003; Walker et al. 2005), which is tundra north of tree line with a mean July temperature of 10°C–12°C. The average annual temperature at the Toolik Field Station is –8°C, but during the summer months of June, July, and August the average temperature may climb above 10°C (Figure 1.6; chapter 2). Winter temperatures average –20°C in the coldest months. While snow may fall on any day of the year, a persistent snow cover begins in mid-to-late September and lasts until late May. Maximum snow depth is around 30 cm, but drifts form behind every obstruction and in every hollow. The sun is above the horizon continuously for more than two months in the summer, and there is no sun for two months in the winter. During June, July, and August of the same period in Figure 1.7, the three-month rainfall averaged 188 mm while total annual precipitation ranged from 250 to 350 mm. Plant photosynthesis begins as soon as the snow melts but by this time, in late May, nearly half of the total solar radiation for the year is already past.

There are only a handful of arctic field stations with climates more extreme than that of Toolik (Figure 1.7). These are located on the northern tip of Alaska (Barrow), in northeast Greenland (Zackenberg), or on the northern coast of Siberia (Samoylov). Despite its far north location at 79° N, the climate of Ny-Ålesund is very similar to that of Toolik because it is located on the Svalbard archipelago at the northern end of the Gulf Stream. All of these field stations are surrounded by tundra vegetation. In contrast, the sites at Bonanza Creek, Abisko, and other warmer and wetter sites are located in the boreal forest.

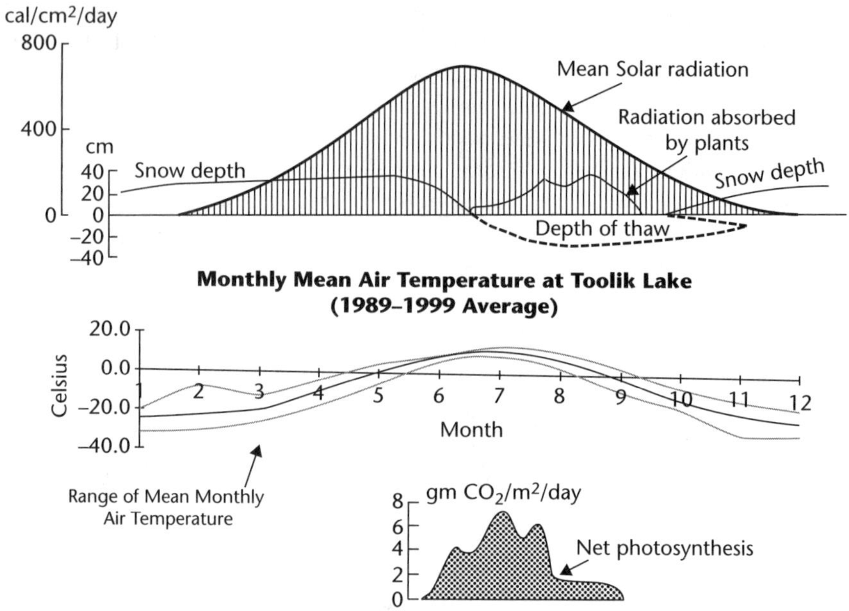

Figure 1.6 Annual course of solar radiation, snow depth, depth of thaw, mean air temperatures, and net photosynthesis at Toolik Lake, Alaska.

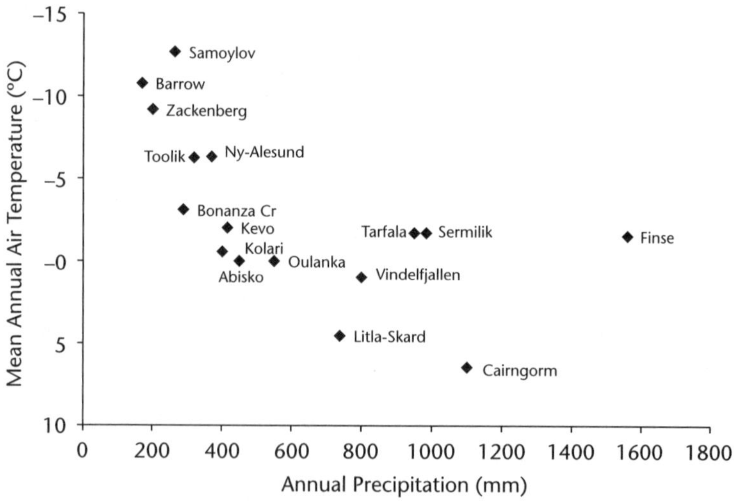

Figure 1.7 Toolik Field Station mean annual air temperature and precipitation in relation to other northern field stations (see www.eu-interact.org for location). The site locations are Samoylov, Lena Delta, Siberia; Barrow, Alaska; Zackenberg, northeast Greenland; Ny-Ålesund, Svalbard; Bonanza Creek, Fairbanks, Alaska; Kevo, Finland; Finse, Norway; Abisko, northern Sweden; Litla-Skard, Iceland.

The low annual temperature means that permafrost, or permanently frozen ground, is present to a depth of about 200 m throughout the LTER site. As noted earlier, permafrost is defined as earth material (soil, rock, ice, and organic material) that remains frozen for at least two consecutive years. This is strictly a temperature-based definition and does not depend on the presence of ice. Nevertheless, permafrost often does contain ice, called "ground ice." This is in several forms: (1) massive ice (e.g., ice wedges and pingo ice), (2) ice contained within microscopic soil pores, or (3) discrete lenses of segregated ice that are highly variable in size and shape. The most common form of massive ice are ice wedges that form as a consequence of fairly recent and regular freeze-thaw cycles when very cold air temperatures lead to vertical cracks in the soil. During snowmelt, water fills the cracks and freezes inside the permafrost. Eventually vertical ice wedges build up and form into honeycomb structures of slightly raised ridges, often with central ponds, that are called "patterned ground." These formations can cover a few hectares to hundreds of square kilometers, especially in flat, wet terrain like the arctic coastal plain.

Ground ice can impart considerable strength and structure to permafrost, but what happens when permafrost thaws? When buildings and roads heat the soil or reduce the insulation of the vegetation, then massive ice inclusions will melt and the soil collapses. When ice in microscope pores in the soil is lost, the soil may lose internal structure and subside unevenly under its own mass. The result of loss of ground ice is called "thermokarst," which refers to uneven ground (karst) caused by thermal changes. Thermokarst formation is a naturally occurring phenomenon that is at present much more important in central Alaska than on the North Slope. For example, Osterkamp et al. (2000) describes the so-called drunken forest, where boreal forest trees grow at strange angles on thermokarst-affected soils.

The large-scale effects of thermokarst formation on streams, nutrient budgets, vegetation, and soil carbon of the arctic landscape are described in detail in chapters 5 and 7. Across the entire Arctic including the North Slope, gradual warming of the permafrost is occurring; temperature changes in permafrost along the Dalton Highway are presented in chapter 2. However, the permafrost temperatures near Toolik Lake are close to –4°C, a long way from the thawing point; widespread thawing is decades or centuries in the future.

Each summer the top layer of soil thaws to a depth of 29–46 cm (range from 1990 to 2000) depending upon how warm and wet the summer happens to be (Hobbie et al. 2003; see also chapter 6). Note that in Figure 1.6 the soil freezing is not complete for some months after the insulating snow cover accumulates. Not only does permafrost restrict the rooting zone of plants to the active layer, it also acts like bedrock and seals the subsoil to water penetration. The result is that water from snowmelt and rain is held in the active layer, especially in the organic matter-rich upper 10–20 cm, and the soils are usually moist despite the low precipitation. When there is enough precipitation to saturate the soil, the resulting runoff is "flashy." That is, there is a quick peak of flow in the streams but there is little water storage in the ground and the recession from the peak is rapid.

The distribution of vegetation is dependent upon topography (dry ridge tops, moist hill slopes, lowland areas of water-saturated soils) and upon the soil chemistry as determined by the age of the soils (see also chapter 3; Hamilton 2003). As shown in Figure 1.8, the oldest soils developed on glacial till from the Sagavanirktok glacial advance (>~300,000 years ago), the next oldest soils on till from the Itkillik I advance (~60,000 years ago), and the youngest soils on till from the Itkillik II advance (~10,000 years ago). As a result of interactions of topography with age of soils, there are four dominant terrestrial ecosystems: moist acidic tussock, moist nonacidic tussock, heath, and wet sedge (see chapter 5). Most of the LTER site is covered with acidic tussock tundra (Figure 1.4). Here, a number of shrubs (willow, dwarf birch) and forbs are always present but the sedge *Eriophorum* dominates. Nitrogen availability limits primary productivity and nitrogen is rapidly recycled. Net ecosystem productivity of the North Slope is ~10–20 g C m^{-2} yr^{-1} (McGuire et al. 2000), while grazing by large and small herbivores is minimal. Nonacidic tussock vegetation grows on the youngest soils and tends to lack dwarf birch. Large caribou herds pass through the site every 5–10 years, and grizzly bear and wolves are the chief predators.

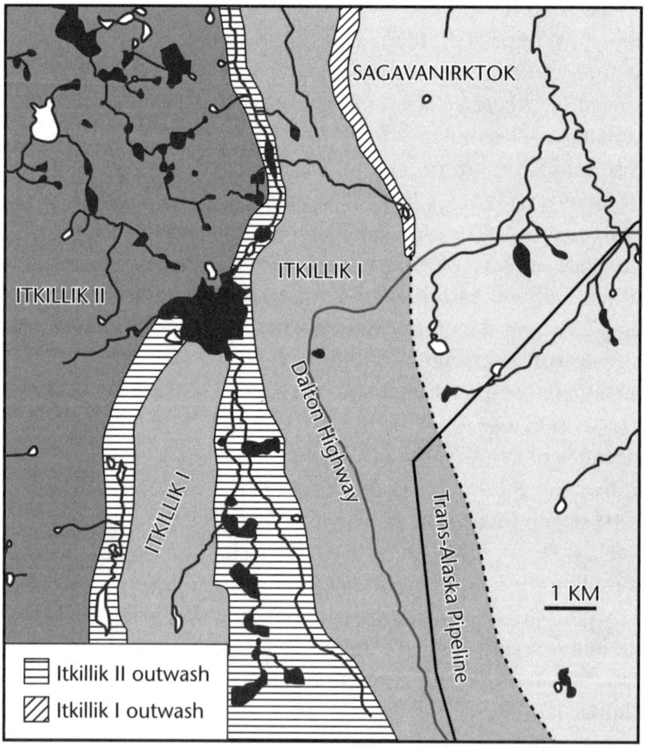

Figure 1.8 The different ages of soils near the Arctic LTER site resulting from glacial advances northward from the valleys of the Brooks Range. Toolik is the large lake in the center of the picture. The Sagavanirktok advance was >300,000 yr BP, the Itkillik I was ~60,000 yr BP, and the Itkillik II was ~10,000 yr BP. Based on map in Hamilton (2003).

The streams in the LTER site are small, 1–10 m across, and relatively shallow. Most streams have bottoms covered with rocks, but some streams have peaty bottoms. Because nutrients are tightly held in the plants and soils, only small amounts move into the stream water and the resulting inorganic N and P concentrations are exceedingly low. Most of the N and P is bound in organic forms as dissolved organic matter (DOM). Only a little of the nutrients in the DOM is available to microbes, perhaps none to algae. As a consequence, stream primary productivity is exceedingly low; most photosynthesis occurs in algae (diatoms) attached either to the stream bottom or to tubes of insect larvae. The food web of the stream is based on the algae, and insect larvae consume the algae and are themselves consumed by other insect larvae or fish. Black fly larvae filter particles from the water; most of their assimilated food is dislodged algae. There is only one species of fish, the arctic grayling, living in the streams. Because the streams freeze completely each fall, the fish must migrate some tens of kilometers to deep lakes where they survive the winter beneath the ice cover.

The only large lake in the LTER site is Toolik Lake, a 26 m deep lake with an area of 1.5 km^2. A number of smaller lakes near Toolik Lake, such as those shown in Figure 1.3, have been used for experiments manipulating nutrients and fish species. All the lakes are ultra-oligotrophic with a net primary productivity around 10 g C m^{-2} y^{-1}. Both N and P have very low concentrations, so much so that either can be limiting to primary production at any given time.

The primary productivity of the water column in the lakes comes mainly from the photosynthesis of nanoplankton, which are mostly small flagellates. Because the algal production and biomass are so low, the zooplankton grazers are also in low numbers and often have little grazing control over their main food. Another major source of primary productivity is benthic algae attached to rocks and sediments in the shallow regions of the lakes. These algae are grazed by snails, which, in turn, are consumed by lake trout or burbot, the top of the food web.

Yet another lake food web, the microbial food web, has its origins in the primary production of lake algae and terrestrial plants (O'Brien et al 1997). In Toolik Lake, however, the numbers and production of bacteria and the numbers of their predator, small colorless flagellates, is much higher than expected for an ultra-oligotrophic lake. The extra organic matter, which may be the major factor affecting bacteria, comes from the dissolved organic matter that originates in peat and soil organic horizons and is transported by streams from land into the lake (chapters 6 and 8).

Vegetation and Human History

The most recent glacial advance, the Itkillik II, retreated from the site around 13,000 years before present (Figure 1.8). Pollen records and radiocarbon dates show that the initial herb tundra of sedge and grass was replaced by a birch-dominated shrub tundra about 12,000 yr BP, and from ~9,500 to 7,000 yr BP alder became established on the North Slope (Brown and Kreig 1983; see also chapter 4). Radiocarbon dates from the Sagavanirktok River valley and elsewhere in the north-central Brooks Range also indicate that peat-forming plants and woody shrubs became relatively widespread during this interval. Bixby (1993) reports evidence for a warming peak

~7,000 yr BP and an unchanged vegetation since that time. Thus the vegetation has been stable at Toolik Lake for the past ~7,000 years.

The LTER site is close to two of the oldest sites in North America for human habitation. One, the Gallagher Flint Station (Brown and Krieg 1983), is some 40 km to the northeast and is radiocarbon dated to 10,400 yr BP. The location is atop a small kame that provides a wide view of the Sagavanirktok River valley. Another site some 200 km to the west, the Mesa Site (Kunz and Reanier 1994), has a similar aspect on top of a small hill, and is radiocarbon dated to between 11,700 and 9,800 yr BP. Both of these sites are evidently places where hunters of the Paleoindian Tradition would sit and scan the valleys for game while reworking flint projectile points and other stone tools. At this time the large herbivores included bison as well as moose and caribou (Kunz and Mann 1997). Archeological evidence, especially the abundant tent rings, indicates that for the past 400 years Nunamiut ("mountain") Iñupiat hunters and fishermen camped on the glacial moraines and other dry locations in the Toolik basin (Huryn and Hobbie 2012).

Oil was discovered at Prudhoe Bay in 1968, and in 1970 a construction camp to build a supply road to Prudhoe was set up at Toolik Lake. This 400-person camp was initially set up by truck transport along a winter road over frozen tundra, the Hickel Highway. In 1974 the gravel haul road, now the Dalton Highway, was completed from Prudhoe Bay to the Yukon River, and the construction camp closed soon after the oil pipeline was completed in 1976. The scientific camp at Toolik was set up in the summer of 1975 and has evolved to become the Toolik Field Station. In recent years the road has been opened to the public. Caribou hunters arrive in late summer for a bow-hunting and rifle season. Tourist buses (one per day) pass Toolik Lake, and there are a number of private vehicles as well. The BLM regulations do not allow camping or off-road vehicles in the Toolik Research Natural Area.

Disturbance

Ecologists now recognize how disturbances shape landscapes (Turner et al. 2003). In fact, ecological systems are seldom in equilibrium; rather, they are often in some phase of recovery from a prior disturbance. Some of these disturbances include fire, flood or drought, hurricanes, insect pests, exotic species, and land use. The landscape of the Arctic LTER site is unusual in that the major disturbance, a glacial advance, occurred some 13,000 years ago, and there has been little change in the vegetation for the past 7,000 years. Fire does affect some tundra systems, and is the key disturbance that has shaped ecosystems of the Bonanza Creek LTER, the other Alaska LTER project near Fairbanks in the boreal forest. For arctic Alaska, Walker (1996) stated that tundra fires are rare and that there was no record of a major fire in arctic tundra—the general belief was the tundra was too moist for a fire to spread. But in 2007 a combination of a dry summer and many lightning strikes on the North Slope created a fire that burned close to 1000 km^2 of tussock tundra northwest of Toolik Lake. This "natural experiment" is under current study to determine the changes in terrestrial carbon lost to the atmosphere and to the streams and lakes (Rocha and Shaver 2011).

The natural changes that take place in a landscape after disturbance by glaciers include the geomorphic softening of landscapes, the leaching and development of soils, the growth of plants, the thickening of soil organic horizons, the erosion of stream channels and formation of water tracks, and the formation of ice wedges. Because the Arctic LTER site includes soils of three different geologic ages, these processes have produced different results across the landscape. For example, the youngest soils have a pH of 6–7 while the soils of the older surfaces are quite acidic (pH 4–5) because of leaching of carbonates and accumulation of dissolved organic matter. Sphagnum moss is only found on the low pH surface.

One natural disturbance that may be increasing as the climate warms is the formation of thermokarst, described earlier. When ice-rich permafrost warms, whole hillsides may shift and new streams may form. Not only is the vegetation disturbed or destroyed, but water erosion may transport soil and nutrients into lakes and rivers, affecting the growth of algae and the survival of freshwater communities. An interesting interaction between fire and thermokarst, which was found in the large 2007 fire described above, is that when fire disturbs the insulating vegetation on the soil surface, deeper soils warm, ice melts, and thermokarst disturbances of the land surface become very common, especially around streams and lakes.

Walker (1996) discussed in detail the possible anthropogenic disturbances in the Arctic, many of which are only found in industrial sites such as the Prudhoe Bay oil fields. Those disturbances of possible importance at Toolik Lake include trash and solid waste, diesel or gasoline spills, thermokarst and thermal erosion, snow drifts from roads and buildings, impoundments, fire, off-road vehicle trails, roads and road dust, gravel borrow pits, acid rain or increased sulfates, airborne pollutants, and climate change. Because of large-scale patterns of air movement, this region of the Arctic receives very low pollutant loads (Ford et al. 1995), but even in this isolated site PCBs and organochlorine pesticides are present in fish (Wilson et al. 1995).

Although there are no major disturbances affecting the landscape around Toolik Lake, there are minor disturbances that must be kept in mind when interpreting biological and chemical records. This region of the Arctic is experiencing warming, which started around 150 years ago. At Barrow, there has been an increase in the annual average air temperature by nearly 2°C in the past 30 years. Details of the record and projections of climate into the future are discussed in chapter 2, and the responses of the Arctic LTER ecosystems to this warming are discussed throughout the book. At Toolik Lake there is evidence that some species of plants now have more biomass and are more abundant, and that the alkalinity of streams and lakes has increased in the last 30 years. This change in alkalinity is very likely caused by increased weathering of previously frozen glacial till, but the exact process remains unknown.

Another disturbance was the establishment of research (1975) and construction camps (1970) at Toolik Lake. The construction camp was sited at the north end of the lake and any drainage of wastewater entered a river downstream from the lake. The scientific camp (now the Toolik Field Station) always had a strict policy that all wastes are removed from the site and taken for disposal to Prudhoe Bay. The field station is located at the site of a former gravel borrow pit.

The building of the road was, however, a measurable disturbance (Walker 1996). One reason was that seven gravel borrow pits were excavated within the Toolik Lake drainage basin. In a typical borrow pit, the upper several meters of a glacial kame, a small hill of water-sorted sand and gravel, was removed for road construction. The resulting surface has no organic soil and is very dry. As a result, there is little revegetation, even in some cases where grasses were seeded and the sites fertilized. More important, the underlying permafrost thaws over the years and exposes fresh glacial till and gravels to erosion and weathering. Hobbie et al. (1999) describe the stream flowing at the edge of one of these pits in which the stream supplies 5% of the water entering Toolik Lake but 35% of the phosphate.

A second disturbance from the road is road dust. Since the road opened in 1974, each truck that passes creates a dust plume, especially in the summer. The dust is deposited downwind in a log-decay relationship, where the amount of dust deposited decreases logarithmically with the log of distance from the road. Everett (1980) measured the dust load near Toolik over 96 days to be 200 g m^{-2} at 10 m and 1.5 g m^{-2} at 1000 m from the road. One effect is an earlier snowmelt by days and weeks along the road corridor caused by a lowered albedo of the dust-covered snow. Another effect is a change in reflectance of the road corridor that is visible in summer satellite pictures. While some environmental changes near the road have occurred (see Leadley et al. 1996), the overall importance of dust deposition is difficult to determine; however, there is no evidence so far of any widespread major changes in vegetation or surface-water chemistry.

Research Approach of the Arctic LTER Project

Observations of Year-to-Year and Long-Term Variability

There are a large number of routine measurements made at regular intervals. These measurements include year-round climate measurements at several sites; stream and lake temperatures; the transparency and oxygen content of lakes; primary productivity of tundra, streams, and lakes; nutrient content of streams and lakes; stream flow; the algal biomass (chlorophyll *a*) of stream rocks and lake waters; zooplankton abundance and species; and fish abundance and growth. Nearly 12,000 grayling have been tagged in the Kuparuk River.

With these measurements, we can describe the environmental variables at the Arctic LTER site; this baseline is beginning to allow scientists to distinguish changes from the noise of year-to-year variability. Some long-term changes have already been noted such as a change in the alkalinity of streams and small changes in the dominant vegetation types (Hobbie et al. 1999). The year-to-year variation also gives information on the controls of plant flowering and grayling growth. For example, adult grayling grow best in summers with cold temperatures and high water flow, while young grayling grow best in low water flows of warm and dry summers (Deegan et al. 1999; see also chapter 7).

Experiments

Large-scale experiments—that is, larger than plot level—are an important tool for developing and testing our ecological understanding. For example, Chapin et al. (1995) report on experiments on the tundra that use greenhouses with nutrient additions and covered with shade cloth or with clear plastic. The Arctic LTER experiments have included:

- Tundra: heat soil, reduce light, fertilize, exclude grazers (Figure 1.9)
- Land-water: increase soil moisture
- Stream: fertilize, add or remove predators
- Lake: fertilize, add or remove top predator (lake trout), add predator (sculpin)

Response to these manipulations is often slow; we found that the results of greenhouse-heating of tundra were more different after nine years than after three or six years. Furthermore, response is often unexpected; mosses in a fertilized reach of stream came to dominate plant biomass, but only after a decade of low-level fertilization. Some of the large-scale experiments of the Arctic LTER were begun before the start of the LTER project. For example, stream fertilization began in 1983.

Figure 1.9 Long-term experiments in moist acidic tundra near Toolik Lake. The experimental treatments at this site are fertilization, heating, shading, and grazer exclusion.

Process Studies

The measurements of processes include growth of organisms, primary productivity of stream and lake algae, changes in biomass, predation, rates of product appearance in chemical reactions, uptake rates of stable and radioactive elements (^{14}C, ^{13}C, and ^{15}N), respiration rates, nitrification, and the growth of microbes. In many cases, the controls of the rates can be measured over a range of conditions (e.g., temperature, nutrients, light) that are manipulated in our large-scale experiments.

Synthesis

One approach to synthesis is the construction of a budget of element fluxes and transformations based on field and experimental measurements. This approach can be extended by bringing together process information to describe fluxes and cycles in whole ecosystems, catchments, and river basins. Modeling can provide another means of synthesis—for example, in the model of the gross primary productivity for the entire Kuparuk River basin (Williams et al. 2001). Data from several field plots were initially used to construct a very detailed physiologically based model with a half-hourly time step and plant responses to temperature, light, CO_2, N content of leaves, and water. Next a simplified version of the model was developed using satellite NDVI to relate leaf area index to leaf N content, and then scaled up to the entire Kuparuk River basin at an hourly time step throughout the summer. The results have been tested against eddy correlation studies of net carbon flux at a number of sites in the basin, and a map was developed of the spatial distribution of gross primary production. This model illustrates the power of mathematical modeling for synthesizing observations and measurements made from very small to very large scales, which has proved useful in the integration and understanding of our many studies conducted by the LTER, as described in detail in several chapters of this book.

References

Bixby, R. J. 1993. The paleolimnology of two arctic lakes: Regional and local changes in climate. M.S. thesis, University of Cincinnati, OH.

Britton, M. E. 1973. Introduction. Pages 9–15 in M. E. Britton, editor, *Alaskan Arctic Tundra*. Arctic Institute of North America Technical Paper No. 25. Washington, DC.

Brown, J., P. C. Miller, L. L. Tieszen, and F. L. Bunnell. 1980. An Arctic Ecosystem. *In* J. Brown, P. C. Miller, L. L. Tieszen, and F. L. Bunnell, editors. Dowden, Hutchinson, and Ross, Stroudsburg, PA.

Brown, J., and R. A. Kreig. 1983. *Elliott and Dalton Highways, Fox to Prudhoe Bay, Alaska: Guidebook to permafrost and related features.* Fourth International Conference on Permafrost, University of Alaska, Fairbanks.

Chapin, F. S., III, G. R. Shaver, A. E. Giblin, K. J. Nadelhoffer, and J. A. Laundre. 1995. Responses of arctic tundra to experimental and observed changes in climate. *Ecology* 76:694–711.

Deegan, L., H. Golden, C. J. Harvey, and B. J. Peterson. 1999. The influence of environmental variability on the growth of age 0 and adult arctic grayling. *Transactions of the American Fisheries Society* 128:1163–1175.

England, J. M. 1982. *A patron for pure science: The National Science Foundation's Formative Years, 1945–57*. Washington, DC: National Science Foundation.

Everett, K. R. 1980. Distribution and properties of road dust along the northern portion of the Haul Road. US Army Cold Regions Research Engineering Laboratory, report 80–19:101–128.

Flanagan, P. W., and A. K. Veum. 1974. Relationships between respiration, weight loss, temperature and moisture in organic residues on tundra. Pages 249–277 in A. J. Holding, O. W. Heal, S. F. MacLean, Jr., and P. W. Flanagan, editors, *Soil organisms and decomposition in tundra: Proceedings of the microbiology, decomposition and invertebrate working groups meeting, Fairbanks, Alaska, August 1973*. Stockholm: International Biological Programme Tundra Biome Steering Committee.

Ford, J., D. Landers, D. Kugler, B. Lasora, S. Allen-Gil, E. Crecelius, and J. Martinson. 1995. Inorganic contaminants in arctic Alaskan ecosystems: Long-range atmospheric transport or local point sources? *Science of the Total Environment,* special issue: Ecological effects of arctic airborne contaminants, 160/161:323–335.

Hamilton, T. D. 2003. Glacial geology of the Toolik Lake and upper Kuparuk River regions. *Biological Papers of the University of Alaska* 26.

Hershey, A. E., W. B. Bowden, L. A. Deegan, J. E. Hobbie, B. J. Peterson, G. W. Kipphut, G. W. Kling, M. A. Lock, M. C. Miller and J. R. Vestal. 1997. The Kuparuk River: A long-term study of biological and chemical processes in an arctic river. Pages 107–129 in A. Milner and M. W. Oswood, editors, *Alaskan Freshwaters.* New York: Springer-Verlag.

Hobbie, J. E. 1964. Carbon-14 measurements of primary production in two Alaskan lakes. *Verhandlungen der Internationalen Vereinigung Limnologie* 15:360–364.

Hobbie, J. E. 1980. Limnology of tundra ponds, Barrow, Alaska. In J. E. Hobbie, editor, US/IBP Synthesis Series no. 13. Stroudsburg, PA: Dowden, Hutchinson and Ross, Inc.

Hobbie, J. E. 1997. History of limnology in Alaska: Expeditions and major projects. Pages 45–60 in A. M. Milner and M. W. Oswood, editors, *Freshwaters of Alaska: Ecological syntheses.* New York: Springer-Verlag.

Hobbie, J. E., B. J. Peterson, N. Bettez, L. A. Deegan, W. J. O'Brien, G. W. Kling, and G. W. Kipphut. 1999. Impact of global change on biogeochemistry and ecosystems of an arctic freshwater system. *Polar Research* 18:207–214.

Hobbie, J. E., G. Shaver, J. Laundre, K. Slavik, L. A. Deegan, J. O'Brien, S. Oberbauer, and S. MacIntyre. 2003. Climate forcing at the Arctic LTER Site. Pages 74–91 in D. Greenland, D. Goodin and R. Smith, editors, *Climate variability and ecosystem Response at Long-Term Ecological Research (LTER) sites*. New York: Oxford University Press.

Huryn, A. D., and J. E. Hobbie. 2012. *Land of extremes: A natural history of the arctic North Slope of Alaska.* Fairbanks: University of Alaska Press.

Kling, G. W., W. J. O'Brien, M. C. Miller, and A. E. Hershey. 1992. The biogeochemistry and zoogeography of lakes and rivers in arctic Alaska. *Hydrobiologia* 240:1–14.

Kunz, M. L., and D. H. Mann. 1997. The Mesa Project: Interactions between early prehistoric humans and environmental change in arctic Alaska. *Arctic Research of the United States* 11:55–62.

Kunz, M. L., and R. E. Reanier. 1994. Paleoindians in Beringia: Evidence from arctic Alaska. *Science* 263:660–662.

Leadley, P. W., H. Li, B. Ostendorf, and J. F. Reynolds. 1996. Road-related disturbances in an arctic watershed: Analyses by a spatially explicit model of vegetation and ecosystem processes. Pages 387–415 in J. F. Reynolds and J. D. Tenhunen, editors, *Landscape function and disturbance in the arctic tundra.* Ecological Studies 120. Berlin: Springer-Verlag.

Livingstone, D. A., K. Bryan, Jr., and R. G. Leahy. 1958. Effects of an arctic environment on the origin and development of freshwater lakes. *Limnology Oceanography* 3:192–214.

McGuire, A. D., J. Clein, J. M. Melillo, D. W. Kicklighter, R. A. Meier, C. J. Vorosmarty, and M. C. Serreze. 2000. Modeling carbon responses of tundra ecosystems to historical and projected climate: The sensitivity of pan-arctic carbon storage to temporal and spatial variation in climate. *Global Change Biology* 6:S141–S159.

McNamara, J. P., D. L. Kane, J. E. Hobbie, and G. W. Kling. 2008. Hydrologic and biogeochemical controls on the spatial and temporal patterns of nitrogen and phosphorus in the Kuparuk River, arctic Alaska. *Hydrological Processes* 22:3294–3309.

Muller, S. V., D. A. Walker, F. E. Nelson, N. Auerbach, J. Bockheim, S. Guyer, and D. Sherba. 1998. Accuracy assessment of a land-cover map of the Kuparuk River basin, Alaska: Considerations for remote regions. *Photogrammetric Engineering and Remote Sensing* 64:619–628.

O'Brien, W. J., M. Bahr, A. E. Hershey, J. E. Hobbie, G. W. Kipphut, G. W. Kling, H. Kling, M. McDonald, M. C. Miller, P. A. Rublee, and J. R. Vestal. 1997. The limnology of Toolik Lake. Pages 61–106 in A. M. Milner and M. W. Oswood, editors, Freshwaters of Alaska. Ecological Studies 119. Berlin: Springer-Verlag.

Oechel, W. C. 1989. Nutrient and water flux in a small arctic watershed: An overview. *Holarctic Ecology* 12:229–237.

Osterkamp, T. E., L. Viereck, U. Shur, M. T. Jorgenson, C. Racine, A. Doyle, and R. D. Boone. 2000. Observations of thermokarst and its impact on boreal forests in Alaska, U.S.A. *Arctic, Antarctic, and Alpine Research* 32:303–315.

Reynolds, J. F., and J. D. Tenhunen, editors. 1996. *Landscape function and disturbance in arctic tundra*. Ecological Studies 120. Berlin: Springer-Verlag.

Rocha, A. V., and G. R. Shaver. 2011. Postfire energy exchange in arctic tundra: The importance and climatic implications of burn severity. *Global Change Biology* 17:2831–2841.

Shaver, G. R. 1996. Integrated ecosystem research in northern Alaska, 1947–1994. Pages 19–33 in J. F. Reynolds and J. D. Tenhunen, editors, *Landscape function and disturbance in arctic tundra*. Ecological Studies 120. Berlin: Springer-Verlag.

Tiwari, J. L., J. E. Hobbie, J. P. Reed, D. W. Stanley, and M. C. Miller. 1978. Some stochastic differential equation models of an aquatic ecosystem. *Ecological Modeling* 4:3–27.

Turner, M. G., S. Collins, A. Lugo, J. Magnuson, S. Rupp, and F. Swanson. 2003. Long-term ecological research on disturbance and ecological response. *BioScience* 53:46–56.

van Everdingen, R. 1998, revised May 2005. *Multi-language glossary of permafrost and related groundice terms*. Boulder, CO: National Snow and Ice Data Center/World Data Center for Glaciology. http://nsidc.org/fgdc/glossary/.

Walker, D. A. 1996. Disturbance and recovery of arctic Alaskan vegetation. Pages 35–71 in J. F. Reynolds and J. D. Tenhunen, editors, *Landscape function and disturbance in arctic tundra*. Ecological Studies 120. Berlin: Springer-Verlag.

Walker, D. A., M. K. Raynolds, F. J. A. Daniëls, E. Einarsson, A. Elvebakk, W. A. Gould, A. E. Katenin, S. S. Kholod, C. J. Markon, E. S. Melnikov, N. G. Moskalenko, S. S. Talbot, B. A. Yurtsev, and CAVM Team. 2005. The circumpolar arctic vegetation map. *Journal of Vegetation Science* 16:267–282.

Williams, M., E. B. Rastetter, G. R. Shaver, J. E. Hobbie, E. Carpino, and B. L. Kwiatkowski. 2001. Primary production of an arctic watershed: An uncertainty analysis. *Ecological Applications* 11:1800–1816.

Wilson, R., S. Allen-Gil, D. Griffin, and D. Landers. 1995. Organochlorine contaminants in fish from an arctic lake in Alaska, USA. *Science of the Total Environment,* special issue: Ecological effects of arctic airborne contaminants, 160/161:511–519.

Zhang, T., T. E. Osterkamp, and K. Stamnes. 1996. Some characteristics of the climate in northern Alaska. *Arctic and Alpine Research* 28:509–518.

2

Climate and Hydrometeorology of the Toolik Lake Region and the Kuparuk River Basin

Past, Present, and Future

Jessica E. Cherry
Stephen J. Déry
Yiwei Cheng
Marc Stieglitz
Amy S. Jacobs
Feifei Pan

Introduction to the Toolik Field Station and the Kuparuk River Basin

The Arctic Long Term Ecological Research site is located at the Toolik Field Station (Toolik) in the northern foothills of the Brooks Range, Alaska, and near the headwaters of the Kuparuk River basin (KRB) at 68°38' N, 149°43' W with an elevation of 719 m. Soils in the Toolik region are glacial till and outwash from glacial advances approximately 10,000 and 60,000 yr BP (see chapters 1 and 3). Soils are typically moist and vegetation is underlain by an organic mat and silty mineral soil with a maximum thaw depth of ~0.25–1.0 m in summer (Hinzman et al. 1991; Zhang et al. 1996). Vegetation at the site is dominated by tussock-forming sedges (mainly *Eriophorum vaginatum*), deciduous shrubs (mainly *Betula nana*), ericaceous evergreens (mainly *Ledum palustre* and *Vaccinium vitis-idaea*), and mosses (mainly *Sphagnum* spp.) (see chapters 3 and 5). The temperature regime for this region has been relatively stable for the last 7,000–7,500 years and favorable for peat accumulation, while prior to ~10,000 yr BP conditions were much colder and drier (Kaufman et al. 2004; see also chapter 4).

The KRB extends northward from the divide of the Brooks Range to the Arctic Ocean, about 250 km. The river basin is centered near 70° N, 150° W and covers ~9000 km^2 in area with an average elevation of 263 m above mean sea level. The Kuparuk River spans three principal physiographic regions: the mountainous headwaters (elevation ~1000 m), the foothills (600–1000 m), and the flat arctic coastal plain near the river's delta at Prudhoe Bay (below 100 m). The basin is underlain by continuous permafrost that reaches depths as great as 600 m in the north; the frozen ground acts as an impermeable boundary that generally hinders water infiltration.

The climate of the KRB is dominated by the cold season, during which snow covers the surface for about two-thirds of the year. Snowmelt is between mid-May and mid-June and provides up to 80% of the river's annual runoff into the Arctic Ocean. Total snowfall amounts are relatively low, with a typical value of 120 mm snow water equivalent (SWE) in any given year. Yet, snowfall (measured as SWE) accounts for about half of the annual precipitation input into the catchment and may occur at any time of the year. High-wind conditions and blowing-snow events are also common in the basin, especially near the arctic coastline and in mountainous regions. Similar to other arctic regions, the KRB has experienced climate change and variability during the past few decades. Throughout the 1980s and 1990s, near-surface air temperatures warmed by about 2°C in this region (Serreze et al. 2000), a trend that slowed somewhat in the early 2010s with the initiation of the "cold" phase of the Pacific decadal oscillation. In response to increasing air temperatures and to increases in snow depth, a significant increase in permafrost temperature (up to 1.5°C at a 20-m depth over a 15-year period) was reported for the area by Osterkamp and Romanovsky (1996).

Climate Datasets

As part of the North Slope Hydrology Research Project of the Water and Environmental Research Center (WERC) at the University of Alaska, Fairbanks (UAF), meteorological information has been recorded throughout the KRB since 1985. Meteorological stations operated by the WERC typically consist of a 3-m or 10-m mast with a suite of instruments that measure similar meteorological parameters as those measured at Toolik. Four Wyoming snow gauges along the Dalton Highway from Atigun Pass to Prudhoe Bay, operated by the Natural Resources Conservation Service (NRCS), supply essential information on wintertime snowfall.

Excluding Toolik, data from 15 meteorological sites within the KRB are summarized in Table 2.1. For these 15 sites, the period chosen begins January 1, 1989, and ends on December 31, 2010 (22 years). Meteorological data that are sparse (e.g., radiation measurements during winter) are not described here. All standard meteorological variables are available on an hourly basis (Table 2.1), including conditions of air temperature (T), relative humidity (RH), wind speed (U), surface atmospheric pressure (P_s) and precipitation (P).

Table 2.1 List of meteorological stations in the vicinity of the Toolik LTER and the Kuparuk River basin of Alaska, their geographical coordinates, elevation, and associated meteorological fields (even if only partially available throughout the period 1988–2004)

Station	Latitude (° N)	Longitude (° W)	Elevation (m)	Available Fields
Atigun Pass	68.13	149.48	1463	Sn
Betty Pingo	70.27	148.88	12	R, RH, T, U
East Headwater	68.57	149.30	909*	R, T, U
Franklin Bluffs	69.83	148.60	78	RH, T, U
Green Cabin Lake	68.53	149.22	975*	R, T, U
Imnavait Creek	68.62	149.28	910	Ps, R, RH, Sn, T, U
Lower Kuparuk	70.28	148.97	78	R, RH, T, U
North Headwater	68.60	149.42	944*	R, T, U
North Prudhoe Bay	70.35	148.52	23**	RH, T
Sagwon	69.42	148.75	299	R, RH, Sn, T, U
Toolik Lake	68.63	149.60	720	Ps, R, RH, T, U
Upper Headwater	68.52	149.33	1003*	R, T, U
Upper Kuparuk	68.63	149.40	774	R, RH, T, U
West Dock	70.37	148.55	8	R, RH, T, U
West Headwater	68.55	149.40	1018*	R, T, U
West Kuparuk	69.42	150.33	160	Ps, R, RH, T, U

* Data taken from the "Elevation Finder," available online at http://216.27.181.157/Elevation.asp.

** Data taken from Zhang et al. (1996).

The measurements of surface air temperature (T), relative humidity with respect to water (RH), wind speed (U), surface atmospheric pressure (Ps), snowfall (Sn) and rainfall (R) rates are listed where available. The data are taken from Kane and Hinzman (2007), the Toolik LTER database (http://ecosystems.mbl.edu/arc), and the Natural Resources Conservation Service (http://www.wcc.nrcs.usda.gov/).

In addition to measurements made throughout the KRB, instruments at Toolik have provided meteorological information from June 1988 to the present. Meteorological instruments deployed on a 5-m tower measured air temperature (T), relative humidity (RH), wind speed (U) and direction (φ), surface atmospheric pressure (P_s), precipitation (P), soil temperature (T_s), and various radiation parameters. The Toolik data chosen for study begin on January 1, 1989, and end on December 31, 2010 (21 years).

Snow data discussed in detail here include data from the Kane and Hinzman group (Kane and Hinzman 2007) based at UAF, the NRCS Snotel Station network, and the gridded weekly snow cover estimates from Armstrong and Brodzik (2005), based on D. Robinson's work on digitizing NOAA/NWS charts (Robinson et al. 2001). Correction factors for gauge-based data in the National Climate Data Center (NCDC) Global Summary of the Day archive are provided by D. Yang (Yang et al. 2005).

Analyses of Historical Climate Data from the Arctic LTER site

Climate data collected at Toolik by the LTER project and the Toolik Field Station staff were used to examine seasonal values and long-term trends of incoming solar radiation, surface air temperature, ground temperatures (surface and 20 cm), precipitation, wind speed and wind direction at 5 m, and growing-season length. Hourly meteorological data recorded at Toolik for the period 1989–2010 were checked, summed, and averaged to derive monthly or annual values. Periods of less than two weeks with missing values were filled in through linear interpolation. For precipitation, frequent periods of missing data greater than two weeks were filled in with data from the nearby Imnavait Creek watershed. General descriptions of the measurements and climatology will be provided, followed by a discussion of trends in each component of the climate system.

Solar Radiation and Thermal Regime

At Toolik a pyranometer (Kipp and Zonen CMP-6) currently measures global solar radiation (sun + sky, 400–1100 nm). Solar radiation data collected from 2009 to 2010 show clear seasonal variation in solar radiation received at Toolik (Figure 2.1). Between mid-November and the end of January, the incident solar radiation is insignificant (<0.005 kW m^{-2}). Beginning in February, solar radiation increases, peaks around the summer solstice, and then decreases to negligible intensity around mid-November. The incoming solar radiation per unit area is strongly affected by the low solar angle (maximum of ~45° at summer solstice). Consequently, Toolik

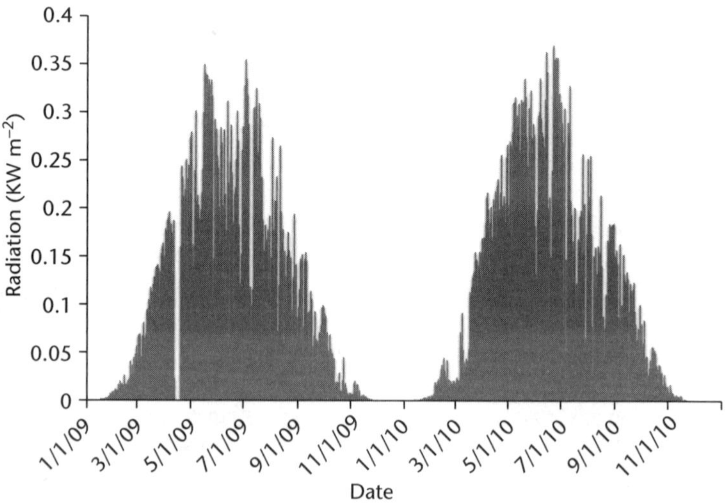

Figure 2.1 Daily averaged incoming solar radiation at the Toolik LTER from January 1, 2009, to December 31, 2010.

Table 2.2 Annual and monthly means of surface air temperature (°C) at 5–m height above the land surface at Toolik LTER from 1989 to 2010

Year	Monthly Mean												Annual Mean
	Jan	Feb	Mar	Apr	May	Jun	Jul	Aug	Sept	Oct	Nov	Dec	
1989	−30.73	−9.22	−20.84	−10.06	−4.06	8.00	12.38	11.51	1.33	−11.05	−26.25	−19.58	−8.21
1990	−30.71	−30.45	−14.08	−8.03	4.73	9.06	12.71	9.24	−0.81	−10.48	−20.56	−23.41	−8.57
1991	−22.47	−24.64	−20.85	−9.16	3.29	8.64	9.18	5.73	1.20	−8.28	−23.74	−24.59	−8.81
1992	−24.97	−27.02	−15.24	−14.31	−5.21	8.83	13.12	8.26	−8.41	−13.00	−16.62	−23.08	−9.80
1993	−21.21	−16.52	−14.22	−6.27	2.65	9.11	13.85	6.15	−0.73	−6.66	−16.07	−21.61	−5.96
1994	−22.34	−22.01	−24.01	−16.13	−0.11	6.49	13.73	9.34	−2.91	−15.32	−22.35	−23.70	−9.94
1995	−20.80	−20.21	−22.45	−8.38	2.15	7.79	11.44	6.61	3.60	−11.73	−19.17	−23.80	−7.91
1996	−20.72	−23.17	−15.85	−17.68	0.32	8.11	10.59	4.07	−4.72	−16.99	−14.70	−19.92	−9.22
1997	−23.36	−20.28	−26.26	−9.74	−0.07	9.40	11.51	9.14	2.67	−15.86	−12.80	−28.30	−8.66
1998	−30.61	−25.60	−13.11	−4.66	1.26	9.87	14.43	7.98	1.77	−9.01	−17.87	−21.54	−7.26
1999	−25.24	−26.10	−26.05	−14.59	0.46	9.66	11.08	9.38	0.82	−16.97	−21.96	−30.17	−10.81
2000	−22.27	−17.99	−20.79	−14.97	−6.56	9.39	10.06	5.10	−1.72	−13.62	−15.69	−11.68	−8.40
2001	−21.06	−17.11	−27.16	−14.90	−9.65	8.10	9.85	7.29	1.64	−16.24	−17.83	−23.15	−10.02
2002	−22.15	−22.39	−16.89	−13.07	0.73	6.22	9.38	4.29	3.62	−7.16	−12.61	−17.08	−7.26
2003	−18.34	−20.09	−22.52	−10.14	−3.28	8.62	8.74	5.88	−3.36	−6.72	−17.49	−19.71	−8.20
2004	−24.59	−28.52	−23.21	−9.46	1.60	12.31	12.59	10.68	−3.40	−10.01	−15.95	−23.47	−8.45
2005	−19.15	−21.16	−18.84	−14.48	0.41	9.15	8.64	8.68	1.81	−11.64	−27.63	−21.82	−8.84
2006	−30.68	−14.47	−22.13	−19.97	−0.59	7.68	10.33	5.74	4.74	−5.38	−18.43	−19.35	−8.54
2007	−22.19	−21.56	−29.11	−10.88	−3.14	11.60	13.43	10.22	3.11	−12.21	−11.99	−19.54	−7.69
2008	−25.46	−23.31	−25.80	−10.39	−2.30	9.58	10.62	4.32	−1.45	−14.90	−18.91	−18.83	−9.73
2009	−23.01	−20.02	−25.96	−12.78	−1.28	8.29	12.71	5.55	−0.99	−8.08	−22.71	−15.65	−8.66
2010	−25.58	−22.04	−22.32	−10.08	−2.40	8.96	10.30	9.18	2.01	−12.96	−15.14	−26.93	−8.92

receives far less solar radiation than tropical locations; peak summer radiation is ~0.35 kW m^{-2}.

Based on the record from 1989 to 2010, annual surface air temperature at 5 m (SAT 5 m) at Toolik fluctuated within the range –10.6°C (1999) to –5.8°C (1993) (Table 2.2). The average annual SAT 5 m during this period was –8.5°C. During this period, surface air temperatures at 5 m reached their minimum typically between January and March and their maximum in July, with the exception of 2005, when the maximum was in June.

The relationship of SAT to the ground surface temperature (GST) at Toolik is heavily influenced by snow. As Stieglitz et al. (2003) demonstrated, changes in ground temperature are not a direct expression of changes in the overlying air temperatures but also depend on changes in snow depth. As an effective insulator, snow attenuates the daily SAT signals—the deeper the snow pack, the higher the attenuation of the daily SAT signals. The Toolik GST record showed significant attenuation of the winter SAT signal (Figure 2.2).

Due to the insulation of snow, GST represents an attenuated SAT but with higher temperatures in winter. During the same period as the SAT measurements, the annual ground surface temperature (GST) at Toolik fluctuated within the range –4.5°C (1994) to –0.1°C (1998) (Table 2.3); the annual average was –2.4°C. During this period, GST reached minimum values typically between January and March and maximum values between June and July.

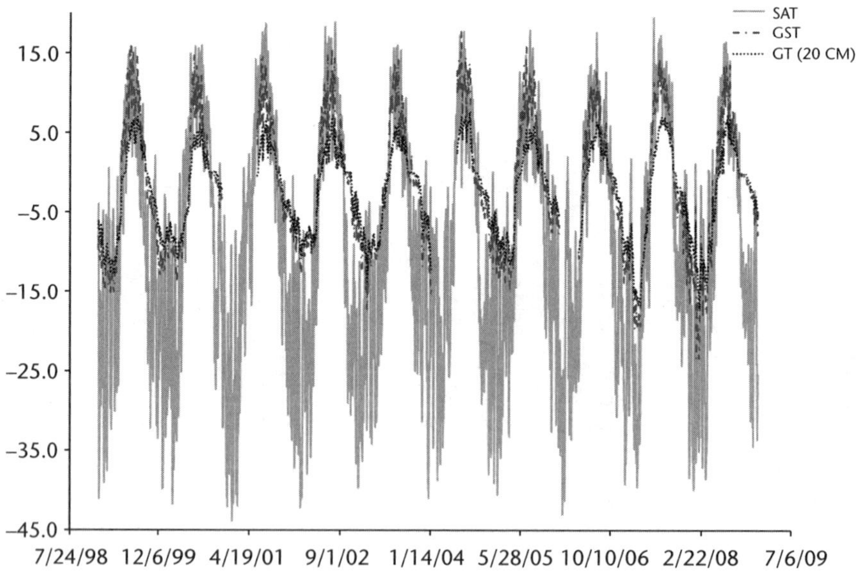

Figure 2.2 Daily average surface air temperature (SAT at 5 m, solid line), ground surface temperature (dot-dash line), and ground temperature at 20 cm depth (dotted line) at Toolik LTER from January 1, 1999, to December 31, 2008 (x-axis labeled as month/day/year).

Table 2.3 Annual and monthly mean of ground surface temperature (°C) at Toolik LTER from 1989 to 2008

Year	Monthly mean												Annual mean
	Jan	Feb	Mar	Apr	May	Jun	Jul	Aug	Sept	Oct	Nov	Dec	
1989	−10.24	−8.18	−10.35	−5.91	1.00	6.33	8.92	5.04	−0.76	−5.63	−10.32	−11.28	−2.79
1993	−14.93	−12.37	−9.23	−6.51	5.56	9.67	7.97	1.73	−0.56	−3.70	−11.68	−10.95	−3.05
1994	−12.10	−16.30	−11.24	−7.60	0.35	7.23	11.19	2.71	−0.34	−1.31	−8.39	−7.26	−4.46
1995	−13.94	−11.18	−9.43	−3.32	3.65	9.68	8.17	3.07	−0.30	−1.49	−3.35	−5.21	−2.16
1996	−6.73	−6.94	−7.62	−7.01	5.57	6.17	7.07	1.03	−0.28	−0.72	−6.29	−6.57	−1.96
1997	−8.09	−6.82	−9.35	−4.16	0.01	9.36	8.98	8.50	−1.79	−2.74	−4.11	−6.78	−1.41
1998	−8.81	−8.64	−6.72	−2.19	7.03	18.44	10.60	3.50	−0.28	−3.41	−6.13	−9.77	−0.12
1999	−13.74	−13.91	−14.90	−6.13	2.55	13.17	9.98	4.60	−1.56	−4.57	−9.73	−11.57	−2.75
2000	−10.16	−10.23	−9.23	−6.32	0.23	10.78	11.15	3.44	−0.20	−1.77	−4.00	−7.00	−2.37
2002	−11.38	−8.08	−10.12	−0.74	6.52	7.75	12.32	3.03	0.03	−0.23	−2.72	−11.43	−1.63
2003	−17.20	−9.27	−9.71	−3.04	4.12	13.45	5.50	3.82	0.68	−0.23	−10.06	−6.12	−2.24
2005	−11.34	−12.50	−14.57	0.79	3.79	10.98	8.26	0.13	−0.67	−4.91	−3.94	−6.82	−1.86
2007	−7.28	−18.20	−13.33	−7.54	6.49	11.30	9.09	5.79	−1.77	−4.67	−5.75	−11.87	−3.36
2008	−15.50	−17.85	−9.75	−5.94	3.87	9.60	4.03	3.91	−0.36	−0.87	−2.90	−7.93	−2.94

* More than two months of data missing in the following years: 1990, 1991, 1992, 2001, 2004 and 2006.

As the GST signal propagates through the subsurface, the signal is attenuated and phase shifted. However, at the shallow subsurface (20 cm) the phase shift was insignificant, and the ground-temperature signal at 20 cm (GT 20 cm) tracked the GST signal closely. During the same period, the annual GT 20 cm fluctuated within the range –4.8ºC (1994) to –1.3ºC (1998) (Table 2.4). The minimum and maximum temperatures occurred in the same years as those of the GST. The average annual GT 20 cm during this period was –2.9ºC. The seasonal trend of GT 20 cm was similar to that of the GST and reached a minimum typically between January and March and a maximum typically in the month of July.

Precipitation

During the period 1989 to 2008, annual precipitation at Toolik ranged from 249 mm yr^{-1} (1994) to 407 mm yr^{-1} (1996) (Table 2.5). Average annual precipitation during this period was 312 mm; 60% of the precipitation fell during summer months (June through August), and 40% fell during winter months. Data were typically collected with an OTT Pluvio gauge, although missing data were filled in using the nearby Imnavait record as described above.

Growing Season Length

Growing season length is defined here as the number of days in a year that the average daily SAT is above freezing. This liberal definition of growing season is used here because arctic plants are highly frost tolerant. The length of the growing season in the Arctic has been projected by some to increase due to earlier snowmelt (Foster et al. 1992; Stone et al. 2002), but warming temperatures could also lead to a deeper snow pack and a later snowmelt. Analysis of the Toolik growing-season data reveal that from 1989 to 2010 the mean growing-season length was 123 (±3) days, with a mean start date of May 15 and a mean end date of September 15. Linear trend analysis shows no statistically significant trend in the length of the growing season during this period (Figure 2.3).

Examining the first and second decades of this record separately (1989–1999, 2000–2010), the mean growing-season length for the period 1989–1999 was 128 (±4) days, with a mean start date of May 9 and mean end date of September 14 (Table 2.6). The mean growing-season length for the period 2000–2010 was shorter than that of the previous decade, 119 (±3) days, with a later mean start date of May 20 (11 days later) and a later mean end date of September 16 (2 days later). At Toolik, a delay in the start of the growing season contributed significantly to the decrease of the growing-season length.

Growing-season length calculated as functions of GST and GT 20 cm also shows no apparent trend from 1989 to 2008. When calculated as a function of GST, the mean growing-season length was 126 (±4) days with a mean start date of May 14 and mean end date of September 17. When calculated as a function of GT 20 cm, the mean growing-season length was 119 (±4) days with a mean start date of May 23 and mean end date of September 18.

Table 2.4 Annual and monthly mean of ground temperature (°C) at 20-cm depth at Toolik LTER from 1989 to 2008.

Year	Monthly Mean												Annual Mean
	Jan	Feb	Mar	Apr	May	Jun	Jul	Aug	Sept	Oct	Nov	Dec	
1989	−9.63	−8.04	−9.82	−6.05	−0.90	3.82	5.79	3.78	−0.59	−4.07	−8.98	−10.01	−3.06
1993	−13.48	−11.65	−9.49	−7.34	0.76	5.52	6.08	2.24	−0.23	−2.73	−10.24	−9.54	−3.50
1994	−12.29	−15.18	−11.29	−7.99	−0.34	3.63	7.76	1.55	−0.18	−0.95	−7.41	−6.79	−4.82
1995	−12.99	−11.17	−9.46	−3.85	0.74	5.73	5.86	1.96	0.43	−1.20	−2.76	−4.76	−2.62
1996	−6.08	−7.44	−7.36	−6.96	0.16	2.95	3.68	1.12	−0.15	−0.48	−5.22	−5.65	−2.40
1997	−7.49	−6.45	−8.77	−5.87	−0.85	2.52	4.40	3.12	0.11	−0.67	−2.07	−4.09	−2.01
1998	−7.46	−7.56	−6.03	−2.40	2.37	6.50	7.34	2.84	0.17	−1.47	−4.15	−6.62	−1.33
1999	−10.08	−11.57	−12.84	−7.41	−0.04	4.52	4.62	2.89	0.01	−2.78	−5.75	−8.99	−3.26
2000	−8.32	−8.65	−9.27	−6.79	−0.69	3.22	4.92	1.41	−0.01	−0.74	−3.16	−6.00	−2.95
2002	−9.79	−7.80	−9.64	−2.63	1.42	2.95	4.89	1.99	0.22	0.01	−2.38	−9.14	−2.44
2003	−13.14	−8.81	−9.22	−3.20	−0.40	5.15	3.10	3.32	0.04	−0.28	−5.56	−5.23	−2.84
2005	−8.67	−9.91	−11.41	−2.10	−0.04	5.37	3.79	1.69	−0.04	−2.42	−3.16	−5.18	−2.66
2007	−7.50	−15.17	−13.37	−7.37	−0.14	5.05	6.38	4.24	−0.07	−2.69	−4.92	−8.11	−3.71
2008	−11.94	−12.58	−10.63	−6.02	−0.08	5.07	4.09	3.09	−0.09	−0.27	−2.23	−6.07	−3.12

* More than two months data missing in the following years: 1990, 1991, 1992, 2001, 2004, and 2006.

Table 2.5 Annual and monthly means of precipitation (mm yr^{-1}) at Toolik LTER from 1989 to 2008

Year	Month												Annual
	Jan	Feb	Mar	Apr	May	Jun	Jul	Aug	Sept	Oct	Nov	Dec	
1989	20.2	15.1	2.5	22.9	12.6	40.7	50.1	55.7	48.3	2.5	5.1	7.6	283.3
1990	0.0	0.0	50.8	20.2	8.4	56.1	12.5	49.5	17.1	5.0	10.2	0.0	229.8
1991	25.3	32.9	2.5	17.8	22.8	44.1	45.6	53.4	14.1	40.4	5.0	17.7	321.6
1992	5.0	12.6	10.1	7.6	17.6	57.2	48.5	117.9	21.6	22.7	38.0	15.2	374
1993	22.8	12.6	7.6	5.0	11.9	34.8	66.3	56.6	25.9	0.0	0.0	0.0	243.5
1994[a]	5.0	15.1	5.0	7.6	20.2	40.7	96.2	48.9	38.0	0.0	0.0	0.0	276.7
1995[b]	0.0	0.0	0.0	0.0	17.8	59.5	118.8	47.5	20.4	12.7	12.7	28.0	317.4
1996	5.1	22.9	12.7	15.3	30.0	55.0	61.3	44.8	60.8	17.5	10.0	7.5	342.9
1997	2.5	12.5	7.6	2.5	21.1	29.6	141.8	99.3	58.1	30.2	15.3	7.5	428
1998	2.5	5.1	2.5	0.0	10.4	61.1	43.4	68.4	38.6	11.2	1.5	21.7	266.4
1999	2.4	9.4	1.2	0.9	16.5	49.6	125.5	57.7	71.3	27.4	0.0	0.0	361.9
2000	3.1	16.5	7.6	5.4	26.0	32.1	68.7	88.5	41.8	15.9	7.4	17.6	330.6
2001	5.0	5.0	2.5	5.1	2.5	40.1	90.5	63.9	8.5	6.9	17.5	0.3	247.8
2002	5.2	16.6	8.2	20.8	17.4	54.4	70.4	120.7	24.8	41.5	2.3	4.1	386.4
2003	9.8	12.1	11.8	28.7	24.6	20.4	154.2	101.9	33.1	34.9	0.0	31.0	462.5
2004[c]	0.0	0.0	0.0	0.0	0.0	22.7	163.7	40.0	20.7	2.4	7.6	6.6	263.7
2005	21.5	4.0	12.4	12.4	5.1	22.5	66.6	17.3	18.6	6.7	15.8	12.7	215.6
2006	0.0	7.5	5.0	2.9	28.7	60.9	85.0	64.8	31.3	18.8	75.6	8.2	388.7
2007[d]	12.8	4.6	16.7	10.1	6.2	26.9	64.5	35.0	3.1	4.5	10.7	5.6	200.7
2008	13.1	10.3	9.5	11.5	23.9	77.5	116.2	17.7	17.6	48.0	10.0	32.8	388.1

[a] Missing data: October 1, 1994–December 31, 1994.
[b] Missing data: January 1, 1995–May 27, 1995.
[c] Missing data: January 22, 2004–May 24, 2004, May 31, 2004–June 8, 2004.
[d] Missing data: October 19, 2007–October 26, 2007.

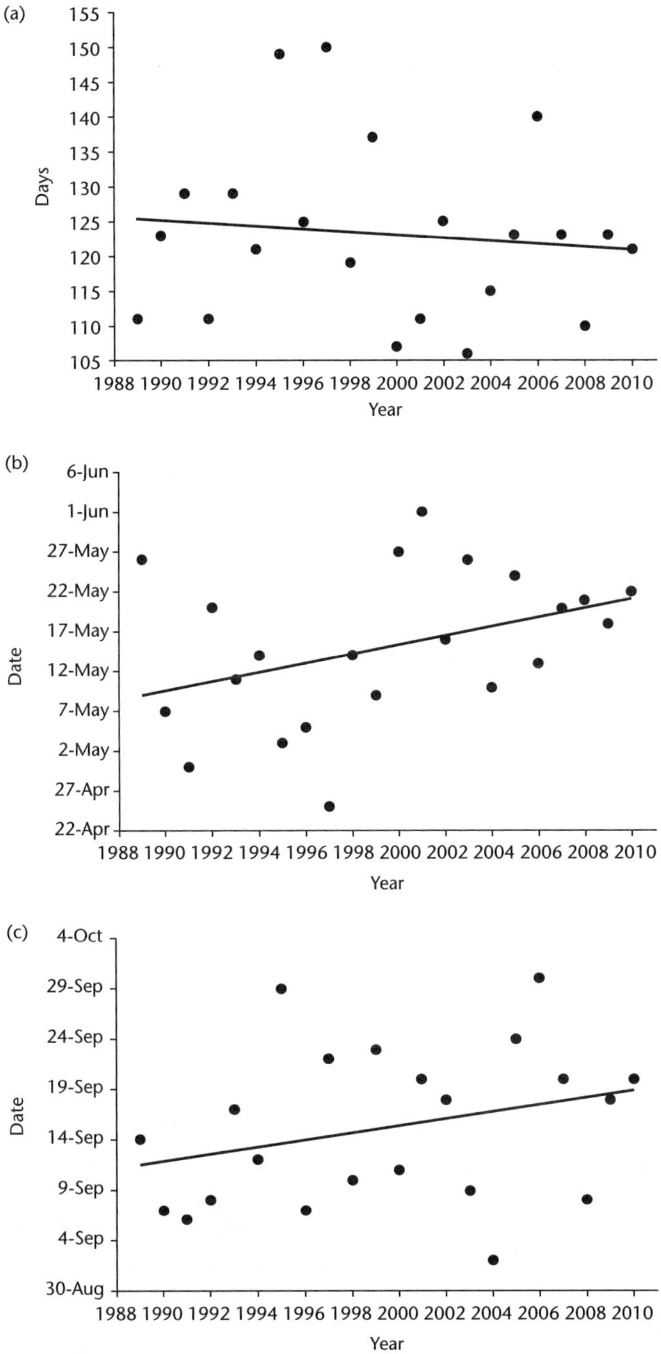

Figure 2.3 (A) Length of growing season from 1989 to 2010; $R^2 = 0.013$, $p = 0.61$ fails student's t-test at confidence interval (CI) = 95%. (B) Start date of growing season from 1989 to 2010; $R^2 = 0.149$, $p = 0.08$ fails t-test at CI = 95%. (C) End date of growing season from 1989 to 2010; $R^2 = 0.085$, $p = 0.19$ fails t-test at CI = 95%.

Table 2.6 Start date, end date, and duration (number of days) of growing season from 1989 to 2010

Growing Season (1989–2010)			
Year	Start	Duration	End
1989	26-May	111	14-Sep
1990	7-May	123	7-Sep
1991	30-Apr	129	6-Sep
1992	20-May	111	8-Sep
1993	11-May	129	17-Sep
1994	14-May	121	12-Sep
1995	3-May	149	29-Sep
1996	5-May	125	7-Sep
1997	25-Apr	150	22-Sep
1998	14-May	119	10-Sep
1999	9-May	137	23-Sep
2000	27-May	107	11-Sep
2001	1-Jun	111	20-Sep
2002	16-May	125	18-Sep
2003	26-May	106	9-Sep
2004	10-May	115	2-Sep
2005	24-May	123	24-Sep
2006	13-May	140	30-Sep
2007	20-May	123	20-Sep
2008	21-May	110	8-Sep
2009	18-May	123	18-Sep
2010	22-May	121	20-Sep
Mean	15-May	123	15-Sep
Standard error	2	3	2

During this period, mean start date, duration, and end date were May 15 (± 2 days standard error), 123 days (± 3 days) and September 15 (± 2 days).

Wind Speed and Direction

Monthly wind speed and direction at 5 m were averaged for the period 1989–2010 to obtain the monthly mean wind speed and direction (Met-One sensors). The monthly means provide insight into the seasonal trend of wind speed and direction at Toolik. Monthly mean wind speed ranges from 2.48 m s^{-1} (May) to 4 m s^{-1} (February) (Table 2.7). On average, wind comes mainly from the south—that is, 180° (Table 2.8). During December and January, winds have a weak easterly component and come from south-southeast (170° in December and 167° in January). In contrast, from April to July, winds have a westerly component. In May, winds predominantly come from the southwest. Seasonal wind shifts are the result of large-scale changes in atmospheric circulation (especially

Table 2.7 Monthly means of wind speed (m s^{-1}) at 5-m height at Toolik LTER from 1989 to 2010

Year	Months											
	Jan	Feb	Mar	Apr	May	Jun	Jul	Aug	Sep	Oct	Nov	Dec
1989	5.27	8.29	2.59	2.72	2.01	3.28	2.30	2.58	3.49	2.51	3.96	1.76
1990	6.44	5.11	2.92	2.12	2.84	3.17	3.26	3.62	—	—	—	—
1991	—	—	—	—	2.46	3.42	3.97	3.38	4.41	6.31	1.60	1.50
1992	2.33	6.02	2.66	1.90	2.04	3.71	—	—	0.70	3.03	6.78	7.81
1993	—	—	—	—	—	—	4.01	3.44	7.19	2.52	3.24	6.30
1994	3.25	1.42	1.89	2.20	2.49	3.36	2.76	5.31	3.77	5.48	5.29	4.82
1995	1.46	5.49	2.13	1.97	2.89	2.86	2.33	2.38	0.84	3.01	9.20	8.09
1996	5.25	5.93	4.14	1.40	1.53	2.66	2.23	2.38	1.77	2.92	0.96	3.29
1997	4.68	1.16	1.76	1.65	3.14	2.85	3.25	2.66	2.29	4.63	1.60	1.04
1998	0.88	1.12	5.09	1.90	2.76	3.34	3.62	4.43	3.24	2.99	1.00	1.10
1999	1.06	1.18	1.16	4.46	2.18	3.41	5.96	2.00	6.88	1.62	2.27	2.98
2000	2.65	4.24	1.50	1.59	1.76	2.07	2.67	4.62	4.74	2.97	0.50	—
2001	—	—	—	—	—	3.60	2.22	2.62	1.81	0.45	3.03	0.95
2002	1.17	5.87	4.40	3.47	—	2.85	2.55	3.10	4.30	5.20	5.69	0.99
2003	1.57	2.63	2.50	4.07	2.22	2.35	1.93	3.08	2.18	2.95	3.10	3.40
2004	1.52	—	—	—	—	3.28	4.18	3.53	2.97	1.24	4.60	5.01
2005	2.68	1.35	1.33	2.23	2.90	3.44	3.12	3.09	0.90	1.25	1.83	0.48
2006	—	—	—	—	—	3.52	2.41	3.61	2.76	1.27	7.61	1.15
2007	6.50	3.35	3.92	1.76	2.33	2.95	1.96	3.24	3.97	1.91	—	—
2008	2.05	2.56	5.80	2.99	2.25	2.24	4.37	2.76	1.76	3.66	1.37	3.66
2009	3.35	10.04	2.39	3.73	3.28	3.00	2.47	5.91	3.12	2.91	4.84	1.00
2010	1.72	1.52	4.44	2.28	3.04	2.69	3.35	2.40	2.83	2.01	3.90	2.26
Monthly mean	2.99	3.96	2.98	2.50	2.48	3.05	3.09	3.34	3.14	2.90	3.62	3.03

Dashes (—) denote months with more than 15 days of missing data.

Table 2.8 Monthly means of wind direction (degrees) at 5-m height at Toolik LTER from 1989 to 2010

Year	Months											
	Jan	Feb	Mar	Apr	May	Jun	Jul	Aug	Sep	Oct	Nov	Dec
1989	—	—	—	—	—	254	182	179	164	137	160	168
1990	102	162	161	175	271	166	226	—	—	—	—	—
1991	—	—	—	—	317	215	157	181	167	168	196	170
1992	168	180	153	223	171	—	—	—	171	169	186	197
1993	—	—	—	—	—	—	195	235	163	151	154	141
1994	163	177	208	220	217	205	215	183	246	179	190	168
1995	126	201	170	138	198	210	228	198	210	174	261	182
1996	211	193	219	256	227	181	177	242	224	247	190	172
1997	175	172	206	217	125	231	207	143	204	174	112	117
1998	172	162	181	207	198	170	234	131	156	183	166	164
1999	191	224	243	160	241	135	183	231	181	165	162	169
2000	148	167	167	269	190	188	209	164	247	162	—	—
2001	—	—	—	—	—	119	199	219	211	147	164	85
2002	118	186	173	98	—	250	202	196	183	180	204	182
2003	176	166	171	285	152	219	206	278	186	228	180	192
2004	—	—	—	—	—	130	276	197	232	175	188	186
2005	197	246	162	191	242	115	147	64	195	168	282	138
2006	—	—	—	—	—	213	222	49	198	125	183	191
2007	190	236	172	226	308	160	200	112	175	—	—	—
2008	199	172	188	69	249	190	274	153	187	178	251	179
2009	175	194	134	179	190	110	195	—	—	—	—	—
2010	—	—	—	—	—	—	223	175	124	173	176	258
Monthly mean	167	189	180	194	220	182	208	175	191	173	189	170

Dashes (—) denote months with more than 15 days of missing data.

the Beaufort Sea pressure high that dominates during the warm season) and smaller-scale processes such as katabatic winds from the mountains or differential heat fluxes from the ocean and land that can result in a pressure gradient and corresponding winds.

Recent Temperature and Precipitation Trends at Toolik

Over the latter half of the twentieth century, arctic air temperatures increased 2°C–3°C and the majority of this warming was in winter. Subsurface warming was also observed in borehole measurements. There is evidence of both SAT warming and snow-depth increases in the long-term record at Barrow (ACIA 2004). Trends during the early 21st century have continued upward at the same approximate rate for winter and autumn, but temperatures have cooled in spring at Barrow (author's analysis from NWS data). Taken as a whole, the North Slope has been the site of rapid climate change during the past few decades and near-surface air temperatures have warmed more than 3°C over the past 60 years (Shulski and Wendler 2007).

In contrast to the warming observed in many arctic regions during the past few decades, a linear trend analysis of the LTER datasets from 1989 to 2010 reveals no significant trend in SAT 5 m (Figure 2.4) or precipitation (Figure 2.5). This is largely an artifact of the shorter period of the Toolik record than of the Barrow record; the Toolik record is consistent with the Barrow temperature record over their period of overlap. Barrow shows a long-term warming trend over its 110-year record. The air temperature sensor used at Toolik, and its lower boundary temperature, have also changed over time. Only 20 km to the south at Galbraith Lake, permafrost temperatures at a depth of 20 m in a borehole have increased significantly by up to 0.8°C over the past 20 years (Smith et al. 2010), consistent with long-term warming in the area. It should be noted that the near-surface air temperature measurements are quite variable from year to year; therefore, small changes are difficult to discern at a decadal scale. On the other hand, the permafrost temperature changes measured at 20 m depth are by their nature an integrated measure of climate over time. In fact, measurements are taken at 20 m because at this depth the seasonal changes are damped out, and it is easier to determine small changes in the trend of climate.

In this book, several other indicators of a warming climate in the vicinity of the Arctic LTER site at Toolik Lake are presented. These are in addition to the warming permafrost temperatures and include changes in stream and lake chemistry, indicating an increase in weathering, and changes in the amount of plant biomass (chapters 5, 6, and 10).

Analyses of Historical Climate Data from the Kuparuk River Basin

Spatial Patterns of Meteorological Parameters

A climatology of observed meteorological parameters for the Arctic LTER site and the KRB over the period 1989–2010 used all available stations listed in Table 2.1.

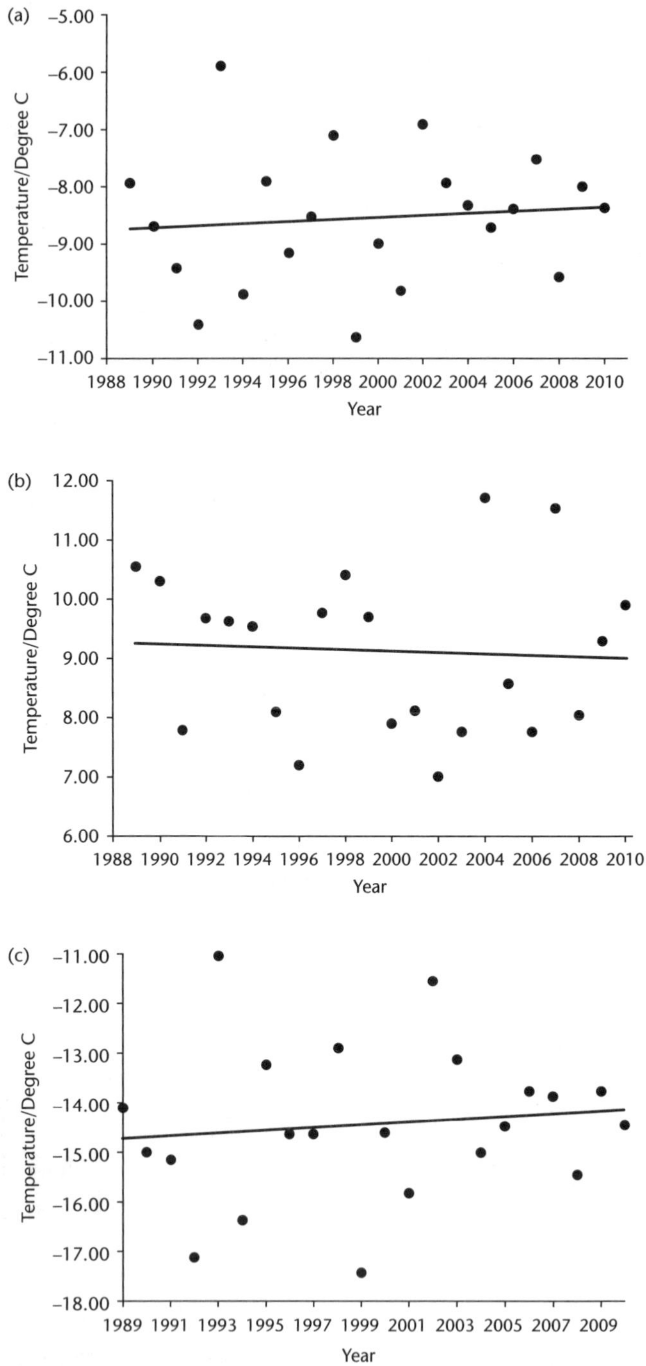

Figure 2.4 (A) Annual mean Toolik LTER surface air temperature (SAT) at 1-m height from 1989 to 2010; $R^2 = 0.009$, $p = 0.66$ fails student's t-test at confidence interval (CI) = 95%. (B) Mean summer Toolik LTER SAT 1 m from 1989 to 2010; $R^2 = 0.003$, $p = 0.79$ fails t-test at CI = 95%. (C) Mean winter Toolik LTER SAT 1 m from 1989 to 2010; $R^2 = 0.013$, $p = 0.61$ fails t-test at CI = 95%.

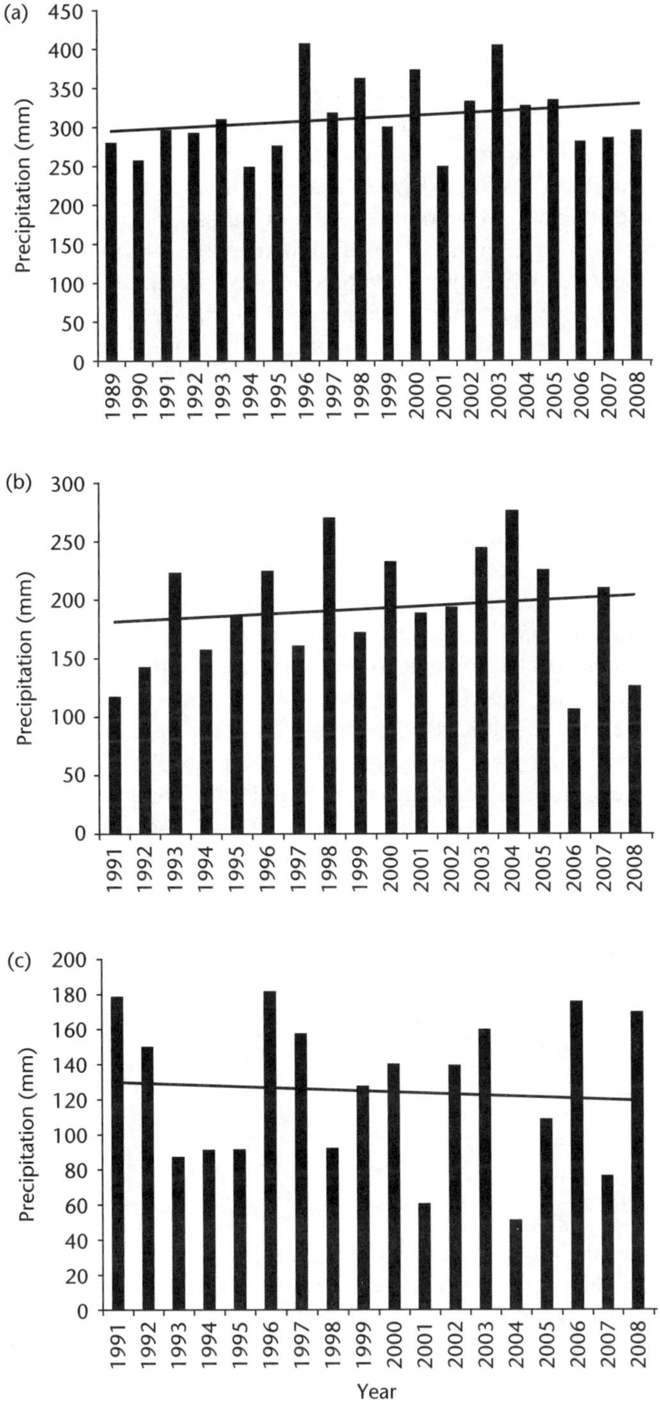

Figure 2.5 (A) Annual mean Toolik LTER precipitation (mm) from 1989 to 2008; $R^2 = 0.053$, $p = 0.33$ fails student's t-test at confidence interval (CI) = 95%. (B) Mean summer precipitation (mm) from 1989 to 2008; $R^2 = 0.021$, $p = 0.25$ fails t-test at CI = 95%. (C) Mean winter precipitation (mm) from 1989 to 2008; $R^2 = 0.006$, $p = 0.77$ fails t-test at CI = 95%.

Hourly meteorological data recorded at the stations were summed or averaged to provide monthly or annual climatologies of air temperature, relative humidity, atmospheric pressure, rainfall, wind speed, and wind direction. Missing data were filled in by linearly weighting the corresponding meteorological parameters from neighboring stations inversely to the square of distance between stations. Those stations listed in Table 2.1 with nonexistent or missing data at the same time did not contribute to the weighting scheme and more emphasis was given to other locations. This yields a continuous, hourly time series of meteorological parameters for the stations that were then summed or averaged to provide monthly or annual totals and means, which were used to construct spatial distribution maps.

Wind speeds measured 5 m above the surface (U_5) are adjusted to a standard reference height of 10 m (U_{10}) through the log-law wind relationship (Oke 1987), such that:

$$U_{10} = U_5 \ln(z_{10}/z_0)/\ln(z_5/z_0) = 1.08U_5$$

where z_{10} = 10 m and z_5 = 5 m above the surface, and where z_0 (taken here as 0.001 m) denotes the aerodynamic roughness length. Thus the adjusted wind speeds may be compared to those recorded at similar heights throughout the KRB.

Obtaining accurate records of precipitation on the North Slope of Alaska remains an arduous task. Gauge undercatch, frequent trace events, and other complicating factors lead to large uncertainties in precipitation measurements in the region, especially for snowfall. Fortunately, there exists a substantial network of rainfall gauges across the KRB to estimate liquid precipitation amounts. For the purpose of this study, precipitation recorded when $T > 0°C$ is considered as rainfall. There are sparse measurements of snowfall such that a gauge-based estimate of solid precipitation is not readily available and will be discussed further in the snow subsection.

An interpolation scheme was adopted to construct spatially complete data over the entire domain of the KRB. A grid with 0.25° × 0.25° horizontal resolution covering 68°N to 71°N and 148°W to 151°W was first developed. Time series of meteorological data at each grid point were interpolated from all available stations using an inverse least squares approach as described previously. This provides the time series to construct the climatology for the KRB.

Results of Analyses

Air temperature

The mean annual air temperature at Toolik, based on the 1989–2010 data, was relatively warmer than that in regions closer to the Arctic Ocean (Figure 2.6). There was a range of about 3.5°C in mean annual air temperatures from north to south across the KRB, equivalent to a gradient of about 1.5°C per degree of latitude. This temperature gradient can be attributed to factors such as incoming solar radiation, height above sea level of the station, and distance to the Arctic Ocean, which are latitudinally dependent.

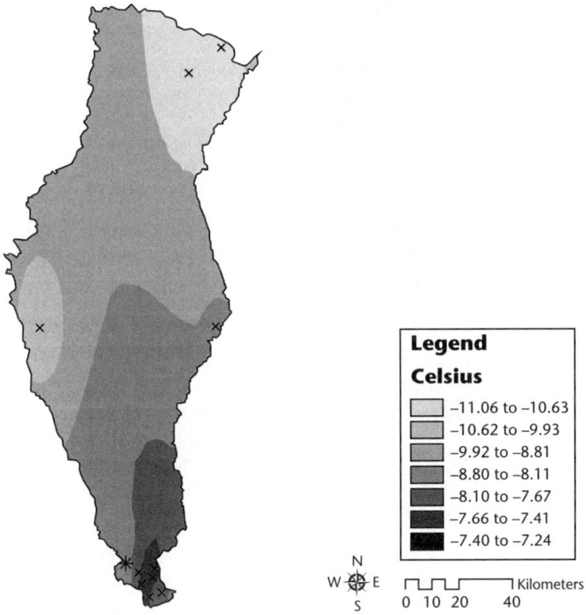

Figure 2.6 Mean annual isotherms (°C) in the Kuparuk River basin (bold outline), 1989–2012. The Toolik LTER is denoted by "*" and other meteorological stations by "×."

Relative Humidity

The relative humidity (RH) at Toolik was near 75% on average. Its values show not only a pronounced annual cycle but also a strong diurnal one in summer. On a monthly average, the highest relative humidity (85–87%) occurred early in the morning (0400 local time) during summer, whereas the driest conditions (RH = 64–67%) occurred at any time of day during January or in late afternoons (1600 local time) in summer. There was a modest gradient in mean annual RH across the KRB from about 75% in the Upper Kuparuk to 85% at the coast.

Precipitation

Similar to mean annual air temperatures, mean annual precipitation shows a strong latitudinal gradient ranging from about 130 mm yr^{-1} along the arctic coast and increasing southward to reach up to 380 mm yr^{-1} (Figure 2.7A). However, few stations report precipitation year round and these numbers are highly uncertain. More stations report summer rainfall; these numbers range from 50 mm yr^{-1} along the coast to 240 mm yr^{-1} closer to Toolik (Figure 2.7B). This sharp gradient in precipitation may be attributed in part to orographic effects of the Brooks Range, which enhance rainfall by mechanical lifting of air masses over the mountains.

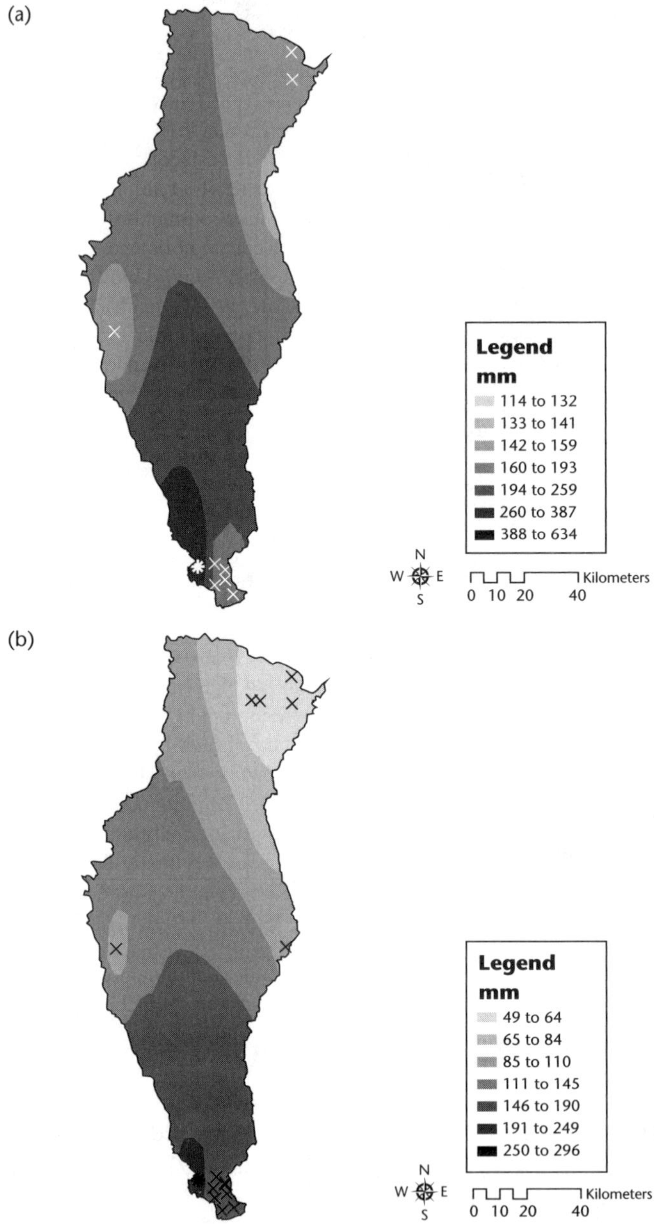

Figure 2.7 Mean annual (A) and mean summer (B) isohyets (lines of constant precipitation, mm yr^{-1}) in the Kuparuk River basin (bold outline), 1989–2010. The Arctic LTER is denoted by "*" and other meteorological stations by "×."

Winds

The North Slope of Alaska is subjected to strong winds during the entire year; mean monthly 10-m height wind speeds peaked near 3.8 m s^{-1} in both summer and winter (Table 2.9). Peak winds registered at Toolik were in excess of 25 m s^{-1} and occurred during intense winter storms. A frequency histogram of wind speeds illustrates the strong tendency for winds in the range of 1–4 m s^{-1}, with a declining probability of stronger winds (Figure 2.8). Calm conditions (wind speeds <1 m s^{-1}) occurred approximately 10% of the time at Toolik. Mean annual wind speeds increased from about 3.3 m s^{-1} in the southern parts of the KRB to values >4.0 m s^{-1} along the arctic coast (Figure 2.9).

Winds are predominantly from the south (~180°) throughout the year except during summer when winds exhibit a more northwesterly component (Table 2.9), making the region somewhat different than Toolik. Regions closer to the coast experience winds preferentially from the southeast compared to those further inland that experience southerly flows (Figure 2.10).

Water Balance Calculations

Water balance over time is a particularly interesting measure of change in the hydroclimate system. The water budget for the KRB may be expressed as:

$$P = R + E + \Delta S$$

where P is precipitation, R is river runoff, E is evapotranspiration, and ΔS is a change in storage (positive for a net gain of water). The evapotranspiration rate can be further partitioned into three distinct components: evapotranspiration with snow-free conditions (including bare soils, lakes, and vegetation), surface sublimation from the snowpack, and wind-related sublimation during periods of blowing snow. All terms are expressed here in units of mm yr^{-1}. McNamara et al. (1998) provide values of P = 262 mm yr^{-1} and of R = 139 mm yr^{-1} based on observations collected from 1993 to 1995 and the USGS gauge at the mouth of the Kuparuk. If no change in soil moisture or subsurface storage (i.e., $\Delta S = 0$) is assumed over this time period, this implies that E = 123 mm yr^{-1}. Kane et al. (2004) report these numbers to be P = 359 mm yr^{-1}, R = 181 mm yr^{-1}, E = 178 mm yr^{-1} for the Imnavait Creek watershed for 1985–2003 and assume no change in storage. For the KRB from 1991 to 2001, Déry et al. (2005) report P = 293 mm yr^{-1} (including 130 mm yr^{-1} of snowfall in SWE) and R = 159 mm yr^{-1} for the Kuparuk River. They also calculated model-based estimates of R = 169 mm yr^{-1}, E = 127 mm yr^{-1} and change in storage (ΔS) of 1 mm yr^{-1}, over the same period 1991–2001. More recently, investigators are using gravity-based measurements to look at changes in groundwater storage. Muskett and Romanovsky (2011) using data from the GRACE satellite report increases in groundwater storage of about 1.15 km^3 yr^{-1} for the period 1999–2009 for the entire arctic coastal plain.

Table 2.9 Climatic norms for the Kuparuk River basin (including the LTER) from 1989 to 2010 for temperature (T), atmospheric pressure (P), relative humidity (RH), wind speed (WS) and wind direction (WD)

Data	Jan.	Feb.	Mar.	Apr.	May	June	July	Aug.	Sep.	Oct.	Nov.	Dec.	Year
Mean T (°C)	−23.6	−21.7	−21.9	−12.3	−2.1	7.7	9.9	6.6	0.0	−10.4	−17.4	−21.2	−8.7
Mean Max.T (°C)	−13.6	−6.9	−10.2	−4.7	5.6	14.1	16.1	13.4	5.1	−3.3	−4.4	−13.4	−5.8
Mean Min. T (°C)	−34.3	−34.6	−30.9	−20.8	−10.4	0.4	3.0	1.8	−9.8	−18.7	−28.2	−30.7	−13.0
Record Highest T (°C) and Year	2.0 (2007)	2.3 (2007)	2.9 (1990)	8.0 (2005)	16.8 (1990)	22.6 (1991)	24.7 (1993)	23.5 (1994)	18.1 (1989)	11.6 (2003)	5.1 (1996)	0.8 (2001)	24.7 (1993)
Record Lowest T (°C) and Year	−55.6 (1989)	−46.1 (1992)	−47.3 (1992)	−41.8 (1992)	−26.6 (1992)	−9.7 (2006)	−4.0 (2002)	−8.0 (1996)	−25.4 (1992)	−32.5 (1996)	−39.1 (1991)	−41.9 (1995)	−55.6 (1989)
Atm. P (hPa)	919.0	920.1	921.9	922.0	925.3	926.1	925.9	922.4	923.2	919.5	918.0	917.7	920.0
RH (%) at Hour 04	75.6	75.8	75.6	83.6	90.2	90.8	90.9	91.5	91.1	87.1	81.2	77.6	84.5
RH (%) at Hour 10	75.7	75.7	75.4	81.1	84.6	79.7	80.0	84.1	87.9	87.0	81.3	77.6	81.6
RH (%) at Hour 16	75.6	75.8	75.4	79.9	82.2	74.0	72.3	76.1	81.8	85.7	81.2	77.6	79.0
RH (%) at Hour 22	75.7	75.9	76.1	84.1	88.0	82.6	81.9	87.4	90.2	87.1	81.2	77.6	82.9
Rainfall (mm)	0.0	0.0	0.0	0.0	7.9	34.0	65.2	44.9	17.3	0.0	0.0	0.0	148.8
WS ($m\ s^{-1}$)	2.8	3.2	3.2	3.2	3.6	3.5	3.5	3.3	3.3	3.1	3.0	2.6	3.2
WD (degrees)	267.2	286.3	276.0	296.8	18.2	58.9	59.6	49.1	316.1	289.0	283.9	284.9	133.1
Peak WS ($m\ s^{-1}$), WD (degrees), and Year	18.213 (2002)	15.0178 (1989)	14.715 (2005)	21.714 (2002)	16.8 28 (1991)	11.4 74 (2007)	11.6 34 (1998)	11.530 (1997)	13.117 (1995)	17.0 65 (2006)	17.64 (1999)	16.4 65 (2007)	21.714 (2002)

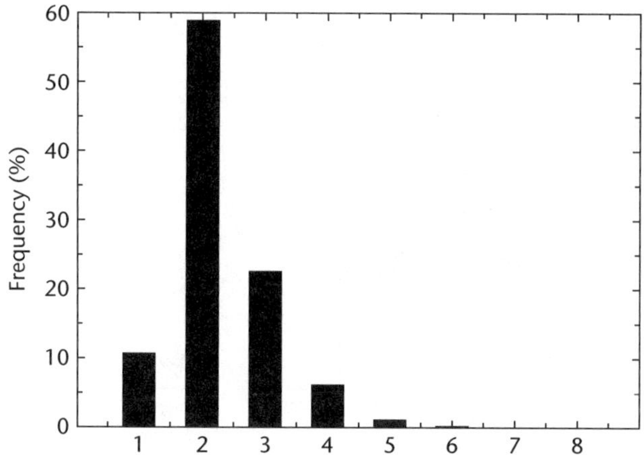

Figure 2.8 Frequency histogram of wind speeds observed at the Arctic LTER, 1989–2010. The wind speed categories are as follows: 1 = U < 1 m s^{-1}; 2 = 1 m s^{-1} < U < 4 m s^{-1}; 3 = 4 m s^{-1} < U < 7 m s^{-1}; 4 = 7 m s^{-1} < U < 10 m s^{-1}; 5 = 10 m s^{-1} < U < 13 m s^{-1}; 6 = 13 m s^{-1} < U < 16 m s^{-1}; 7 = 16 m s^{-1} < U < 19 m s^{-1}; 8 = U > 19 m s^{-1}.

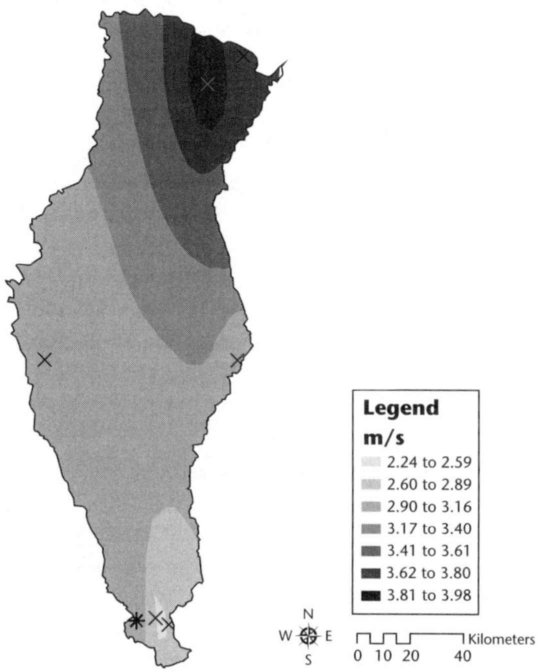

Figure 2.9 Mean annual isotachs (lines of constant windspeed, m s^{-1}) in the Kuparuk River basin (bold outline). The Arctic LTER is denoted by "*" and other meteorological stations by "x".

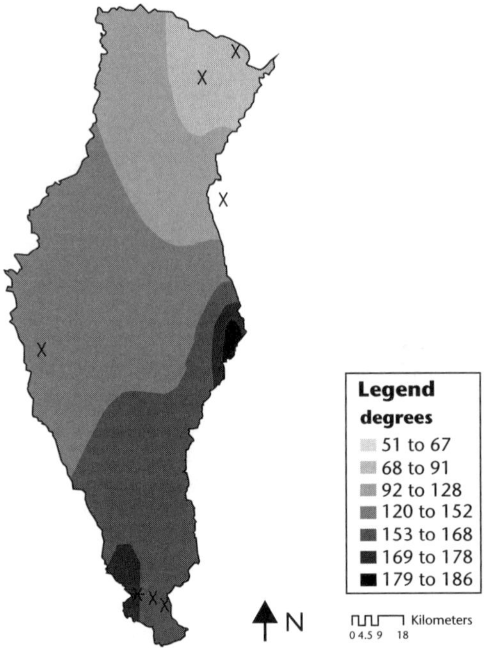

Figure 2.10 Mean annual isogons (lines of constant wind direction, °) in the Kuparuk River basin (bold outline). The Arctic LTER is denoted by "*" and other meteorological stations by "x".

Spatial Variability in Temperature and Precipitation

Determining spatial variability in the Arctic is difficult because strong microclimates can exist even in gently sloping topography, due in part to the extreme angles of the sun. Sparse data throughout the Arctic are another challenge to estimating spatial variability. Figures 2.6–2.10 show spatial variability on the scale of the KRB. Relative to the pan-Arctic, Alaska is colder than Scandinavia in winter but not as cold as central Siberia. In summer, there is less spatial variability across the Arctic and sub-Arctic, but a more maritime climate prevails in the coastal areas as sea ice retreats to the high Arctic. In general, Alaska receives less precipitation than Scandinavia but more than central Siberia.

There is an entirely different scale of spatial variability over Alaska (Figures 2.11 and 2.12). These plots are climatologies for the period 1961–1990 and are the output from an analysis (combination of station and model data) produced by the PRISM Climate Group at Oregon State University and made available by the Spatial Climate Data Analysis Service. This statistical methodology for interpolating station and model data over complex topography is described in Daly et al. (1994) and Simpson et al. (2005). The mean annual precipitation on a 4-km grid shows a dominant north-south gradient, as also happens at the KRB and pan-Arctic scales, but with a more obvious influence of topography and proximity to the ocean (Figure 2.11). The North Slope shows considerably less variability because the variation in the topography is smaller than the grid cell and data here are very sparse (WERC sites were not assimilated).

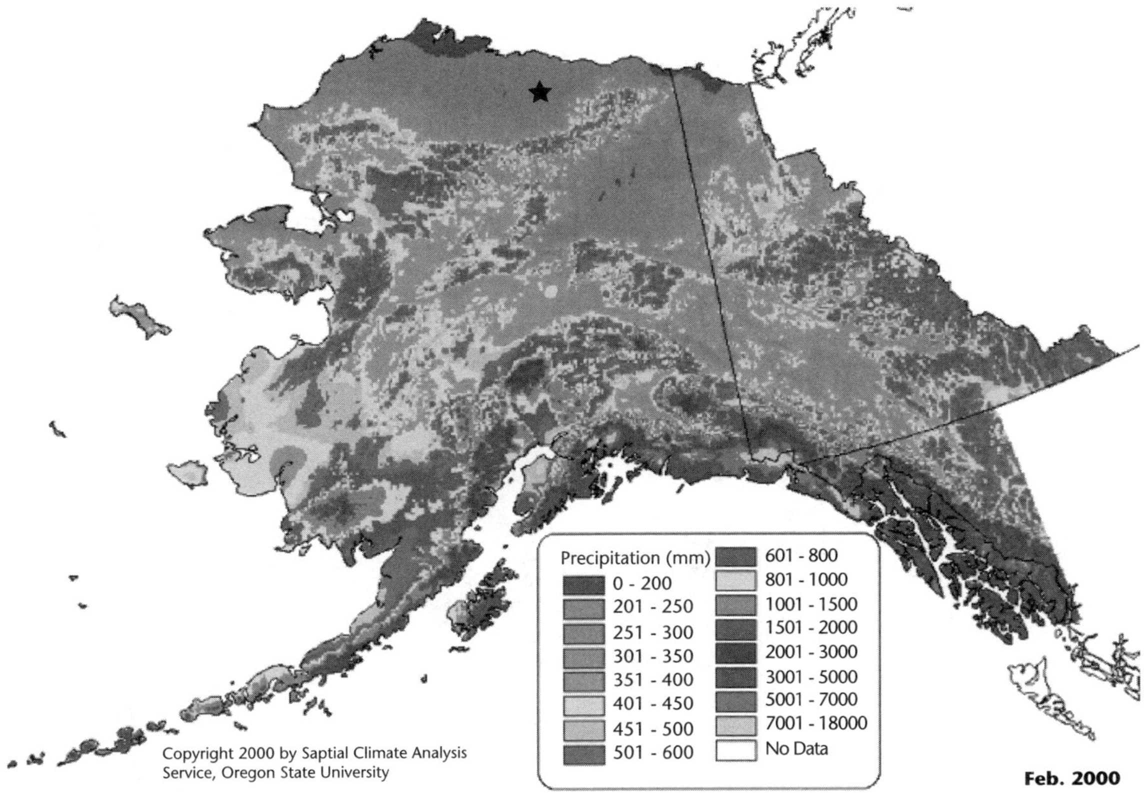

Figure 2.11 Mean annual precipitation for 1961–2000. Approximate location of the Toolik Field Station has been marked with a star. Used with permission. Copyright © 2000, PRISM Climate Group, Oregon State University. [REFER COLOR MAP]

A small deviation can be seen at Imnavait Creek where the data from the NRCS Wyoming snow gauge were assimilated. Toolik and the North Slope have a cold continental climate during winter, similar to that of the ice-locked coast, but a much warmer climate during summer, quite different than that of the arctic coast (Figure 2.12).

In-Depth Analysis of Snowfall and Snow Cover of the Kuparuk River Basin

Snow-Climate Interactions

Snow seasonally covers up to 40% of the Northern Hemisphere landmass and constitutes the most prominent transient feature across continental surfaces. With its high albedo and high thermal emissivity, snow cover leads to a suppression of near-surface air temperatures. The low thermal conductivity of snow insulates the underlying ground from the relatively cold atmosphere. Through surface and blowing sublimation processes, a snow cover constitutes a sink of energy near its surface while providing a source of atmospheric moisture (Déry et al. 1998). During the spring transition period, snow retained in this temporary reservoir is quickly released, yielding high runoff rates that contribute up to 80% of the yearly discharge of some arctic streams and rivers. Snow also has significant interactions with biogeochemical cycles and vegetation, which are discussed throughout this book.

Spatial variability in snow cover is found on regional (10 km to 1000 km), local (100 m to 10 km), and small (1 cm to 100 m) scales (McKay and Gray 1981; Pomeroy and Gray 1995). Atmospheric processes and patterns that govern storm tracks and the formation of solid precipitation determine the spatial distribution of snow on a regional scale (Stewart et al. 1998). Altitudinal control over precipitation type and intensity (by enhanced orographic lifting), in addition to ambient air temperatures, contribute to regional and local nonuniformity in the snow cover. At local scales, the elevation, slope, and aspect regulate the amount of incoming radiation available for melt. Snow redistribution during high-wind events further enhances the heterogeneous aspect of the snow cover at local to small scales. The presence of trees and shrubs can also influence snow depth at small scales.

Measurement Problems

Of all climatic variables, snowfall may be the most difficult to measure well. The simplest bucket gauge and the most sophisticated shielded gauge share similar systematic biases because they disrupt the normal flow of snow in the atmospheric boundary layer, and snow preferentially falls away from the gauge. In addition, there are many unsystematic biases including unrepresentative gauge siting, mechanical failures, and snow freezing around the gauge (Goodison et al. 1998).

Previous efforts to estimate "true" precipitation, liquid and frozen, include correcting gauges according to the amount of precipitation they catch relative to a well-tested reference gauge or a pit. This correction is typically a regression function based on meteorological parameters such as wind speed and temperature and

(a)

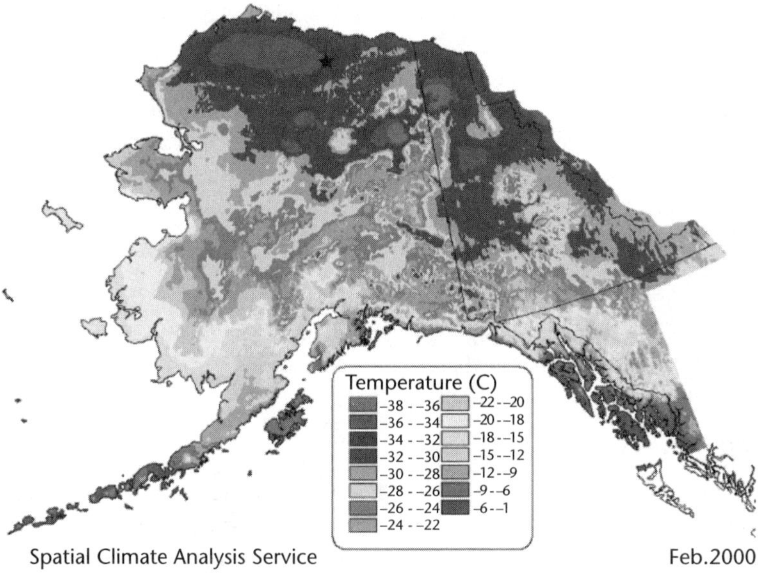

(b)

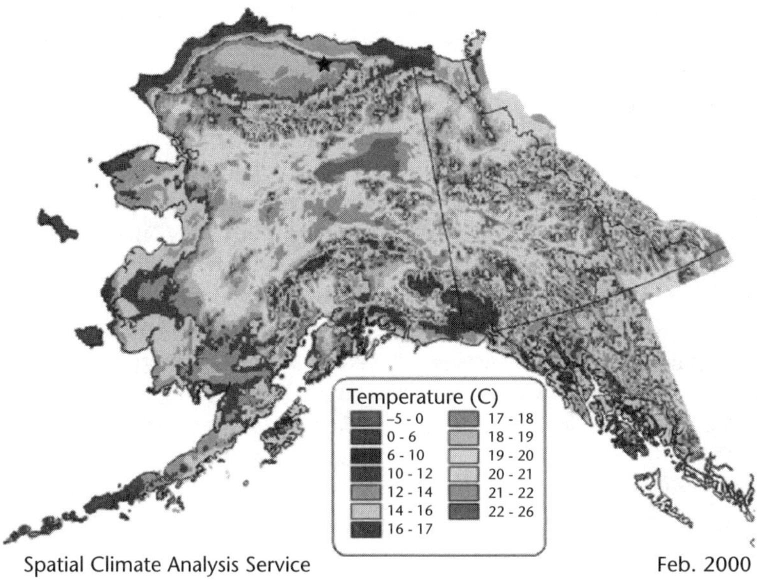

Figure 2.12 (A) January mean minimum temperature for 1961–2000. (B) July mean maximum temperature for 1961–2000. Approximate location of the Toolik Field Station has been marked with a star. All used with permission. Copyright © 2000, PRISM Climate Group, Oregon State University. [REFER COLOR MAP]

has shown gauge errors of up to 200% for snowfall (Yang et al. 2005). Biases for liquid precipitation are smaller by an order of magnitude and will not be addressed here. Adam and Lettenmaier (2003) published an excellent summary of gauge corrections. Problems associated with the statistical approach may be that it does not account for gauge siting and other unsystematic biases (Cherry et al. 2005).

Other efforts to estimate solid and liquid precipitation include radar measurements, merged gauge-satellite analysis products, budgetary approaches, and global climate re-analyses. Disadvantages to these approaches include a relatively short satellite era, errors associated with the indirect measurement of precipitation from satellites, inconsistent levels of assimilated observations in the re-analyses, and errors associated with interpolating or assimilating sparse observations of a heterogeneous substance to large grid cells.

Spring Snowmelt

The emphasis of both the WERC and the NRCS surveys has been to capture the timing and water equivalent of snowmelt during the spring. There are scientific as well as practical reasons for this. The timing of snowmelt is critical for the start of the growing season, represents a regime shift in land-atmosphere coupling, and is strongly controlled by the temperature of regional air masses and the radiation budget at the surface. When the snow pack contains liquid water that evaporates and condenses in clouds, downwelling longwave radiation is a particularly powerful mechanism for snow ablation (lead author's field observations). The melting of snow and breakup of lakes and rivers also starts the active hydrologic season in which there is far more exchange of heat and moisture between the land, atmosphere, and ocean than in the winter. The other reason for emphasizing observations during snowmelt is that manual snow surveys are costly and time consuming. Because the active hydrologic season starts in the spring, it is simply the most efficient time to focus observational efforts.

A time series of the day of the year since 1985 when the average snow depth in Imnavait basin goes to zero (i.e., melts completely) is shown in Figure 2.13. While average surface air temperatures on the North Slope have certainly increased over the past twenty years, there is not a strong trend in the timing of snowmelt. This plot shows a modest trend toward later snowmelt (caused by cooler springs or deeper snow), but this trend is not statistically significant. Figure 2.14 shows snow pack depletion curves for Imnavait basin since 1985. While the final spring snow depth maximum is not always captured, in general deep snow packs tend to be associated with later snowmelt years.

Estimates of Annual Snowfall over Time

Because of the challenges of measuring solid precipitation, particularly in regions of high winds, the snowfall budget on the North Slope is poorly constrained. The NRCS's Snotel Station network includes six stations in northern Alaska: (south to north) Gobblers Knob, Coldfoot, Atigun Pass, Imnavait, Sagwon River, and Prudhoe Bay. Wyoming snow gauges at these stations may have less undercatch than typical

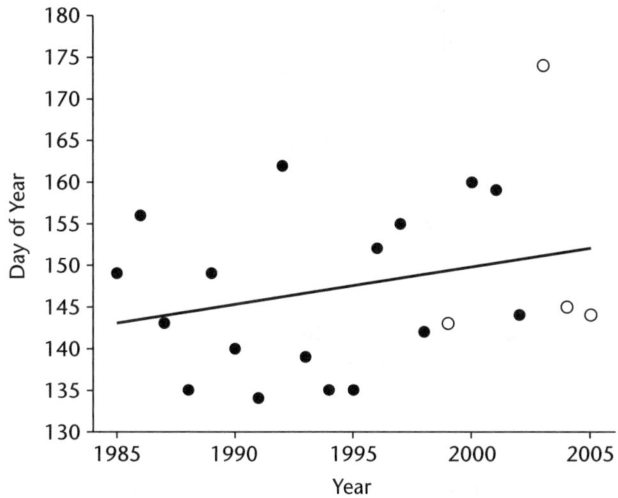

Figure 2.13 The day of year on which snow melted away to bare ground in the Imnavait basin. Data are from D. Kane and L. Hinzman (personal communication). Closed circles are end-of-snow season dates that were observed precisely. Open circles are inferred from observations taken within a few days of final melt. While these data show a trend of +0.45 days later melt per year, it is not statistically significant ($R^2 = 0.26$, $p = 0.25$ fails student's t-test at confidence interval = 90%).

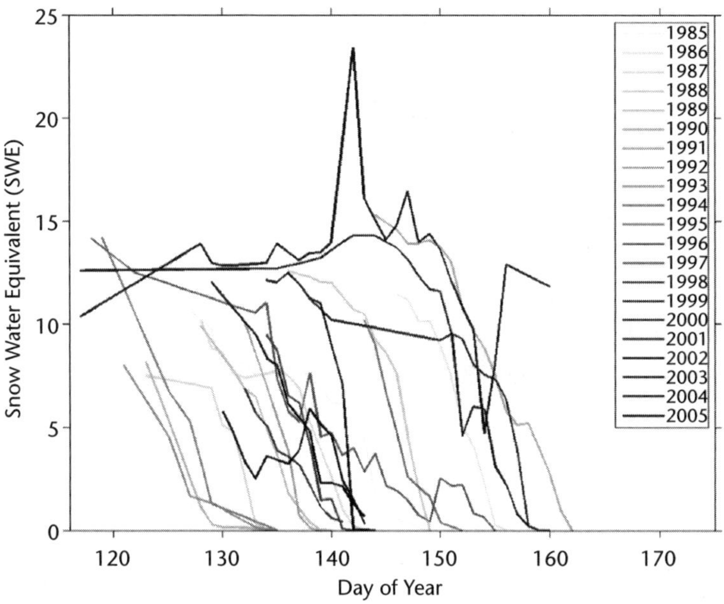

Figure 2.14 Snowpack depletion curves (snow water equivalent versus day of year) for the Imnavait basin. Data are from Kane and Hinzman (2007). These basin averages are calculated from fixed transects.

precipitation gauges, but they are still known to have losses of up to 70% or more. In the same way as for liquid precipitation, moisture appears to be transported from south to north. Unfortunately, the data record is less than a decade long for all but three of these stations, and there are several missing years. However, analysis of the time series suggests that Imnavait is the only station to show an upward trend, although Atigun, the next station to the south, shows a slight decrease in total precipitation over the same period. These data do not have a correction factor applied, which requires a wind measurement that is all too often missing because of mechanical failures or animal disturbances. Correction factors for solid precipitation at National Weather Service stations on the North Slope vary from about 1.00 (no correction) to about 1.75 according to Yang et al. (2005), but only for periods when the anemometer was functional and only when calculated over a year or less.

Snow Redistribution by Wind

While snowfall is difficult to estimate, the redistribution and sublimation of snow is even more difficult to measure. For the KRB, Liston and Sturm (2002) used a blowing snow model and end-of-winter snow depth observations to determine the sublimated losses. Expressed as the fraction of winter precipitation returned to the atmosphere by blowing-snow sublimation, they estimated ~15% in the south (near Toolik) up to 34% in the north, near the coast, and a basin average of 21%. These results are based on three years of observations (1995–1997) including snow depth, water equivalent, and interpolated meteorological observations such as wind. Déry and Yau (2002) took a different approach for the pan-Arctic. They looked at the large-scale mass balance effects of both blowing snow and surface sublimation and used the surface wind, temperature, and humidity fields from the European Centre for Medium-Range Weather Forecasts (ECMWF) Re-Analysis project for 1979–1993. They estimated sublimation of 25–50 mm (SWE) per year, of which 2–5 mm or ~10% is due to blowing snow. From the wind fields they calculated a mass flux of snow to be 1–5 mm, or no more than 10% of the total sublimation. This mass flux increased from south to north. Both studies suggest that the blowing-snow sublimation had a much bigger influence over the water budget on the medium to large scale than does redistribution by wind.

Spatial Variability of Snow

Another challenge to the observation of snow is its tremendous variability over all spatial scales. Measurements collected from a transect in the Imnavait basin by the WERC in 2007 show snow depth variation on scales from 1 cm to 1 km. Modest depth differences are seen over the small scale, representative of tussock-scale topographic change up to the significant differences in distribution even in this gently sloping landscape. Liston and Sturm (2002) show large variability in depth distributions from 100 m to larger spatial scales, driven by larger-scale topography and vegetation trapping. Representative sampling in a data-sparse region is particularly difficult. The depth sampling that occurs during the WERC snowmelt sampling includes averaging of 100 measurements, 1 m apart, split between two transects, 90 degrees apart. Density shows less variability and is time consuming to measure, so five measurements are taken at each site and averaged. Berezovskaya et al. (2008)

describe average snow density values ranging from 214 kg m^{-3} (in the mountains) to 248 kg m^{-3} (on the coast) during that year's survey. While these surveys and others have expanded in recent years, the long-term observations of snow depth on the North Slope are generally limited to the small (2.2 km^2) Imnavait basin. As the North Slope is approximately the size of Nebraska, there is likely to be large-scale variability that is simply not captured by the existing observational programs.

Remote sensing products derived from MODIS and AMSR-E have much greater spatial coverage. Their limitation is that the field of view is relatively large (100s of meters for MODIS and 1000s of meters for AMSR-E). For MODIS, the time series of snow-covered area (SCA) is short and it depends on visible wavelengths (of no use in the Arctic during midwinter or on cloudy days); for AMSR-E, there are known to be large errors in the estimate of SWE and it is no longer operating at the time of writing. Whenever comparing data from a pixel or grid cell with a ground-truth station, it is important to take into account that, with few exceptions, meteorological stations are representative of an area much smaller than the typical pixel or grid cell. Comparisons of SCA over space and time for (1) the pixel nearest the Arctic LTER site, (2) the North Slope, and (3) Alaska for both May and October for 1982–2006 show no significant trends over time. These data are from remote sensing-based snow maps digitized by D. Robinson and gridded by Armstrong and Brodzik (2005). However, the 25-km pixel resolution does not capture the small-scale melt initiation or possible shifts in ablation time on the order of days. This can be seen in the Kane and Hinzman data record from Imnavait basin (Figures 2.13 and 2.14). On the other hand, records such as the one at Imnavait may not capture the large-scale regional trends evidenced in the satellite-based records.

Recent Climate Trends of the Kuparuk River Basin

The magnitude and spatial extent of high latitude warming during the past century is well documented and expected to continue into the next century. Over the last half-century, air temperatures in the Arctic have increased 2°C–3°C; the majority of this warming has been winter warming. In many arctic regions this warming is associated with increased precipitation, a longer growing season, a change in the distribution of plant species, an increase in the spatial extent of woody species, and higher aboveground photosynthetic rates. Subsurface warming has been observed in borehole measurements (Figure 2.15). The impact of climate changes on ground temperatures is nonlinear; for example, as demonstrated by Stieglitz et al. (2003), changes in ground temperature are not a direct expression of changes in the overlying air temperatures but also depend on changes in snow depth. Their study shows that changes in the permafrost temperatures from 1983 to 1998 on the North Slope of Alaska are consistent with decadal-scale variability in snow cover. Specifically, increase in snow cover contributes as much to the increase in permafrost temperature as does the increase in near-surface air temperature. Trends in arctic cloudiness and shortwave radiation observed from satellites show increased cloudiness in spring and summer, decreased cloudiness in winter, and decreased surface albedo during all seasons with the strongest decreases in fall and winter. Changes in cloudiness

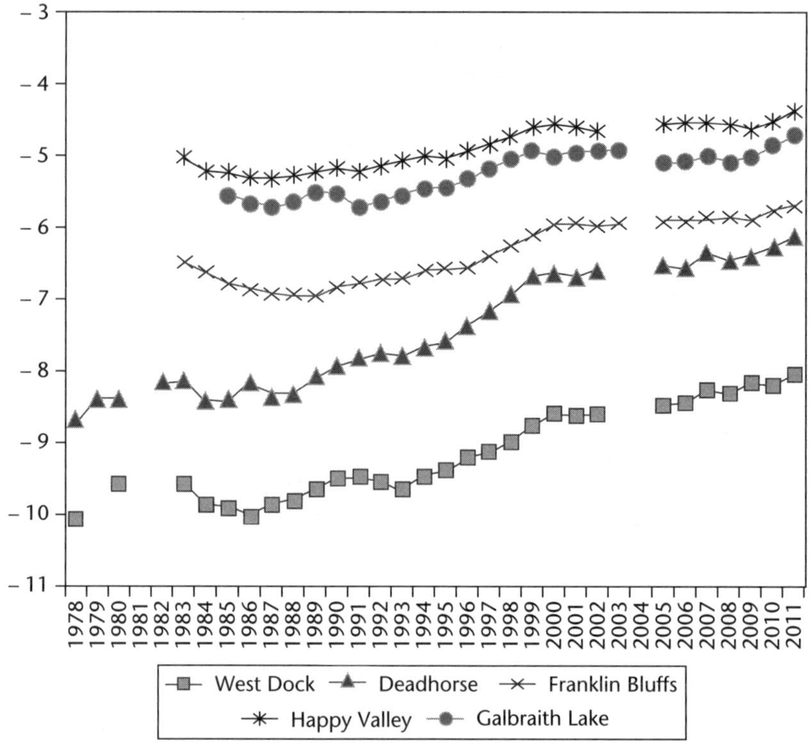

Figure 2.15 Time series of permafrost temperatures (°C) measured in Alaska showing warming trends between 1983 and 2011. Measurements are at 20-m depths at several sites across the continuous permafrost zone. The Arctic LTER lies between Happy Valley (to the north) and Galbraith Lake (to the south). Source: Romanovsky and Osterkamp (2012).

are associated with increased cooling in the summer, fall, and winter, suggesting that clouds in these seasons may be damping warming trends (Wang and Key 2003).

In contrast to ground and air temperatures, changes in the distributions of precipitation are less clear. White et al. (2007) show from satellite images that there were increases in precipitation in most but not all regions. Sparse ground observations in the Arctic lead to considerable uncertainty about the amount and distribution of precipitation. The problem of sparse observations is further exacerbated when station records are interpolated to fit a grid. Much of the uncertainty regarding precipitation and other arctic freshwater trends is tied to the difficulty in measuring solid precipitation. Furthermore, according to estimates by Aagaard and Carmack (1989), the pan-Arctic landmass is the single greatest contributor to the arctic freshwater budget. The uncertainty associated with arctic precipitation makes it difficult to interpret observed changes in surface runoff as being caused by changes in subsurface storage of water, land use, or precipitation distribution.

A confounding factor for attribution of trends in the Arctic, and in the KRB in particular, is the strong influence of large-scale modes of variability such as the Pacific decadal oscillation (PDO), the Arctic oscillation (AO), and the El Niño southern oscillation (ENSO). Of these modes, the PDO has the most dominant impact on this region (Papineau 2001) and its effects are strongest in winter. Few records in arctic Alaska capture more than two or three decadal shifts in the PDO, making the distinction between natural climate variability and long-term change difficult to make. This highlights the importance of long-term monitoring and is discussed further in the final section.

Climate Projections for the Kuparuk River Basin

General circulation models (GCMs) have been used to conduct climate simulations spanning the period 1800–2100 by the Intergovernmental Panel on Climate Change (IPCC). First, a control integration simulation using present-day CO_2 concentrations was conducted to evaluate the ability of the models to simulate today's climate and to determine the stability of the simulated climate. Thereafter, a warming simulation was conducted using an effective greenhouse forcing corresponding to that observed from 1900 to 1990 and a forcing thereafter which corresponds to an increase of CO_2 along particular economic development pathways known as emissions scenarios (IPCC 2007; ACIA 2004). While mean annual air temperatures are expected to increase by 4ºC–6ºC, most of the warming is projected to occur during the winter. In coastal regions, nearby ice-free ocean conditions that prevail in summer months act to diminish the overriding warming through the high thermal inertia of the ocean. A decrease in the interannual variability of air temperature seems to be associated with an enhanced period of ice-free ocean conditions. With higher annual air temperatures, the atmospheric water holding capacity increases and precipitation increases by 10% over the next 100 years. While some GCM simulations do produce slightly more warming, the drying effects of enhanced warming are offset by enhanced precipitation. All simulations, however, yield more winter warming than summer warming, as has been observed.

To determine what the long-term climate projections are for the Arctic LTER site and the surrounding KRB, the output from five GCMs under the B1 emissions scenario (SRES) was analyzed. Spatial distributions, climatologies, and projected time series were plotted for surface air temperature, precipitation, relative humidity, and wind speed and direction in order to correspond with the historical data. The GCMs were chosen from the full suite of IPCC AR4 WG1 models because they are known to agree well with past observations in Alaska. These models, their boundaries for the KRB, and the number of grid cells used are shown in Table 2.10.

Surface Air Temperature

The multi-model mean of projected annual mean surface air temperature for the decade 2090–2099 shows a warming of the KRB of approximately 0.2°C in the south to 3°C in the northern end of the basin (compare Figure 2.16 with Figure 2.6). Consequently, the north to south temperature gradient has decreased by 1.5°C. The projected time series of annual mean surface air temperature over the KRB for the

Table 2.10 Characteristics and boundaries of GCMs used for climate predictions

Model	Grid Cells	Latitude—Longitude boundaries
CCC	2 × 3 = 6 cells	lat1 = 68.652; lat2 = 72.362; lon1 = 206.25; lon2 = 210; lon3 = 213.75
CSM	7 × 8 =5 6 cells	lat1 = 65.135; lat2 = 66.536; lat3 = 67.9367; lat4 = 69.337; lat5 = 70.738,lat6 = 72.139; lat7 = 73.539; lat8 = 74.940; lon1 = 205.313; lon2 = 206.719; lon3 = 208.125; lon4 = 209.531; lon5 = 210.938; lon6 = 212.3438; lon7 = 213.75
ECH	5 × 5 = 25 cells	lat1 = 66.216; lat2 = 68.081; lat3 = 69.946; lat4 = 71.811; lat5 = 73.676; lon1 = 206.25; lon2 = 208.125; lon3 = 210.0; lon4 = 211.875; lon5 = 213.75
GFD	4 × 4 = 16 cells	lat1 = 67.0; lat2 = 69.0; lat3 = 71.0; lat4 = 73.0; lon1 = 206.25; lon2 = 208.75; lon3 = 211.25; lon4 = 213.75
HAD	5 × 3 = 15 cells for SAT, Pr, Rh	lat1 = 65.0; lat2 = 67.5; lat3 = 70.0; lat4 = 72.5; lat5 = 75.0; lon1 = 206.25; lon2 = 210.0; lon3 = 213.75
HAD	4 × 2 = 8 cells for wind	lat1 = 66.25; lat2 = 68.75; lat3 = 71.25; lat4 = 73.75; lon1 = 208.125; lon2 = 211.875

For each General Circulation Model (GCM) used here, the number of grid cells and latitude and longitude boundaries are described.

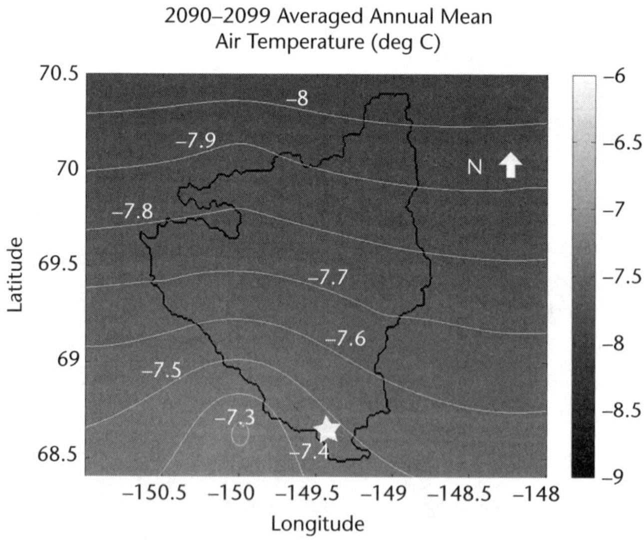

Figure 2.16 Average annual mean surface air temperature for the Kuparuk River basin between 2090 and 2099, predicted by 5 GCMs. The location of the Arctic LTER is marked by a star.

21st century for each model and the multi-model mean show strong interannual variability, with an approximately linear long-term warming of ~0.4°C per decade and a slight slowing of warming after 2070 (Figure 2.17). However, Figure 2.17 also shows a cold bias in the current temperature mean of the KRB in the model mean of ~3°C.

Precipitation

The projected multi-model mean of total annual precipitation (mm) averaged for 2090–2099 indicates a much wetter future on the North Slope with ~1.5 times the amount of current precipitation in the southern end of the basin and ~3 times the amount in the north (compare Figure 2.18 with Figure 2.7A). As was the case for temperature, these results also show a substantially decreased precipitation gradient from ~160 mm south to north to a projected 50 mm south to north. The projected time series of annual precipitation totals over the KRB for the 21st century show that, as is the case for temperature, there is strong interannual variability (stronger for individual models than for the mean) and an approximately linear increase in precipitation of ~10 mm per decade (Figure 2.19). However, as for temperature, the multi-model mean shows an overall bias that is wet relative to the historical observations.

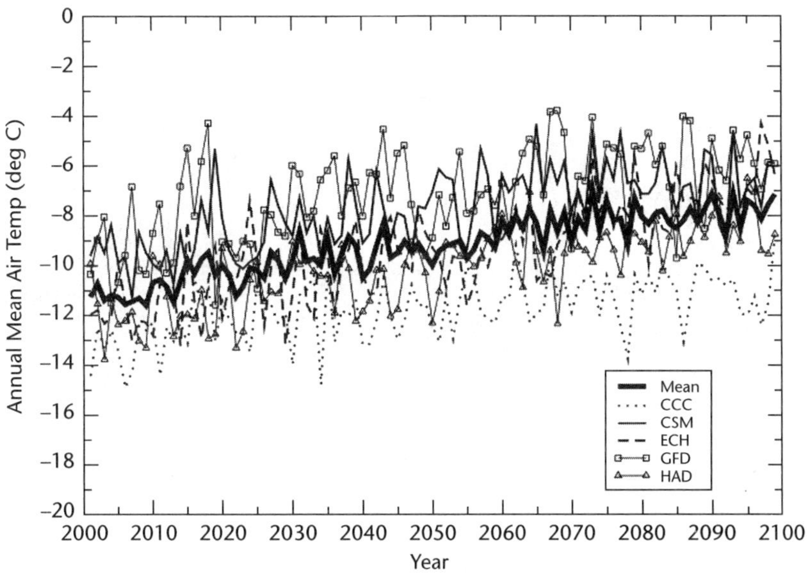

Figure 2.17 Time series of annual mean surface air temperature over the Kuparuk River basin between 2001 and 2099, simulated by five GCMs (CCC, CSM, ECH, GFD, and HAD) and the multi-model mean.

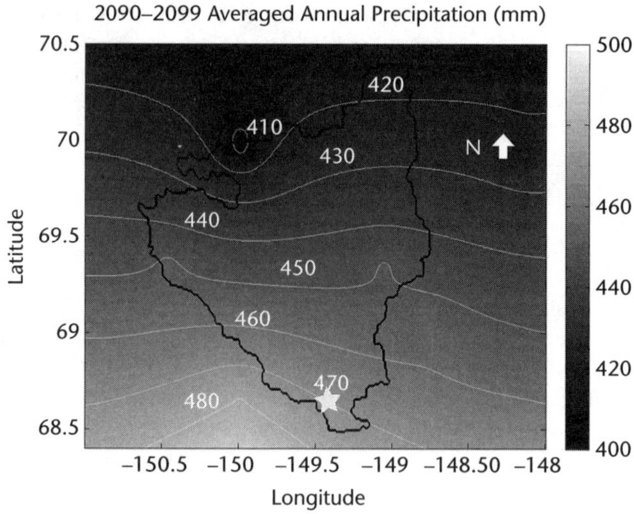

Figure 2.18 Averaged annual precipitation (mm) between 2090 and 2099 predicted by five GCMs (CCC, CSM, ECH, GFD, and HAD). The location of the Arctic LTER is marked by a star.

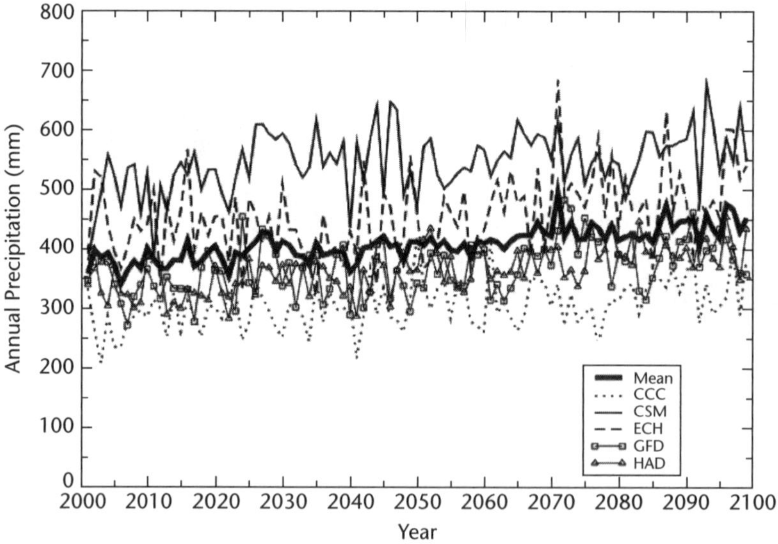

Figure 2.19 Time series of annual precipitation predicted for the Kuparuk River basin between 2001 and 2099, simulated by five GCMs (CCC, CSM, ECH, GFD, HAD) and the multi-model mean.

Relative Humidity

Relative humidity is projected to remain stable over the next century, although the models show a much higher humidity than do the observations. The model projections show a reversed humidity gradient from that currently existing, with decreasing humidity from the south to the north. The time series shows no significant long-term trend; both the mean and time series are calculated for a subset of the GCMs that output relative humidity.

Wind Speed and Direction

Future wind speed is lower over the KRB, with a weaker gradient from south to north, although from the time series it can be seen that the models are simply projecting too low a wind speed, even for the present (data not shown). The time series actually shows a slight increase in wind over the course of the 21st century. The models suggest a future with a somewhat different wind direction (115° over most of the KRB) than the present gradient of 155–175° north to south. The time series also shows a modest shift from around 115° to more like 120° by end of the century, also with strong interannual variability. Again the direction of the wind in the models seems to be misrepresented, even for the present. This may be indicative of problems in the atmospheric pressure values and other dynamics for the Alaska and high arctic domain.

Discussion and Conclusions

This chapter focuses on the climate past, present, and future for Toolik and its environs within the KRB. The arctic climate is highly variable on interannual and longer time scales. While at the time of this writing 20 years of observations exist at Toolik, this may be insufficient for significant trends to emerge in this highly variable data-set. It is easier to separate the "signal" from the "noise" in air temperature, snow depth, and ground temperatures at Barrow, for example, which has temperature records for over 100 years and snow depth data for more than 50 years. The advantage that the Toolik site has over the Barrow site is the wealth of process-oriented data described throughout this book. Certainly the various biologic, hydrologic, and biogeochemical processes described in the other chapters are strongly impacted by interannual variability in climate and, over the longer term, climate change.

The second important caveat to the past and present observations is the problem of biased measurement that is described throughout the snow subsection. For the case of snow, not only is the interannual variability in amount, timing, and distribution high but the measurement error is greater than the magnitude of long-term change over the past 20 years. This makes it particularly difficult to separate all of the snow-related climate parameters (i.e., precipitation and sublimation) using observations alone. On the other hand, the measurements taken during snowmelt in the KRB are of a spatial resolution unmatched in Alaska and most other northern basins. This is a particularly valuable data set for the Arctic LTER community.

Finally, there are caveats associated with the projections of the future climate at Toolik and throughout the KRB. It was mentioned that for several variables the models and the multi-model mean are simply biased in magnitude even for the present period. There are known shortcomings in the models with regard to the representation of arctic clouds and their feedbacks, which impact the projections of many hydroclimatological processes. Another critical problem is the misrepresentation of atmosphere–land surface feedbacks with respect to snow, ground thermodynamics, and subsurface storage of water in a permafrost-dominated landscape. The development of most GCMs has largely focused on the equatorial dynamics. Regional model development for the Arctic is a field that is rapidly growing to fill this void and is making use of data from process studies such as those at Toolik.

While it seems likely that the future climate throughout the KRB will be warmer and wetter, with more snow, it is not entirely clear how these changes will impact other characteristics such as the ground temperatures and vegetation. The answer to this may depend in part on the seasonality and timing of changes, which may not be accurately represented in the climate model projections. The more information that can be gained from studying the feedbacks of environmental response (e.g., shifts in vegetation) to climate variations, the higher the accuracy of the projection of this response into the future on climate-change time-scales. The Arctic LTER site, in coordination with other observing programs, is well positioned to provide the data to help build this understanding.

References

Aagaard, K., and E. C. Carmack. 1989. The role of sea ice and other fresh water in the arctic circulation. *Journal of Geophysical Research* 94:14485–14498.

Adam, J. C., and D. P. Lettenmaier. 2003. Adjustment of global gridded precipitation for systematic bias. *Journal of Geophysical Research* 108(D9):4257. doi:10.1029/2002JD002499.

ACIA (Arctic Climate Impact Assessment). 2004. *Impacts of a warming Arctic.* New York: Cambridge University Press.

Armstrong, R. L., and M. J. Brodzik. 2005, updated 2007. Northern hemisphere EASE-grid weekly snow cover and sea ice extent, version 3. National Snow and Ice Data Center, Boulder, Colorado USA. Digital media.

Berezovskaya, S. L., J. E. Derry, D. L. Kane, R. E. Gieck, M. R. Lilly, and D. M. White. 2008. Snow survey data for the Kuparuk foothills hydrology study: Spring 2008. Report INE/WERC 08.14:40. University of Alaska, Fairbanks, Water and Environmental Research Center.

Cherry, J. E., L. B. Tremblay, S. J. Déry, and M. Stieglitz. 2005. Reconstructing solid precipitation from snow depth measurements and a land surface model. *Water Resources Research* 41:W09401. doi:10.1029/2005WR003965.

Daly, C., R. P. Neilson, and D. L. Phillips. 1994. A statistical-topographic model for mapping climatological precipitation over mountainous terrain. *Journal of Applied Meteorology* 33:140–158.

Déry, S. J., and M. K. Yau. 2002. Large-scale mass balance effects of blowing snow and surface sublimation. *Journal of Geophysical Research* 107:4679. doi: 10.1029/2001JD001251.

Déry, S. J., P. A. Taylor, and J. Xiao. 1998. The thermodynamic effects of sublimating, blowing snow in the atmospheric boundary layer. *Boundary Layer Meteorology* 89:251–283.

Déry, S. J., M. Stieglitz, A. K. Rennermalm, and E. F. Wood. 2005. The water budget of the Kuparuk River basin, Alaska. *Journal of Hydrometeorology* 6:633–655.

Foster, J. L., J. W. Winchester, and E. G. Dutton. 1992. The date of snow disappearance on the arctic tundra as determined from satellite, meteorological station and radiometric in situ observations. *IEEE Transactions on Geoscience and Remote Sensing* 30:793–798.

Goodison, B. E., P. Y. T. Louie, and D. Yang. 1998. *WMO solid precipitation intercomparison, final report.* WMO/TD-872:212. Geneva: World Meteorological Organization.

Hinzman, L. D., D. L. Kane, R. E. Gieck, and K. R. Everett. 1991. Hydrologic and thermal-properties of the active layer in the Alaskan Arctic. *Cold Regions Science and Technology* 19:95–110.

IPCC (Intergovernmental Panel on Climate Change). 2007. Summary for policymakers. Pages 7–22 in M. L. Parry, O. F. Canziani, J. P. Palutikof, P. J. van der Linden, and C. E. Hanson, editors, Climate change. 2007: *Impacts, adaptation and vulnerability. Contribution of Working Group ii to the fourth assessment report of the Intergovernmental Panel on Climate Change.* Cambridge: Cambridge University Press.

Kane, D. L., and L. D. Hinzman. Accessed 2007. Climate data from the North Slope hydrology research project. University of Alaska, Fairbanks, Water and Environmental Research Center, Fairbanks, AK. http://www.uaf.edu/water/projects/NorthSlope/

Kane, D. L., R. E. Gieck, D. C. Kitover, L. D. Hinzman, J. P. McNamara, and D. Yang. 2004. Hydrological cycle on the North Slope of Alaska. Pages 224–236 in D. L. Kane and D. Yang, editors, *Northern research basins water balance*. International Association of Hydrologic Sciences, publication 290. Wallingford: IAHS Press.

Kaufman, D. S., T. A. Ager, N. J. Anderson, P. M. Anderson, J. T. Andrews, P. J. Bartlein, L. B. Brubacker, L. L. Coats, L. C. Cwybar, M. L. Duvall, A. S. Dyke, M. E. Edwards, W. R. Eisner, K. Gajewski, A. Geirsdottir, F. S. Hu, A. E. Jennings, M. R. Kaplan, M. W. Kerwin, A. V. Lozhkin, G. M. MacDonald, G. H. Miller, C. J. Mock, W. W. Oswald, B. L. Otto-Bliesner, D. F. Porinchu, K. Ruhland, J. P. Smol, E. J. Steig, and B. B. Wolfe. 2004. Holocene thermal maximum in the Western Arctic (0–180°W). *Quaternary Science Review* 23:529–560.

Liston, G. E., and M. Sturm. 2002. Winter precipitation patterns in arctic Alaska determined from a blowing snow model and snow-depth observations. *Journal of Hydrometeorology* 3:646–659.

McKay, G. A., and D. M. Gray. 1981. The distribution of snowcover. Pages 153–190 in D. M. Gray and D. H. Male, editors, *Handbook of Snow.* Toronto: Pergamon Press.

McNamara, J. P., D. L. Kane, and L. D. Hinzman. 1998. An analysis of streamflow hydrology in the Kuparuk River basin, arctic Alaska: A nested watershed approach. *Journal of Hydrology* 206:39–57.

Muskett, R., and V. E. Romanovsky. 2011. Alaskan permafrost groundwater storage changes derived from GRACE and ground measurements. *Remote Sensing* 3:378–397. doi:10.3390/rs3020378

National Snow and Ice Data Center (NSIDC). Accessed April 30, 2008. Arctic climatology and meteorology. http://nsidc.org/arcticmet/basics/arctic_climate.html

Oke, T. R. 1987. *Boundary layer climates.* 2nd edition. Methuen Press, London.

Osterkamp, T. E., and V. E. Romanovsky. 1996. Characteristics of changing permafrost temperatures in the Alaskan arctic, U.S.A. *Arctic and Alpine Research* 28:267–273.

Papineau, J. M. 2001. Wintertime temperature anomalies in Alaska correlated with ENSO and PDO. *International Journal of Climatology* 21:1577–1592.

Pomeroy, J. W., and D. M. Gray. 1995. *Snowcover accumulation, relocation and management.* NHRI Science Report 7:144. Saskatoon, SK, Canada.

Robinson, D. A., A. Bamzai, and B. Ramsay. 2001. Evaluating northern hemisphere snow cover during the satellite era: Variations in extent and associations with temperature. Pages 36–39 in *Proceedings of the 12th Symposium on Global Change and Climate Variations*. Albuquerque, NM: American Meteorological Society.

Romanovsky, V.E. and T. Osterkamp. Accessed 2012. Online data and references. http://www.permafrostwatch.org/

Serreze, M. C., J. E. Walsh, F. S. Chapin, III, T. Osterkamp, M. Dyurgerov, V. Romanovsky, W. C. Oechel, J. Morison, T. Zhang, and R. G. Barry. 2000. Observational evidence of recent change in the northern high-latitude environment. *Climatic Change* 46:159–207.

Shulski, M., and G. Wendler. 2007. *The Climate of Alaska*. Fairbanks, AK: University of Alaska Press.

Simpson, J. J., G. L. Huffford, C. Daly, J. S. Berg, and M. D. Fleming. 2005. Comparing maps of mean monthly surface temperature and precipitation for Alaska and adjacent areas of Canada produced by two different methods. *Arctic* 58:137–161.

Smith, S. L., V. E. Romanovsky, A. G. Lewkowicz, C. R. Burn, M. Allard, G. D. Clow, K. Yoshikawa, and J. Throop. 2010. Thermal state of permafrost in North America: A contribution to the international polar year. *Permafrost and Periglacial Processes* 21:117–135.

Stewart, R. E., K. K. Szeto, R. F. Reinking, S. A. Clough, and S. P. Ballard. 1998. Midlatitude cyclonic cloud systems and their features affecting large scales and climate. *Reviews of Geophysics* 36:245–273.

Stieglitz, M., S. J. Déry, V. E. Romanovsky, and T. E. Osterkamp. 2003. The role of snow cover in the warming of arctic permafrost. *Geophysical Research Letters* 30:1721. doi:10.1029/2003GL017337.

Stone, R. S., E. G. Dutton, J. M. Harris, and D. Longenecker. 2002. Earlier spring snowmelt in northern Alaska as an indicator of climate change. *Journal of Geophysical Research* 107:4089. doi:10.1029/2000JD000286.

Wang, X., and J. R. Key. 2003. Recent trends in arctic surface, cloud, and radiation properties from space. *Science* 299:1725–1728.

White, D., L. Hinzman, L. Alessa, J. Cassano, M. Chambers, K. Falkner, J. Francis, W. J. Gutowski, Jr., M. Holland, R. M. Holmes, H. Huntington, D. Kane, A. Kliskey, C. Lee, J. McClelland, B. Peterson, T. S. Rupp, F. Straneo, M. Steele, R. Woodgate, D. Yang, K. Yoshikawa, and T. Zhang. 2007. The arctic freshwater system: Changes and impacts. *Journal of Geophysical Research* 112:G04S54. doi:10.1029/2006JG000353.

Yang D., D. Kane, Z. Zhang, D. Legates, and B. Goodison. 2005. Bias corrections of long-term (1973–2004) daily precipitation data over the northern regions. *Geophysical Research Letters* 32:L19501. doi:10.1029/2005GL024057.

Zhang, T., T. E. Osterkamp, and K. Stamnes. 1996. Some characteristics of the climate in northern Alaska, U.S.A. *Arctic and Alpine Research* 28:509–518.

3

Glacial History and Long-Term Ecology in the Toolik Lake Region

Donald A. Walker
Thomas D. Hamilton
Hilmar A. Maier
Corinne A. Munger
Martha K. Raynolds

Introduction

An understanding of the glacial history of the Toolik Lake region is key to understanding the present-day variety of landforms, soils, and vegetation that occupy the extraordinarily beautiful landscapes that surround the lake. Landscape evolution following deglaciation proceeds much slower in the Arctic than in more temperate regions. Glacial surfaces with ages that span hundreds of thousands of years are present within a few kilometers of the Toolik Field Station. These surfaces have not been altered by agriculture or other anthropogenic influences, so the region is also an excellent laboratory to study the effects of glacial age on arctic ecosystem function. Such studies can help us understand how arctic systems change over long periods of time and provide insights about how they might change in the future. Here we synthesize information from maps and analysis of glacial geology, landforms, and vegetation of the 823-km^2 area of Toolik Lake and the upper Kuparuk River region (Figure 3.1).

Glacial Geology

The complicated topography of the upper Kuparuk River region (Figure 3.1) is the result of deposits laid down by valley glaciers that flowed into the region from the south during three major glacial advances that span the period of time between the middle Pleistocene (610,000–132,000 yr BP) and the late Pleistocene (115,000–10,000 yr BP) (Table 3.1; Figure 3.2). The glacial deposits are assigned to the Sagavanirktok (middle Pleistocene), Itkillik I, and Itkillik II (late Pleistocene)

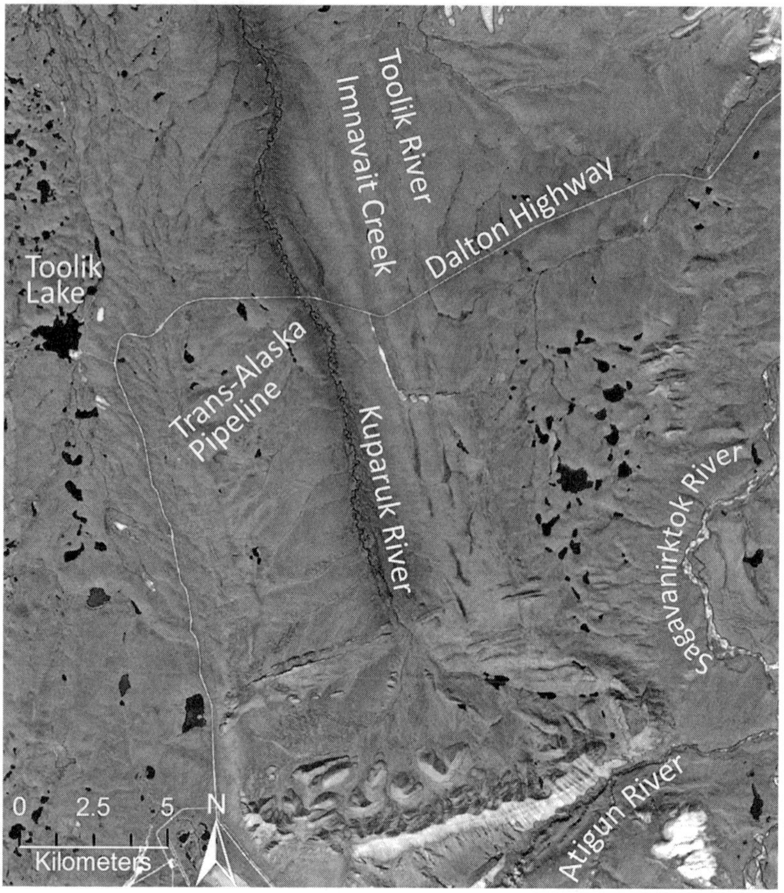

Figure 3.1 False color-infrared SPOT satellite image of the mapped area in the Toolik Lake region. Surfaces glaciated during the Itkillik glaciations have a browner tone and numerous lakes, whereas areas glaciated during the older Sagavanirktok glaciation have a brighter red tone and few lakes (see Figure 3.2 for glacial boundaries). The brightest red areas, e.g., along streams and water tracks, are densely vegetated with shrubs. More intermediate red tones are generally tussock tundra with varying amounts of shrubs. The more brownish tones are less densely vegetated and include dry tundra, nonacidic tundra, and wet tundra. Rocky areas are either light colored, such as the limestone in the lower right corner or greenish gray as in the sandstones and conglomerates along the north side of the Atigun River. Image taken July 28, 1989. [REFER COLOR IMAGE]

glaciations of the central Brooks Range glacial succession (see Hamilton 1986, 2003a, and 2003b for more details on the glacial chronology of the region).

Most of the upper Kuparuk River watershed, including Imnavait Creek, developed on Sagavanirktok-age glacial deposits (light and dark purple, Figure 3.2). The Sagavanirktok River glaciation consisted of several glacial events dating broadly from about 610,000–132,000 yr BP. During the oldest recognized

Table 3.1 Glacial sequence in the central Brooks Range based on Hamilton (1994)

Glacial periods (k yr BP)	Glaciation	Phase
Holocene (12–0)	Neoglaciation	
Late Pleistocene (115–12)	Itkillik II	Latest Itkillik II readvance (12.8–11.4)
		Main Itkillik II advance (25–17)
	Itkillik I	Phase B (>55)
		Phase A
Middle Pleistocene (610–132)	Sagavanirktok	Late Phase
		Main Phase
Early Pleistocene (2,580–610)	Anaktuvuk	
Late Tertiary (>2,580)	Gunsight Mountain	

Timing of ages of glacial periods in thousands of years before present (BP) are based on Richmond and Fullerton (1986) for main chronology and on Gibbard et al. (2009) for the end of the Tertiary. Radiocarbon dates for the Itkillik advances are from Hamilton (2003a).

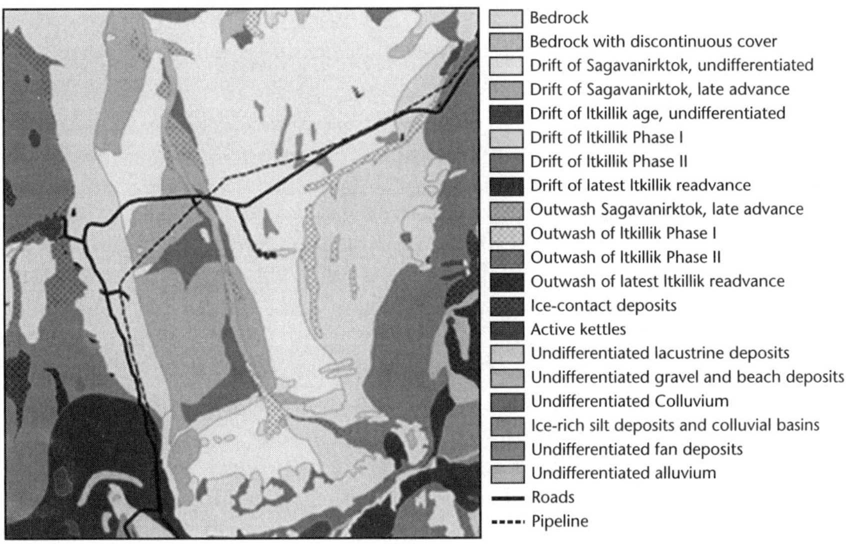

Figure 3.2 Simplified glacial geology map of the upper Kuparuk River region (based on Hamilton 2003a).

(maximum) Sagavanirktok-age advance, large valley glaciers flowed north down the Itkillik, Sagavanirktok, and Kuparuk river drainages and reached their northern limit 50–60 km beyond the northern flank of the Brooks Range. End moraines of the Sagavanirktok River glaciation are about 25 km north of northern edge of the map along the Kuparuk River and about 9 km north along the Toolik River and about 55 km north along the Sagavanirktok River. Several massive lateral moraines of Sagavanirktok age trend SSE to NNW across the central part of the upper Kuparuk River watershed (Figure 3.3, top). These broad, gently sloping moraines are rounded by slope-erosion processes and rise about 100 m from the valley bottoms to their crests (Hamilton 1986, 2003a, 2003b). The glacial till of these moraines is well covered by varying thicknesses of loess and colluvial sediments. Only a few widely dispersed glacial erratics protrude above the tundra surface. The hillcrests and slopes are generally topped by 20–40 cm of peat and tundra vegetation of various types that have been described in studies of a west-facing hill toposequence at Imnavait Creek (Walker and Walker 1996).

A less extensive late Sagvanirktok-age advance (darker purple on Figure 3.2) overflowed the upper Kuparuk drainage from the west and south to form moraines and outwash remnants that are intermediate in appearance between those of the maximum advance and the subsequent Itkillik I moraine succession.

Glaciation of Itkillik I age occurred more than 55,000 calibrated ^{14}C years BP (referred to here as "yr BP"; light green on Figure 3.2) and probably postdates the warmest part of the last interglacial ~120,000 yr BP. These glaciers abutted Sagavanirktok-age moraines on the west side of the upper Kuparuk River drainage, and bedrock ridges on the south side, but overflowed divides on the east and southeast sides of the upper Kuparuk River watershed. Itkillik I moraines are modified by weathering and erosion, but on a much smaller scale than deposits of the Sagavanirktok River glaciations; stony surfaces, small patterned-ground features, and lakes are more common on the Itkillik I surfaces than on the Sagavanirktok surfaces.

The subsequent but much less extensive Itkillik II advances, which date between about 25,000 and 11,400 yr BP (medium green on Figure 3.2 and Figure 3.3, bottom), are contemporaneous with the major late Wisconsin advances of the standard North American glacial succession. Two major advances of Itkillik II age took place between about 25,000 and 17,000 yr BP, forming extensive ice-stagnation features around Toolik Lake. Glacial flow patterns during the Itkillik II advance were generally similar to those of the present-day river drainages. A subsequent Itkillik II re-advance occurred between 12,000 and 11,400 yr BP (dark green on Figure 3.2). Itkillik II age surfaces are generally rocky and gently undulating with small blockfields (areas with >50% cover of moderate to large-sized angular rocks). Crests are slightly flattened, and loess and vegetation cover are locally absent. Kettle lakes with irregular shorelines are common.

Bedrock exposures also occur within the mapped area. These include Cretaceous-age fluvial and shallow marine conglomerates, sandstones, shales, and siltstones (Chandler, Toruk, and Fortress Mountain formations) that outcrop along several WSW–ENE trending ridges (Brosgé et al. 1979). A mountain, composed of Pennsylvanian and Mississippian limestone (Lisburne formation), occurs in the southeast corner of the map (Figure 3.1).

Figure 3.3 Glaciated terrain of Sagavanirktok and Itkillik II age. Top—Sagavanirktok-age moraine, Imnavait Creek vicinity. Note long smooth-flanking slopes and continuous vegetation cover broken only by dispersed large, weathered erratic boulders of resistant lithology. The vegetation is a rather homogeneous cover of tussock tundra. The cottongrass, *Eriophorum vaginatum*, is flowering profusely in this photo. Bottom—Drift and outwash surfaces of Itkillik II age, looking northeast from Jade Mountain, west of Toolik Lake. Note narrow moraine crests, steep-flanking slopes, and discontinuous vegetation cover on the moraine and kame features. Minor as well as major depositional features are well preserved. Photos by D. A. Walker. [REFER COLOR IMAGE]

Surface Geomorphology

The surficial deposits of Pleistocene age within the study area have been modified by a variety of postglacial (Holocene) geomorphologic processes (Figure 3.4). Common periglacial surface geomorphic features within the mapped area include sorted and nonsorted circles (frost boils) and stripes, turf hummocks, gelifluction lobes and terraces, water tracks, ice-wedge polygons, wetland features (strangmoor, aligned hummocks, palsas), and thermokarst features. More detailed descriptions of the various surface geomorphic units are contained in the maps of Walker and Walker (1996) for Imnavait Creek.

The broad hillslope deposits of the Sagavanirktok-age surfaces are more modified by weathering and erosion than are the younger Itkillik-age surfaces. Area analysis of the three dominant glacial surfaces (Figure 3.5) shows that the Sagavanirktok-age surfaces have more well-developed and indistinct hillslope water tracks than the Itkillik I and Itkillik II age surfaces (55%, 13%, and 9%, respectively), fewer lakes (1%, 2%, and 5%), more wetlands (3%, 2%, and 1%), fewer stony surfaces (1%, 8%, and 3%), fewer nonsorted stripes (12%, 13%, and 16%), and fewer areas with greater than 20% cover of nonsorted circles (1%, 1%, and 3%).

Vegetation

The vegetation map of this part of the upper Kuparuk River basin (823 km^2, Figure 3.6) shows 14 map units originally mapped at 1:25,000 scale (Walker and Maier 2008). Within these map units, 33 distinct plant-community types and 19

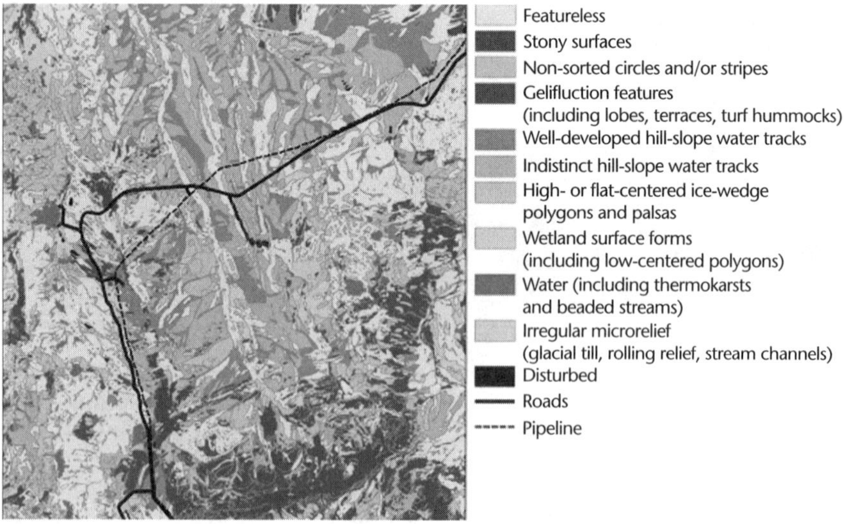

Figure 3.4 Surface geomorphology of the upper Kuparuk River region (based on Walker and Maier 2008). [REFER COLOR MAP]

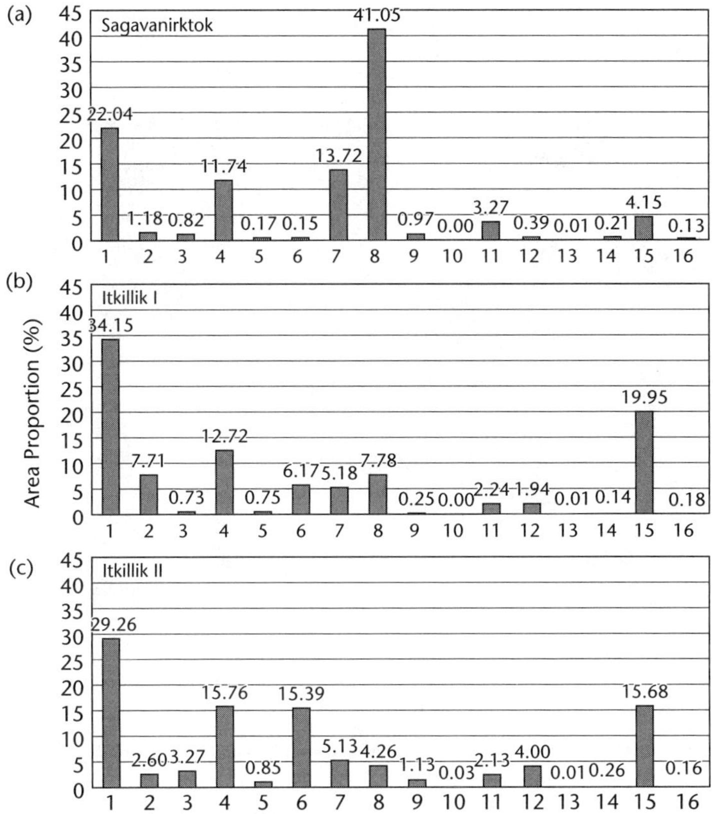

1. Featureless
2. Stony surfaces
3. Non-sorted circles
4. Sorted and non-sorted stripes
5.Turf hummocks
6. Gelifluction features
7. Well-developed hill-slope water tracks
8. Indistinct hill-slope watertracks
9. High-or flat-centered ice-wedge polygons and palsas
10. Low-centered ice-wedge polygons
11. Wetland surface forms
12. Water
13. Pond Complex
14. Thermokarst and beaded streams
15. Irregular microrelief
16. Disturbed

Figure 3.5 Area analysis of surface geomorphology units (Figure 3.4) on each glacial surface (based on Munger et al. 2008).

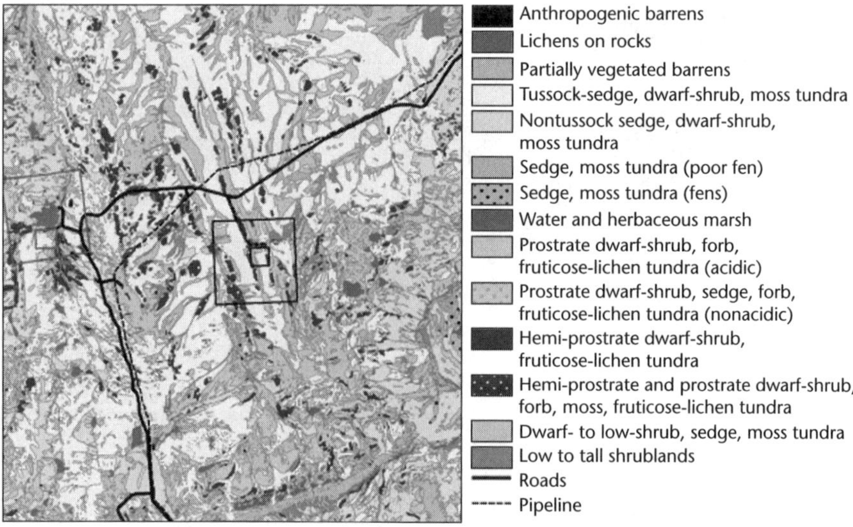

Figure 3.6 Vegetation of the upper Kuparuk River region. Red rectangles are areas of vegetation maps at Toolik Lake LTER research area (Walker and Maier 2008). Black boxes are areas of maps of research areas at Imnavait Creek (described in Walker and Walker 1996). [REFER COLOR MAP]

subtypes are recognized. Seven of the most common plant-community types are shown in Figure 3.3 (top) and Figure 3.7. Most of the published plant-community information in the Toolik Lake region comes from 81 permanent vegetation study plots at Toolik Lake and 73 plots at Imnavait Creek (Walker et al. 1994). The locations of the study plots and details of plant-species cover, soil properties, site factors, and photographs of each study plot are contained in two data reports (Walker et al. 1987; Walker and Barry 1991), which are available at the Arctic Data Coordination Center (ADCC), Boulder, Colorado, at http://adcc.colorado.edu and the Arctic Geobotanical Atlas at http://www.arcticatlas.org/support/.

Description and Environmental Controls

Toolik is within arctic bioclimate subzone E, the southernmost subzone of the circumpolar Arctic (Walker et al. 2005). Mean July temperatures in subzone E are typically 9°C –12°C. The zonal vegetation is dominated by either low-shrub tundra (in areas with warmer soils and thick active layers) or tussock tundra (in areas with ice-rich permafrost, cold wet soils, and shallow active layers).

Jorgenson (1984) first described the contrasting nature of the vegetation on different-age glacial surfaces near Toolik Lake. Several more recent studies have expanded on his observations (e.g., Walker et al. 1995, 1996; Munger et al. 2008).

Sphagno-Eriophoretum vaginati (Walker et al. 1994) (moist acidic tussock tundra; Figure 3.3, top) is the zonal plant association on ice-rich sediments with shallow active layers and low soil pH (3.8–5.5 in the Toolik Lake area); the nomenclature

Figure 3.7 Common plant communities in the Toolik Lake region: Top left—barrens, *Cetraria nigricans-Rhizocarpon geographicum* comm. on block field of Itkillik I–age surface; Top right—Dry acidic tundra, *Selaginello sibericae-Dryadetum octopetalae* on south-facing Itkillik II–age kame west side of Toolik Lake. Middle left—Snowbed, *Carici microchaetae-Cassiopetum tetragonae* on steep north-facing slope of Itkillik II–age till deposit. Middle right—Moist nonacidic tundra, *Dryado integrifoliae-Caricetum bigelowii* on Itkillik II–age hill slope west of Toolik Lake. Lower left—Shrub tundra, *Sphagno-Eriophoretum vaginati betuletosum nanae* in water track margin. Lower right—Wetland, *Carex aquatilis-Carex chordorrhiza* comm. in rich fen complex on Itkillik II surface. Also see Figure 3.3 for photo of tussock tundra (*Sphagno-Eriophoretum vaginati* typicum), the most common and zonal vegetation type in the region. Photos by D. A. Walker. [REFER COLOR IMAGE]

for plant associations follows that of the Braun-Blanquet approach (Westhoff and van der Maarel 1978). This plant association is common throughout much of subzone E of Beringia on older surfaces that have been unglaciated for long periods of time and where ice-rich permafrost and cold wet soils have developed on stable hillslopes. It is found extensively in northern Alaska, northwestern Canada (Lambert

1968), and Chukotka (Alexandrova 1980). Typical taxa in this association include tussock cottongrass (*Eriophorum vaginatum),* a mixture of dwarf shrubs (including *Betula nana, Ledum decumbens, Salix pulchra, Vaccinium* spp.), mosses (including *Sphagnum* spp., *Dicranum* spp., *Aulacomnium* spp., *Polytrichum strictum, Hylocomium splendens*), and lichens (*Peltigera aphthosa, Cladonia* spp., *Dactylina arctica*); the species nomenclature here for the most part follows that used by Walker et al. (1994). Either graminoids or shrubs can be dominant in response to local variations of temperature, moisture, and nutrients. These physiognomic variations within the association can be important from an ecosystem-function perspective. For example, the relative cover and size of deciduous shrubs can strongly affect microclimate, net primary productivity, energy, water and trace-gas fluxes, and animal habitat characteristics (Myers-Smith et al. 2011).

The plant association *Dryado integrifoliae–Caricetum bigelowii* (Walker et al. 1994) (moist nonacidic tundra; Figure 3.7, middle right) occurs on surfaces with somewhat higher soil pH (5.5–7.5) and warmer (in summer) soils, which are typically found on mesic sites of loess deposits, solifluction features, frost-boil complexes, alluvial terraces, and younger glacial surfaces (Walker and Everett 1991; Kade et al. 2005). Characteristic plant species in this association include graminoids (*Carex bigelowii, C. membranacea, C. scirpoidea, Eriophorum triste, Arctagrostis latifolia*), prostrate dwarf shrubs (*Dryas integrifolia, Salix arctica, S. reticulata, Arctous rubra*), forbs (*Bistorta vivipara, Senecio atropurpureus, Chrysanthemum integrifolium, Pedicularis lanata, P. capitata, Tofieldia coccinea, Astragalus* spp., *Oxytropis* spp., *Hedysarum* spp., *Saxifraga oppositifolia*), bryophytes (*Tomentypnum nitens, Hylocomium splendens, Aulacomnium turgidum, Rhytidium rugosum, Ptilidium ciliare*), and lichens (*Thamnolia* spp., *Cetraria* spp., *Peltigera* spp.).

Variations of plant-species composition within moist acidic and moist nonacidic tundra occur along the north-south climate gradient in northern Alaska (Kade et al. 2005) and Russia (Matveyeva 1998). For example, plant communities in bioclimate subzone E and the southern part of subzone D are more species rich and have more erect shrubs than similar communities farther north.

Relationship of Vegetation to Glacial History

Compared to the younger Itkillik-age surfaces, Sagavanirktok-age glacial surfaces have much greater cover of tussock-sedge, dwarf-shrub, and moss tundra (unit 3 in Figure 3.7); much less nontussock-sedge, dwarf-shrub, and moss tundra (unit 4); relatively high cover of erect dwarf-shrub tundra types (sum of units 12, 13, and 14); more poor-fen wetlands (unit 5); fewer rich fens (unit 6); fewer snowbeds (units 10 and 11); and less dry nonacidic tundra (unit 9) (Figure 3.8).

Each glacial surface also has distinctive patterns of biomass (Figure 3.9 and Figure 3.10). The normalized difference vegetation index (NDVI) is a commonly used measure of photosynthetic capacity that is derived from the red and infrared bands of multi-spectral data (Tucker and Sellers 1986). NDVI is strongly correlated with tundra aboveground biomass at the circumpolar scale (Raynolds et al. 2012) and with a variety of tundra biophysical properties including aboveground biomass, leaf area index, and trace-gas flux at the local scale (Shippert et al. 1995;

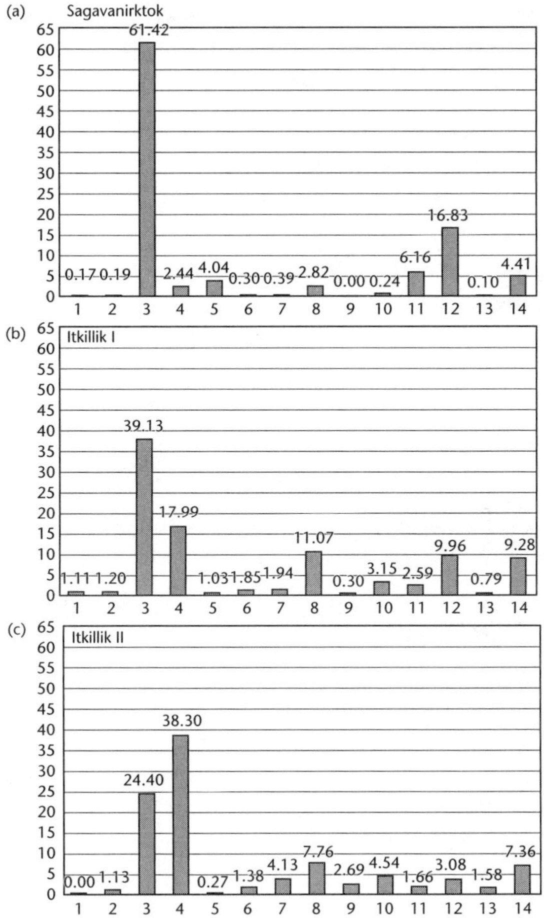

1. Lichens on rocks
2. Partially vegetated barrens and revegetated disturbed areas.
3. Tussock sedge, dwarf-shrub, moss tundra
4. Non-tussock sedge, dwarf-shrub, moss tundra
5. Sedge, moss tundra (poor fens)
6. Sedge, moss tundra (fens)
7. Water and deep water herbaceous marsh
8. Prostrate dwarf-shrub, forb, fruticose-lichen tundra (acidic)
9. Prostrate dwarf-shrub, sedge, forb, fruticose lichen tundra (nonacidic)
10. Hemi-prostrate dwarf-shrub, fruticose-lichen tundra
11. Hemi-prostrate and prostrate dwarf-shrub, forb, moss fruticose-lichen tundra
12. Dwarf-shrub, sedge, moss tundra
13. Dwarf- to low-shrub moss tundra
14. Low to tall shrublands

Figure 3.8 Area analysis of vegetation units on the major glacial units (from Munger et al. 2008.)

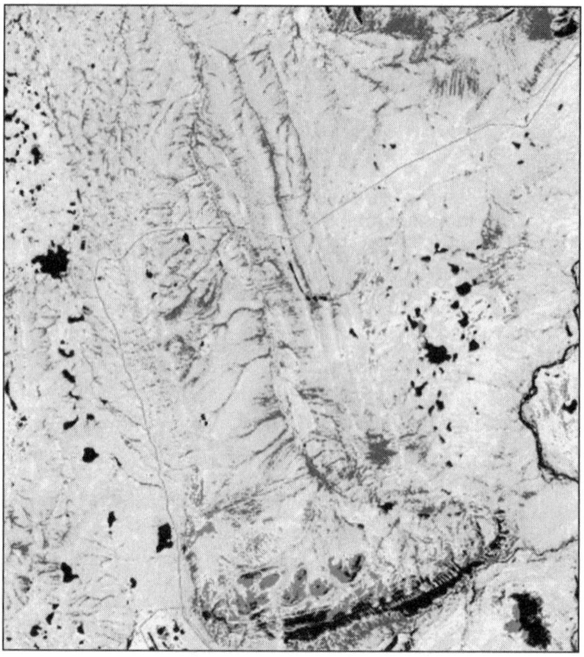

Figure 3.9 NDVI of the upper Kuparuk River region. The image is from a SPOT-1 multispectral image acquired July 29, 1989. NDVI = (NIR – R)/(NIR + R), where R is the reflectance in the red band (630–690 nm) and the NIR is the reflectance in the near infrared band (760–900 nm). Compare the pattern on the NDVI image with the glacial geology map (Figure 3.2) and note greater biomass per unit area on the older Sagavanirktok-age surfaces. Based on Shippert et al. (1995). [REFER COLOR MAP]

Stow et al. 2004). Older landscapes in the Toolik Lake region have higher NDVI (Figure 3.10). The higher NDVI values of the older landscapes are due in part to more-shrubby zonal vegetation. Biomass of the *Sphagno–Eriophoretum vaginati* tussock tundra, which grows on the older acidic surfaces, is about 25% greater than its nonacidic counterpart *Dryado integrifoliae–Caricetum bigelowii*, which is dominant on the younger glacial surfaces (512 g m^{-2} vs. 403 g m^{-2}; see also chapter 5). Older landscapes have more water tracks filled with high-biomass shrub tundra (average of 735 g m^{-2} in the shrubby plots of Walker et al. 1995). The older surfaces also have less area with low NDVI, such as lakes and ponds, nonsorted circles, bare soil, and cobbles.

Several studies have shown broad-scale tundra greening of the Alaskan Arctic in recent decades as documented with AVHRR satellite data at 1- to 8-km pixel resolution (Jia et al. 2003; Verbyla 2008; Bunn et al. 2007; Bhatt et al. 2010). Do the different landscapes also have different rates of vegetation change under the prevailing climate? We might expect change to occur more rapidly on disturbed sites and on younger landscapes with warmer, more nutrient-rich soils. Tundra disturbances—for example, landslides, thermokarst, fire, roadside areas, and off-road

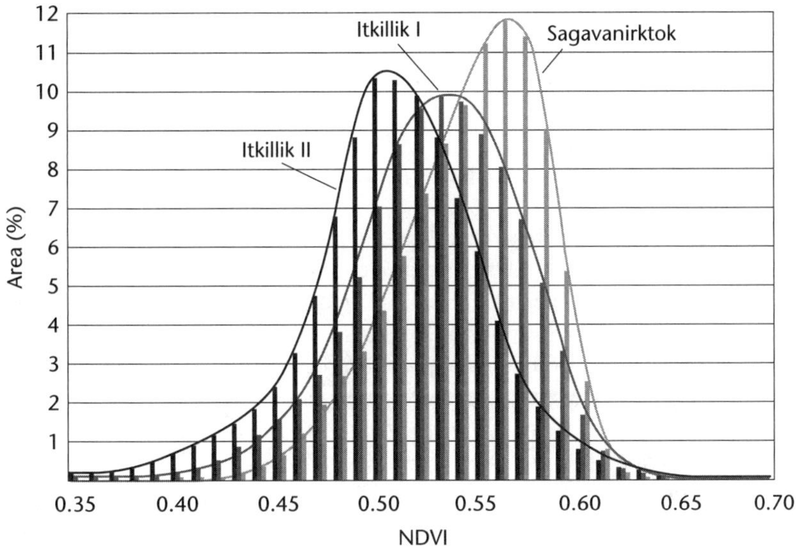

Figure 3.10 Distribution of NDVI on three glacial units in the Toolik Lake region. Curves represent smoothed data for 45 NDVI classes (based on Munger et al. 2008).

vehicle trails—often lead to enhanced plant growth and are likely to be areas of increased productivity and enhanced NDVI (Walker et al. 2009).

An analysis of NDVI changes in the upper Kuparuk region using 30-m resolution Landsat data permitted a closer examination of the parts of the region where significant changes in NDVI have occurred (Raynolds et al. 2010). The most rapid and significant ($p > 0.05$) changes occurred in areas of anthropogenic disturbance such as roadsides and revegetated gravel pads (21% increase). Compared to the Sagavanirktok-age surfaces, areas on the Itkillik surfaces with significant changes in peak-NDVI had larger NDVI increases, and the areas of NDVI increase covered a larger percentage of the vegetated glacial drift and outwash surfaces. The changes on the Itkillik-age surfaces were distributed widely across the surfaces, whereas significant changes on the Sagavanirktok-age surface were concentrated in a few areas, possibly areas of local buried ice or thawing permafrost or other large disturbances. These results suggest that NDVI and biomass changes in the future will also be larger and more widespread on younger surfaces and in areas of disturbance.

Landscape Paludification

Paludification is the long-term accumulation of organic matter (peat) that leads to increased soil moisture and water logging of previously dry landscapes (Gorham et al. 2007). The process has been described extensively in forested landscapes and also has been invoked to describe landscape-evolution in the low Arctic (Walker and Walker 1996). Paleoecological studies from lakes on the Itkillik II and Sagavanirktok-age surfaces near Toolik Lake indicate that during the early part of the Holocene both surfaces likely had tundra with many prostrate dwarf shrub and species indicative of drier, nonpaludified conditions (e.g., *Equisetum, Thalictrum,* Rosaceae, *Encalypta, Selaginella*) (Walker and Walker 1996; Mann et al. 2002; Oswald et al. 2003; see chapter 4). Tundra with plants typically found in acidic tussock tundra (e.g., *Rubus chamaemorus, Sphagnum* spp., Ericales, *Betula nana, Polygonum bistorta*) increased on the older surfaces between 10,000 and 7,500 years ago.

Mosses are critical to this paludification process. The advent of the mosses changes the soil hydrology, soil thermal properties, and soil chemistry, which results in acidic mires in colluvial basins and tussock tundra and extensive water-track development on hillslopes. Observations of the moss carpet at Prudhoe Bay, Sagwon, and Toolik Lake indicate that before *Sphagnum* becomes established, other mosses cause the initial trend toward wetter and more acidic conditions. Early colonizing species include small mosses such as *Encalypta* spp., *Ceratodon purpureus, Distichium* spp., and *Ditrichum flexicaule.* These are followed by larger branching moss species, such as *Tomentypnum nitens*, *Aulacomnium turgidum*, and *Rhytidium rugosum,* which can develop thick moss mats that insulate the soil, trap moisture, reduce the active-layer thickness, and promote the process of paludification. Once the soils are continuously wet, peat buildup and pH reduction occur, permitting the spread of *Sphagnum* spp., *Dicranum* spp., ericaceous shrubs, and other acidophilus species.

Based on the present-day contrasts between the glacial surfaces, we can deduce that enhanced peat formation and reduced thickness of summer thaw layers on the older surfaces led to restricted drainage, a general acidification of the soils, and the introduction of thick moss carpets to hillslopes (Walker and Walker 1996). The landscapes gradually became less diverse with more tussock tundra, more shrub tundra (primarily in water tracks), fewer lakes, less moist and dry nonacidic tundra, fewer snowbeds, and fewer rich fens.

The patterns of vegetation on present-day landscapes that span multiple glaciations do not, however, represent a continuous successional sequence spanning >100,000 years. Cores taken along foothill toposequences near the Mesa archeological site 150 km west of Toolik Lake show that peat accumulation began near the beginning of the Holocene, slowed during the cooling of the Younger Dryas (12,800–11,500 yr BP), and then resumed by about 8,500 yr BP when the organic surface horizons probably had approached their present wide distribution (Mann et al. 2002; see chapter 4 for the local paleoenvironmental interpretation). A critical factor affecting the different rates of soil and vegetation development on the different-aged glacial surfaces during the Holocene was the accumulation of weathered clays and wind-blown, glacially derived silt (loess) on the older landscapes, which promoted higher soil moisture retention, colder soils, more mosses, thicker organic soil horizons, and greater overall vegetation cover.

Relevance to Arctic Ecology

Much of the LTER terrestrial research at Toolik Lake occurs on the south and east sides of the lake on glacial drift surfaces of Itkillik I age (>55,000 yr BP).

Research occurs on the west and north sides of Toolik Lake on drift and outwash deposits of Itkillik II age (<25,000 yr BP; Figure 3.3, bottom). The 1 × 1 km research grid at Toolik Lake (small red square in Figure 3.6) is mainly on an Itkillik I glacial moraine, but includes areas of Itkillik II outwash on the east side of the map. More detailed maps of the Toolik Lake area are contained in Walker and Maier (2008) and the online Toolik-Arctic Geobotanical Atlas (http://www.arcticatlas.org/; Walker et al. 2009).

LTER research is also conducted on surfaces of Sagavanirktok age (>125,000 yr BP) in the upper Imnavait Creek watershed (Reynolds and Tenhunen 1996) (Figure 3.1; Figure 3.3, top). The 1 × 1 km research grid at Imnavait Creek (small black rectangle in Figure 3.6) spans two broad Sagavanirktok-age lateral moraines with a stream and colluvial basin in between the moraines. The landforms and vegetation at Imnavait Creek were mapped as part of Department of Energy R4D studies, which readers should consult for more detailed vegetation and surficial geomorphology maps of this area (Walker and Walker 1996).

The relationship between soil pH and vegetation patterns has been described in association with other types of long-term disturbance such as wind-blown loess at Prudhoe Bay (Walker and Everett 1991), cryoturbation associated with small-scale patterned-ground features in northern Alaska (Kade et al. 2005), and landslides on the Yamal Peninsula, Russia (Walker et al. 2009). A study of ecosystem properties of acidic and nonacidic tundra was conducted at a major soil pH boundary near Sagwon, Alaska (Walker et al. 1998). Compared to acidic tundra at the same location, nonacidic tundra had less gross photosynthesis, respiration, leaf area index, NDVI, average canopy heights, moss cover, and shrub cover; and much greater evapotranspiration, soil heat flux, active layer depths, and cover of nonsorted circles. More recent studies on different-aged glacial surfaces near Toolik have noted that litter decomposition, soil respiration, dissolved organic C production, and net N mineralization are much greater in acidic tundra than in nonacidic tundra (Hobbie et al. 2002; Hobbie and Gough 2004). Both soil pH and calcium ion concentrations appear to affect microbial respiration and dissolved organic carbon dynamics of tundra and different-aged surfaces, and both need to be considered in models of tundra biogeochemistry (Whittinghill and Hobbie 2011).

Plant diversity is also higher in areas of moist nonacidic tundra (Walker et al. 1998; Gough et al. 2000). Kade et al. (2005) showed the southern variant of nonacidic tundra, *Dryado integrifoliae–Caricetum bigelowii* var. *Lupinus arcticus*, has very high diversity within this plant community (alpha diversity) with an average of 30.8 vascular-plant species per 1-m^2 study plot, compared to 13.8 species per plot in the acidic *Sphagno-Eriophoretum* plant community. A total of 155 species, including vascular plants, lichens, and bryophytes, were recorded in the nonacidic plant association. Numerous basiphilous species in this association have distributions restricted to Beringia or western North American (e.g., *Claytonia bostockii, Lagotis glauca, Novosieversia glacialis, Parrya nudicaulis* ssp. *septentrionalis, Potentilla biflora,* and *Saussurea angustifolia).*

Furthermore, plant community diversity (beta diversity) is higher within nonacidic tundra landscapes because of the abundance of nonsorted circles that have a different plant association (*Junco biglumis–Dryadetum integrifoliae;* Kade et al.

2005), which also has high vascular-plant species diversity (20 species per 1-m^2 plot). This greatly increases the amount of species and microhabitat diversity within nonacidic-tundra landscapes (Walker et al. 2011). Exceptional diversity has also been noted in the equivalent vicariate plant association, *Carici arctisibiricae–Hylocomietum alaskani*, on the Taimyr Peninsula in Russia (Matveyeva 1998).

Although the connection between wildlife and nonacidic tundra has apparently not been studied in Alaska, there are numerous characteristics that likely make it important to a wide range of animal species (Walker et al. 2001). The higher floral diversity and greater diversity of microhabitats undoubtedly also affects the local diversity of microbes and invertebrates. Our aerial and ground observations of wildlife during many years of vegetation mapping indicate that many mammals, including ground squirrels, caribou, muskoxen, wolves, and grizzly bears are more commonly found in areas that are rich in nonacidic habitats. The ecosystem properties of moist nonacidic tundra are analogous to those of the hypothesized "mammoth steppe" or steppe tundra of glacial Beringia (Guthrie 1990). Compared to acidic tussock tundra, nonacidic tundra has firm, well-drained, deeply thawed, nutrient-rich soils; high diversity of plant species and habitats; and plants low in secondary protective compounds. Plant-tissue calcium is also much higher in nonacidic tundra (Walker et al. 2001; Hobbie and Gough 2002) and could be a factor affecting wildlife patterns.

Within the southernmost tundras (bioclimate subzone E), nonacidic tundra is not common except on carbonate-rich tills or bedrock, river floodplains, and on late-Pleistocene glacial surfaces. Such areas are likely to be especially important to wildlife. Areas with a juxtaposition of acidic and nonacidic tundra may be particularly valuable to some species, such as caribou, that use different parts of the landscape during their annual migrations as plants change their phenological development.

Conclusion

Most studies that have related vegetation succession to glacial history have taken place on surfaces recently exposed by the retreat of glaciers (e.g., Chapin et al. 1994). The terrain around Toolik Lake offers a unique opportunity to study landscape evolution on glacial surfaces that span much longer periods of time. The analyses at Toolik Lake showed that glacial surfaces all have characteristic vegetation, geomorphology, and patterns of plant production and changes in productivity. Over time, plant succession trends toward peaty, wetter upland surfaces and infilling of lakes in lowland sites. The highest NDVI values occur on those portions of the landscape with abundant shrubs, such as water tracks, on moderate slopes, and on older glacial surfaces. Greening as indicated by NDVI during the period of available satellite imagery occurred heterogeneously across the landscape, with the most rapid change occurring in areas of recent disturbance and on relatively young glacial surfaces. The oldest surfaces show little change except in isolated areas that may be undergoing thawing permafrost. Acidic and nonacidic areas that are characteristic of relatively old versus young landscapes have very different ecosystem properties that are important to aquatic ecosystems (e.g., Kling et al. 1992, 2000) and wildlife, and to their response to climate change.

The surfaces discussed in this chapter span the second (younger) half of the Quaternary. Further north there are glacial deposits from early-Pleistocene and Tertiary times (Table 3.1; Hamilton 1986). Aerial photos and initial ground surveys of these areas indicate that these surfaces have even fewer lakes, more subdued topography, more abundant and larger water tracks, more abundant shrub cover associated with the water tracks, less dry heath vegetation on hillcrests, and even less diverse landscapes than the Sagavanirktok-age surfaces of this study. Thus, each different-aged surface spanning the entire Quaternary period in northern Alaska has distinctive assemblages of periglacial features and vegetation that are legacies of their geomorphic histories. These differences are clear and striking and can be quantified by means of maps derived from aerial photographs and satellite-derived data. Such differences are important with respect to water chemistry of streams and transport of materials into river systems and the general ecology of these regions.

Acknowledgments

NSF Grants OPP-9908829, OPP-0120736, ARC-0425517l, ARC-0531180, and ARC-0902175 provided funds for this manuscript. Numerous other grants provided funds for the original research, the NSF ARCSS FLUX project, the NSF ATLAS project, and the Department of Energy's R4D project. Many people and institutions have contributed to the fieldwork and the analysis of the data over the course of about 30 years, and unfortunately not all can be named here. Particular thanks go to personnel at the Institute of Arctic and Alpine Research, which housed the early GIS work for these projects, the staff and personnel at the Institute of Arctic Biology where the recent work has been done, the Toolik Field Station, VECO, which provided logistical support for much of the field work, and the Alaska Division of Geological and Geophysical Surveys, which supported much of Tom Hamilton's surficial geology mapping.

References

Alexandrova, V. D. 1980. *The Arctic and Antarctic: Their division into geobotanical areas.* Cambridge: Cambridge University Press.

Bhatt, U. S., D. A. Walker, M. K. Raynolds, J. C. Comiso, H. E. Epstein, G. Jia, R. Gens, J. E. Pinzon, C. J. Tucker, C. E. Tweedie, and P. J. Webber. 2010. Circumpolar arctic tundra vegetation change is linked to sea-ice decline. *Earth Interactions* 14:1–20.

Brosgé, W. P., H. N. Reiser, and J. T. Dutro, Jr., and R. L. Detterman. 1979. *Bedrock geologic map of the Philip Smith Mountains Quadrangle, Alaska.* Miscellaneous field studies map MF-879-B, scale 1:250,000. U.S. Geological Survey.

Bunn, A. G., S. J. Goetz, J. S. Kimball, and K. Zhang. 2007. Northern high-latitude ecosystems respond to climate change. *Eos Transactions: American Geophysical Union* 88:333.

Chapin, F. S., III., L. R. Walker, C. L. Fastie, and L. C. Sharman. 1994. Mechanisms of primary succession following deglaciation at Glacier Bay, Alaska. *Ecological Monographs* 64:149–175.

Gibbard, P. L., M. J. Head, M. J. C. Walker, and the Subcommission on Quaternary Stratigraphy. 2009. Formal ratification of the Quaternary System/Period and the Pleistocene Series/Epoch with a base at 2.58 Ma. *Journal of Quaternary Science* 25:96–102.

Gorham, E., C. Lehman, A. Dyke, J. Janssens, and L. Dyke. 2007. Temporal and spatial aspects of peatland initiation following deglaciation in North America. *Quaternary Science Reviews* 26:300–311.

Gough, L., G. R. Shaver, J. Carroll, D. L. Royer, and J. A. Laundre. 2000. Vascular plant species richness in Alaskan arctic tundra: The importance of soil pH. *Journal of Ecology* 88:54–66.

Guthrie, R. D. 1990. *Frozen fauna of the mammoth steppe: The story of Blue Babe.* Chicago: University of Chicago Press.

Hamilton, T. D. 1986. Late Cenozoic glaciation of the Central Brooks Range. Pages 9–49 in T. D. Hamilton, K. M. Reed, and R. M. Thorson, editors, *Glaciation in Alaska: The geologic record.* Anchorage: Alaska Geological Society.

Hamilton, T. D. 2003a. *Surficial geology of the Dalton Highway (Itkillik-Sagavanirktok rivers) area, southern Arctic Foothills, Alaska.* Professional report 121, scale 1:63,360. Alaska Department of Natural Resources, Alaska Division of Geological and Geophysical Surveys.

Hamilton, T. D. 2003b. *Glacial geology of Toolik Lake and the Upper Kuparuk River regions.* Biological papers of the University of Alaska, no. 26, scale 1:100,000. Fairbanks, AK: Institute of Arctic Biology.

Hobbie, S. E., and L. Gough. 2002. Foliar and soil nutrients in tundra on glacial landscapes of contrasting ages in northern Alaska. *Oecologia* 131:453–462.

Hobbie, S. E., and L. Gough. 2004. Litter decomposition in moist acidic and non-acidic tundra with different glacial histories. *Oecologia* 140:113–124.

Hobbie, S. E., T. A. Miley, and M. S. Weiss. 2002. Carbon and nitrogen cycling in soils from acidic and nonacidic tundra with different glacial histories in northern Alaska. *Ecosystems* 5:761–774.

Jia, G. J., H. E. Epstein, and D. A. Walker. 2003. Greening of arctic Alaska, 1981–2001. *Geophysical Research Letters* 30:HLS 3-1~3-4.

Jorgenson, M. T. 1984. The response of vegetation to landscape evolution on glacial till near Toolik Lake, Alaska. Pages 134–141 in *Inventorying forest and other vegetation of the high latitude and high altitude regions, proceedings of an international symposium.* Fairbanks, AK: Society of American Foresters Regional Technical Conference.

Kade, A., D. A. Walker, and M. K. Raynolds. 2005. Plant communities and soils in cryoturbated tundra along a bioclimate gradient in the low Arctic, Alaska. *Phytocoenologia* 35:761–820.

Kling, G. W., W. J. O'Brien, M. C. Miller, and A. E. Hershey. 1992. The biogeochemistry and zoogeography of lakes and rivers in arctic Alaska. *Hydrobiologia* 240:1–14.

Kling, G. W., G. W. Kipphut, M. C. Miller, and W. J. O'Brien. 2000. Integration of lakes and streams in a landscape perspective: The importance of material processing on spatial patterns and temporal coherence. *Freshwater Biology* 43:477–497.

Lambert, J. D. H. 1968. The ecology and successional trends in the low Arctic subalpine zone of the Richardson and British Mountains of the Canadian western Arctic. Ph.D. dissertation, University of British Columbia, Vancouver.

Mann, D. H., D. M. Peteet, R. E. Reanier, and M. L. Kunz. 2002. Responses of an arctic landscape to late glacial and early Holocene climatic changes: The importance of moisture. *Quaternary Science Reviews* 21:997–1021.

Matveyeva, N. V. 1998. Zonation of plant cover in the Arctic. St. Petersburg: Russian Academy of Science.

Munger, C. A., D. A. Walker, H. A. Maier, and T. D. Hamilton. 2008. Spatial analysis of glacial geology, surficial geomorphology, and vegetation in the Toolik Lake region: relevance to past and future land-cover changes. Pages 1255–1260 in D. K. Kane and K. M. Hinkel, editors. *Ninth International Conference on Permafrost.* Fairbanks: Institute of Northern Engineering, University of Alaska.

Myers-Smith, I. H., B. C. Forbes, M. Wilmking, M. Hallinger, T. Lantz, D. Blok, K. D. Tape, M. Macias-Fauria, U. Sass-Klassen, E. Lévesque, S. Boudreau, P. Ropars, L. Hermanutz, A. Trant, L. Siegwart Collier, S. Weijers, J. Rozema, S.A. Rayback, N. M. Schmidt, G. Schaepman-Strub, S. Wipf, C. Rixen, C. B. Ménard, S. Venn, S. Goetz, L. Andreu-Hayles, S. Elmendorf, V. Ravolainen, J. Welker, P. Grogan, H. E. Epstein, and D. S. Hik. 2011. Shrub expansion in tundra ecosystems: Dynamics, impacts, and research priorities. *Environmental Research Letters.* doi:10.1088/1748-9326/6/4/045509.

Oswald, W. W., L. B. Brubaker, F. S. Hu, and G. W. Kling. 2003. Holocene pollen records from the central arctic foothills, northern Alaska: Testing the role of substrate in the response to climate change. *Journal of Ecology* 91:1034–1048.

Raynolds, M. K., D. A. Walker, H. E. Epstein, J. E. Pinzon, and C. J. Tucker. 2012. A new estimate of tundra-biome phytomass from trans-Arctic field data and AVHRR NDVI. *Remote Sensing Letters* 3:403–411.

Raynolds, M. K., D. A. Walker, D. Verbyla, C. A. Munger. 2010. Changes in tundra vegetation over 25 years as measured by Landsat NDVI in the Upper Kuparuk River basin, North Slope Alaska, 1985–2009. Abstract GC43B-0984, American Geophysical Union Meeting, San Francisco, CA, December 13–17.

Reynolds, J. F., and J. D. Tenhunen. 1996. *Landscape function and disturbance in arctic tundra.* Berlin: Springer-Verlag.

Richmond, G. M., and D. S. Fullerton. 1986. Introduction to Quaternary glaciations in the United States of America. *Quaternary Science Reviews* 5:3–10.

Shippert, M. M., D. A. Walker, N. A. Auerbach, and B. E. Lewis. 1995. Biomass and leaf-area index maps derived from SPOT images for Toolik Lake and Imnavait Creek areas, Alaska. *Polar Record* 31:147–154.

Stow, D. A., A. Hope, M. D. Verbyla, J. Gamon, F. Huemmrich, S. Houston, C. Racine, M. Sturm, K. Tape, L. Hinzman, K. Yoshikawa, C. Tweedie, B. Noyle, C. Silapaswan, D. Douglas, B. Griffith, G. Jia, H. Epstein, D. Walker, S. Daeschner, A. Petersen, L. Zhou, and R. Myneni. 2004. Remote sensing of vegetation and land-cover change in arctic tundra ecosystems. *Remote Sensing of Environment* 89:281–308.

Tucker, C. J., and P. J. Sellers. 1986. Satellite remote sensing of primary production. *International Journal of Remote Sensing* 7:1395–1416.

Verbyla, D. 2008. The greening and browning of Alaska based on 1982–2003 satellite data. *Global Ecology and Biogeography* 17:547–555.

Walker, D. A., N. A. Auerbach, J. G. Bockheim, F. S. Chapin, III, W. Eugster, J. Y. King, J. P. McFadden, G. J. Michaelson, F. E. Nelson, W. C. Oechel, C. L. Ping, W. S. Reeburg, S. Regli, N. I. Shiklomanov, and G. L. Vourlitis. 1998. Energy and trace-gas fluxes across a soil pH boundary in the Arctic. *Nature* 394:469–472.

Walker, D. A., N. A. Auerbach, and M. M. Shippert. 1995. NDVI, biomass, and landscape evolution of glaciated terrain in northern Alaska. *Polar Record* 31:169–178.

Walker, D. A., E. Barbour, and H. Maier. 2009. Arctic geobotanical atlas: A plant-to-planet volume of maps and related information produced by the Alaska Geobotany Center. http://www.arcticatlas.org.

Walker, D. A., and N. Barry. 1991. Toolik Lake permanent vegetation plots: Site factors, soil physical and chemical properties, plant species cover, photographs, and soil descriptions. Unpublished data report, Institute of Arctic and Alpine Research, University of Colorado, Boulder.

Walker, D. A., J. G. Bockheim, F. S. Chapin, III, W. Eugster, F. E. Nelson, and C. L. Ping. 2001. Calcium-rich tundra, wildlife, and the "mammoth steppe." Quaternary *Science Reviews* 20:149–163.

Walker, D. A., P. Kuss, H. E. Epstein, A. N. Kade, C. M. Vonlanthen, M. K. Raynolds, F. J. A. Daniëls. 2011. Vegetation of zonal patterned-ground ecosystems along the North America Arctic bioclimate gradient. *Applied Vegetation Science* 14:440–463.

Walker, D. A., and K. R. Everett. 1991. Loess ecosystems of northern Alaska: Regional gradient and toposequence at Prudhoe Bay. *Ecological Monographs* 61:437–464.

Walker, D. A., N. D. Lederer, and M. D. Walker. 1987. Permanent vegetation plots: Site factors, soil physical and chemical properties, and plant species cover. Unpublished data report for R4D Program, U.S. Department of Energy, Institute of Arctic and Alpine Research, University of Colorado, Boulde.

Walker, D. A., M. O. Leibman, H. E. Epstein, B. C. Forbes, U. S. Bhatt, M. K. Raynolds, J. Comiso, A. A. Gubarkov, A. V. Khomutov, G. J. Jia, E. Kaarlejärvi, J. O. Kaplan, T. Kumpula, H. P. Kuss, G. Matyshak, N. G. Moskalenko, P. Orekhov, V. E. Romanovsky, N. K. Ukraientseva, and Q. Yu. 2009. Spatial and temporal patterns of greenness on the Yamal Peninsula, Russia: Interactions of ecological and social factors affecting the arctic normalized difference vegetation index. *Environmental Research Letters.* doi:10.1088/1748-9326/4/4/045004.

Walker, D. A., and H. A. Maier. 2008. *Vegetation in the vicinity of the Toolik Field Station, Alaska*, 1:63,360. Biological papers of the University of Alaska, no. 28. Fairbanks, AK: Institute of Arctic Biology.

Walker, D. A., M. K. Raynolds, F. J. A. Daniëls, E. Einarsson, A. Elvebakk, W. A. Gould, A. E. Katenin, S. S. Kholod, C. J. Markon, E. S. Melnikov, N. G. Moskalenko, S. S. Talbot, B. A. Yurtsev, and CAVM team. 2005. The circumpolar arctic vegetation map. *Journal of Vegetation Science* 16:267–282.

Walker, D. A., and M. D. Walker. 1996. Terrain and vegetation of the Imnavait Creek Watershed. Pages 73–108 in J. F. Reynolds and J. D. Tenhunen, editors, *Landscape function: Implications for ecosystem disturbance, a case study in arctic tundra.* New York: Springer-Verlag.

Walker, M. D., D. A. Walker, and N. A. Auerbach. 1994. Plant communities of a tussock tundra landscape in the Brooks Range foothills, Alaska. *Journal of Vegetation Science* 5:843–866.

Westhoff, V., and E. van der Maarel. 1978. The Braun-Blanquet approach. Pages 287–399 in R. H. Whittaker, editor, *Classification of plant communities.* Den Haag: W. Junk.

Whittinghill, K. A., and S. E. Hobbie. 2011. Effects of pH and calcium on soil organic matter dynamics in Alaskan tundra. *Biogeochemistry* 113:283–305. doi 10.1007/s10533-011-9688-6.

4

Late-Quaternary Environmental and Ecological History of the Arctic Foothills, Northern Alaska

W. Wyatt Oswald
Linda B. Brubaker
Feng Sheng Hu
George W. Kling

Introduction

To understand the modern processes and future dynamics of arctic ecosystems, in particular their response to climatic variability and anthropogenic perturbations, a paleoecological perspective can yield important insights. Records of past environmental change provide numerous natural experiments, including examples of past climate changes and associated ecosystem responses at a variety of temporal and spatial scales. Various types of late-Quaternary paleoenvironmental evidence are available for arctic Alaska, including geomorphology and soils, plant and insect macrofossils, lake-sediment geochemistry, and pollen records from peat deposits and lake sediments. In this chapter we summarize the environmental and ecological history of northern Alaska to characterize the responses of terrestrial and aquatic ecosystems to climatic changes of the past ~30,000 years (calibrated ^{14}C years, referred to here as "yr BP"). Our interpretations of past environmental and ecological changes focus mainly on evidence from the Toolik Lake area, although we also draw upon data from other parts of the arctic foothills and coastal plain. We pay special attention to pollen records of changing vegetation composition, as these data are available from many sites in the arctic foothills, including Toolik Lake itself (Bergstrom 1984).

Glacial Interval (~27,000–15,000 yr BP)

Several types of paleoenvironmental evidence suggest that cold and dry conditions existed in northern Alaska during the late Pleistocene. For example, glaciers advanced from the Brooks Range onto the arctic foothills, including the Toolik

Lake area, during the early phase of the Itkillik II glaciation (~27,000–21,000 yr BP; Hamilton 2003). Severe environmental conditions during the glacial interval are also evidenced by an extensive area of eolian dunes on the arctic coastal plain (Carter 1981) and the formation of sand wedges in the lower Kuparuk River basin by periglacial freezing and thawing (Carter 1983).

In lake-sediment records from sites in the arctic foothills, including Tukuto Lake (Oswald et al. 1999), Ahaliorak Lake (Eisner and Colinvaux 1990), and Oil Lake (also known as Lake E5; see chapter 8, Figure 8.1; Eisner and Colinvaux 1992; Oswald unpublished), Pleistocene pollen spectra are dominated by herbaceous taxa, including Poaceae, Cyperaceae, and *Artemisia* (Figure 4.1, Figure 4.2). Although these major taxa are difficult to interpret because of their broad ecological tolerances, minor taxa such as *Encalypta rhaptocarpa* and *Selaginella siberica* indicate sparse, xeric tundra. Taken together, the geomorphic and paleoecological data suggest cold, dry climatic conditions and xeric vegetation in northern Alaska during the late Pleistocene.

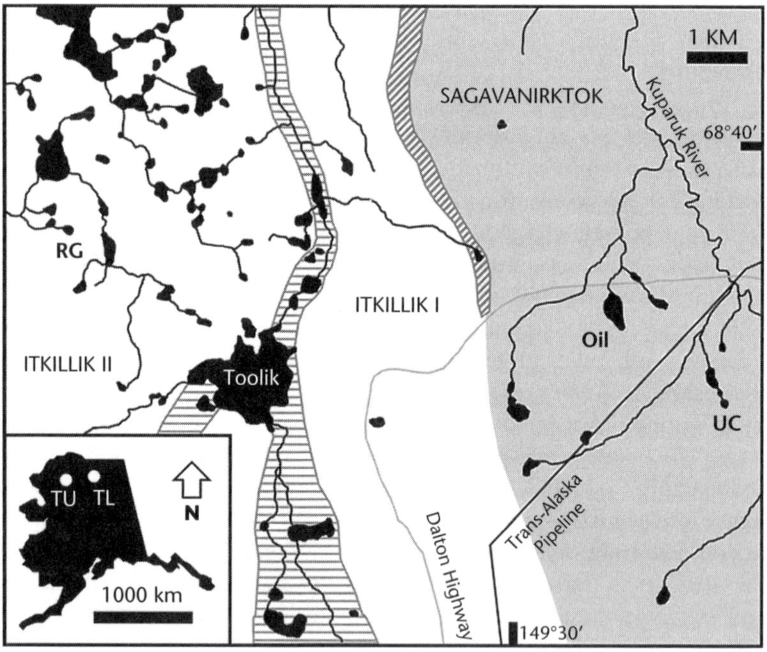

Figure 4.1 Map of glacial geology and study sites in the Toolik Lake area of the arctic foothills, northern Alaska. The Sagavanirktok glaciated surface is shaded gray, the Itkillik I and II surfaces are white, and areas of Itkillik I and II outwash are indicated by diagonal and horizontal lines, respectively (Walker et al. 1995; Hamilton 2003). Paleoecological study sites include Red Green Lake (Oswald et al. 2003b), Toolik Lake (Bergstrom 1984; Bixby 1992), Oil Lake (Eisner and Colinvaux 1992; Oswald, unpublished), and Upper Capsule Lake (Oswald et al. 2003b). The inset map shows the locations of Toolik Lake (TL) and Tukuto Lake (TU; Oswald et al. 1999) in northern Alaska.

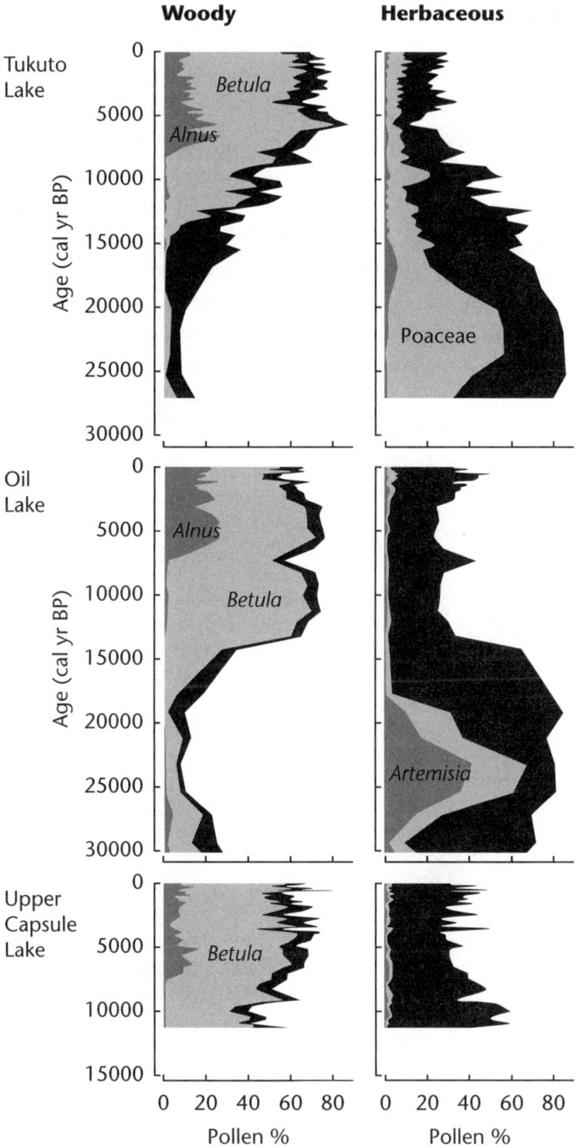

Figure 4.2 Pollen data from Tukuto Lake (Oswald et al. 1999), Oil Lake (Oswald, unpublished), and Upper Capsule Lake (Oswald et al. 2003b). Left-hand graphs show woody taxa; black fill is total woody, light gray is *Betula*, and dark gray is *Alnus*. Right-hand graphs show herbaceous taxa; black fill is total herbaceous, light gray is Poaceae, and dark gray is *Artemisia*.

Late Glacial (~15,000–11,500 yr BP)

During the transition from the Pleistocene to the Holocene, landscapes of arctic Alaska underwent major changes. A summary of late-glacial paleoenvironmental evidence by Mann et al. (2010) suggests that conditions between ~15,000 and 13,000 yr BP were warmer and wetter than during the glacial interval, leading to stabilization of eolian dunes, aggradation of streams, formation of organic peat deposits, higher lake levels, and the presence of scattered *Populus balsamifera* trees in the arctic foothills (Figure 4.3). The pollen record from Tukuto Lake (Oswald et al. 1999) indicates the development of more-continuous vegetation cover in the western arctic foothills, with shrub tundra communities dominated by Cyperaceae, *Salix*, and non-*Sphagnum* moss species and the rising abundance of *Betula* (Figure 4.2). In the Toolik Lake area, the late-glacial interval saw the re-advance of glaciers into the foothills during the Itkillik II late-phase glaciation (~15,000–13,000 yr BP; Hamilton 2003), perhaps due to increased snowfall. In the pollen record from Oil Lake, Poaceae and *Artemisia* pollen percentages decline while Cyperaceae and *Betula* percentages increase during the late-glacial interval, supporting the interpretation of a warmer, moister climate (Oswald unpublished; Figure 4.2).

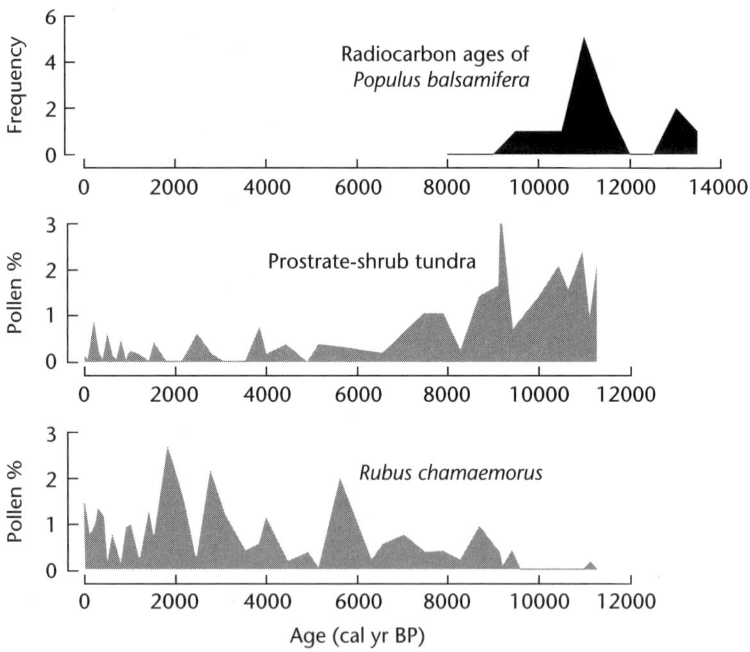

Figure 4.3. Diagrams showing: ^{14}C dates on *Populus* wood occurring beyond its present-day range limit in arctic Alaska (Hopkins et al. 1981; Mann et al. 2002; Kaufman et al. 2004); Upper Capsule Lake pollen-percentage sum for taxa indicative of prostrate-shrub tundra (Oswald et al. 2003a, 2003b); and Upper Capsule Lake pollen percentages for *Rubus chamaemorus*, an indicator of organic soils and tussock tundra (Oswald et al. 2003a, 2003b).

Several lines of evidence indicate that climate changes during the Younger Dryas interval (~13,000–11,500 yr BP) had effects on ecosystems in northern Alaska. Mann et al. (2002, 2010) report that for this period there are geomorphic and paleoecological signs of stream down-cutting, eolian dune activity, reduced peat formation, and a contraction in the range of *Populus balsamifera*, indicating a return to cold and dry conditions. In the pollen record from the Lake of the Pleistocene in the western arctic foothills, the Younger Dryas interval is characterized by increased pollen percentages for xeric tundra taxa, including *Artemisia*, Poaceae, *Selaginella siberica*, and *Juniperus* (Mann et al. 2002). Reduced abundance of *Pediastrum* algal cell nets in the Lake of the Pleistocene record may indicate lower lake levels during this interval. In the Toolik Lake area, the Itkillik II late-phase glacial advance ended around the onset of the Younger Dryas, perhaps in response to drier conditions.

Early Holocene (~11,500–7,500 yr BP)

After the end of the Younger Dryas interval, climate in arctic Alaska appears to have become warmer or wetter or both, presumably in response to a peak in summer insolation between ~11,000 and 9,000 yr BP (Kaufman et al. 2004). Fossil insects from an early-Holocene fluvial deposit in the northern arctic foothills indicate summer temperatures 2°C–3°C warmer than present (Nelson and Carter 1987), although some paleoclimatic records from other parts of Alaska do not feature elevated temperatures during the early Holocene (Clegg et al. 2011). The re-expansion of *Populus balsamifera* north of the Brooks Range (Figure 4.3) has been interpreted as a response to warm climatic conditions, but *Populus balsamifera* may also have established on newly stabilized gravel bars as stream aggradation increased the availability of alluvial deposits (Mann et al. 2002).

Pollen records from several sites in the arctic foothills, including Imnavait Creek, Oil Lake, and Upper Capsule Lake in the Toolik Lake area, suggest that additional ecological changes coincided with the early-Holocene expansion of *Populus balsamifera*. For example, relatively high percentages of *Selaginella siberica*, Polypodiaceae, *Equisetum*, and Rosaceae indicate well-drained soils and prostrate-shrub communities (Eisner 1991; Oswald et al. 1999, 2003b). Itkillik II–age landscapes were likely unstable at this time due to the melting of residual ice and low vegetation cover on coarse-textured soils. However, increasing pollen percentages of Ericaceae, *Rubus chamaemorus*, and other taxa indicative of mesic tundra, along with evidence of increased paludification (Mann et al. 2010), suggest that some moist substrates became available during the early Holocene, especially on fine-textured soils like those of surfaces dating to the Sagavanirktok River glaciation (>132,000 years ago; see chapter 3).

Middle to Late Holocene (~7,500 yr BP to present)

During the middle Holocene there appears to have been a shift from the warmth of the early Holocene to cooler conditions and higher effective moisture. *Populus balsamifera* became rare in the arctic foothills after ~9,000 yr BP (Figure 4.3), and *Alnus* expanded in the Brooks Range and southern arctic foothills between ~8,000 and 7,000 yr BP (Figure 4.2). The pollen record from Upper Capsule Lake, located on the Sagavanirktok surface 7 km east of Toolik Lake, suggests that prostrate-shrub tundra elements disappeared from the vegetation of older glaciated surfaces ~7,500 yr BP (Oswald et al. 2003b), marking the completion of the transition from relatively dry and sparse tundra to tussock tundra dominated by *Betula nana*, *Eriophorum vaginatum*, Ericaceae species, and *Sphagnum* mosses. Similar changes are seen in pollen records from other old-landscape sites in the arctic foothills (Eisner 1991; Oswald et al. 1999). In contrast, the pollen record from Red Green Lake, located on the Itkillik II surface 3 km northwest of Toolik Lake, does not indicate a dramatic change in vegetation composition at 7,500 yr BP. Vegetation cover may have increased somewhat at this time, but the pollen evidence suggests that the Itkillik II surface responded differently than the older surfaces and has supported prostrate-shrub tundra communities since the early Holocene.

Differences in soil texture and geomorphology between the young Itkillik II and old Sagavanirktok glaciated surfaces were likely responsible for the contrasting ecological responses at Red Green and Upper Capsule to the climatic changes of the early to middle Holocene (Oswald et al. 2003b). As effective moisture increased between ~10,000 and 7,500 yr BP, soil moisture became higher on the Sagavanirktok surface than on the Itkillik II surface because its fine-textured soils and flatter landforms retained more water than the better-drained soils of the Itkillik II landscape. Higher soil moisture led to the development of continuous vegetation cover, which led to a series of ecological and hydrological changes leading ultimately to the shift from prostrate-shrub tundra to tussock-tundra communities. Of primary importance were effects of the vegetation on the depth of the permafrost thaw layer. Higher plant coverage, including upright, woody species such as *Betula nana* and *Vaccinium uliginosum*, would have shaded and thus cooled the soil. In addition, these woody species, as well as mosses, would have contributed recalcitrant litter to the soil, thus helping to create a thicker organic layer and more insulation. The deep layer of organic matter, mosses, and Cyperaceae tussocks would have further reduced the amount of solar energy reaching the soil, and together these changes would have reduced soil temperature and depth of seasonal thawing. Shallower thawing reduced soil drainage, leading to anoxic and acidic conditions. The combined effects of cold soil temperatures, increased abundance of slowly decomposing litter, and soil anoxia and acidity would have greatly reduced decomposition rates, contributing to further permafrost aggradation. The outcome of this transition can be seen at present, as tussock-tundra soils have thicker organic horizons and shallower active layers than soils of drier tundra communities (Walker et al. 1995; Bockheim et al. 1998).

In contrast to the evidence suggesting that terrestrial ecosystems on the Sagavanirktok glaciated surface were more sensitive than those of the Itkillik II surface to middle-Holocene climate changes, geochemical data indicate a more-pronounced shift in the aquatic ecosystems on younger landscapes at

~7,500 yr BP than was the case for older glaciated surfaces (Kling 2000). In the record from Red Green Lake, an increase in sediment organic matter (OM) and a drop in the C:N ratio (Hu unpublished) indicate a rise in aquatic productivity at that time. The sediment record from Toolik Lake, which like Red Green is located on a younger landscape, also features an increase in OM during the middle Holocene (Bixby 1993). At Upper Capsule Lake, OM increases gradually throughout the Holocene, whereas C:N exhibits a step increase ~7,500 yr BP (Hu unpublished). This shift suggests a higher influx of terrestrial OM related to the changing vegetation and soils within the watershed. The C:N ratios are consistently higher for Upper Capsule than for Red Green since ~7,500 yr BP, supporting the inference that organic soils have been more abundant or thicker on older than on younger glaciated landscapes since the middle Holocene. Mollusk remains from Red Green Lake sediments also indicate further aquatic ecosystem changes because remains are present in sediments older than ~4,000 yr BP but absent in the youngest, late-Holocene sediments (Oswald 2006). This shift may reflect reduced alkalinity of the lake in response to acidification of the upland ecosystem.

Conclusions

A primary conclusion from paleoecological studies conducted in arctic Alaska is that the composition of tundra plant communities experienced dramatic changes in response to late-Quaternary environmental variations. The cold, dry climate of the glacial interval resulted in xeric tundra dominated by herbaceous taxa, including Poaceae, *Artemisia*, and Cyperaceae. Increasing temperatures during the late-glacial and early-Holocene intervals allowed the expansion of *Betula nana* and other shrub species, and *Populus balsamifera* stands were common in the arctic foothills when a peak in summer insolation at ~11,000–9,000 yr BP fostered warm temperatures. Effective moisture increased after the early-Holocene warm interval, reducing the prevalence of *Populus balsamifera* and allowing the regional expansion of *Alnus* between ~8,000 and 7,000 yr BP. This long-term perspective is consistent with modern studies illustrating the sensitivity of arctic tundra ecosystems to recent variations in climate (e.g., Chapin et al. 1995; Sturm et al. 2001).

The finding that sites with contrasting edaphic characteristics experienced different ecological responses to the middle-Holocene rise in effective moisture also has important implications for understanding ongoing and future changes in arctic tundra ecosystems. Substrate varies greatly across the Arctic, with heterogeneity occurring across spatial scales in response to glacial history, parent material, or eolian deposition (Walker et al. 1998; Walker 2000). The critical role of substrate in the ecosystem history of the Toolik Lake area suggests that this edaphic variability may impart strong spatial heterogeneity on the responses of arctic tundra ecosystems to future climate change.

Acknowledgments

We thank the following colleagues for their contributions to this research: Pat Anderson, Tom Brown, Terry Chapin, Wendy Eisner, Dan Gavin, Laura Gough, Tom Hamilton, Sarah Hobbie, Torre Jorgenson, Gary Michaelson, Chien-Lu Ping, Marilyn Walker, Skip Walker, and the staff of the Toolik Field Station. Two anonymous reviewers provided constructive feedback. This research was funded by National Science Foundation grant OPP-9615947, 9911278 and DEB-9810222.

References

Bergstrom, M. F. 1984. Late Wisconsin and Holocene history of a deep arctic lake, north-central Brooks Range, Alaska. M.S. thesis, Ohio State University, Columbus.

Bixby, R. 1993. The paleolimnology of two arctic lakes: Local and regional changes in climate. M.S. thesis, University of Cincinnati, OH.

Bockheim, J. G., D. A. Walker, L. R. Everett, F. E. Nelson, and N. I. Shiklomanov. 1998. Soils and cryoturbation in moist nonacidic and acidic tundra in the Kuparuk River basin, arctic Alaska, U.S.A. *Arctic and Alpine Research* 30:166–174.

Carter, L. D. 1981. A Pleistocene sand sea on the Alaskan Arctic coastal plain. *Science* 211:381–383.

Carter, L. D. 1983. Fossil sand wedges on the Alaskan arctic coastal plain and their paleoenvironmental significance. Pages 109–114 in *Permafrost: 4th International Conference Proceedings, Fairbanks Alaska.* Washington, DC: National Academy Press.

Chapin, F. S., III, G. S. Shaver, A. E. Giblin, K. J. Nadelhoffer, and J. M. Laundre. 1995. Responses of arctic tundra to experimental and observed changes in climate. *Ecology* 76:694–711.

Clegg, B. F., R. Kelly, G. H. Clarke, I. R. Walker, and F. S. Hu. 2011. Nonlinear response of summer temperature to Holocene insolation forcing in Alaska. *Proceedings of the National Academy of Sciences* 108:19299–19304.

Eisner, W. R. 1991. Palynological analysis of a peat core from Imnavait Creek, the North Slope, Alaska. *Arctic* 44:279–282.

Eisner, W. R., and P. A. Colinvaux. 1990. A long pollen record from Ahaliorak Lake, arctic Alaska. *Review of Paleobotany and Palynology* 63:35–52.

Eisner, W. R., and P. A. Colinvaux. 1992. Late Quaternary pollen records from Oil Lake and Feniak Lake, Alaska, USA. *Arctic and Alpine Research* 24:56–63.

Hamilton, T. D. 2003. *Surficial geology of the Dalton Highway (Itkillik-Sagavanirktok Rivers) area, southern arctic foothills, Alaska.* Professional report 121, scale 1:63,360. Alaska Department of Natural Resources, Alaska Division of Geological and Geophysical Surveys.

Hopkins, D. M., P. A. Smith, and J. V. Matthews, Jr. 1981. Dated wood from Alaska and the Yukon: implications for forest refugia in Beringia. *Quaternary Research* 15:217–249.

Kaufman, D. S., T. A. Ager, N. J. Anderson, P. M. Anderson, J. T. Andrews, P. J. Bartlein, L. B. Brubaker, L. L. Coats, L. C. Cwynar, M. L. Duvall, A. S. Dyke, M. E. Edwards, W. R. Eisner, K. A. Gajewski, A. Geirsdottir, F. S. Hu, A. E. Jennings, M. R. Kaplan, M. R. Kerwin, A. V. Lozhkin, G. M. MacDonald, G. H. Miller, C. J. Mock, W. W. Oswald, B. L. Otto-Bliesner, D. F. Porinchu, K. Rühland, J. P. Smol, E. J. Steig, and B. B. Wolfe. 2004. Holocene thermal maximum in the western Arctic (0 to 180° W). *Quaternary Science Reviews* 23:529–560.

Kling, G. W. 2000. A lake's life is not its own. *Nature* 408:149–150.

Mann, D. H., D. M. Peteet, R. E. Reanier, and M. L. Kunz. 2002. Responses of an arctic landscape to late glacial and early Holocene climatic changes: The importance of moisture. *Quaternary Science Reviews* 21:997–1021.

Mann, D. H., P. Groves, R. Reanier, M. L. Kunz. 2010. Floodplains, permafrost, cottonwood trees, and peat: What happened the last time climate warmed suddenly in arctic Alaska. *Quaternary Science Reviews* 29:3812–3830.

Nelson, R. E., and L. D. Carter. 1987. Paleoenvironmental analysis of insects and extralimital *Populus* from an early Holocene site on the arctic slope of Alaska. *Arctic and Alpine Research* 19:230–241.

Oswald, W. W. 2006. Mollusks in a Holocene lake-sediment core from the arctic foothills of northern Alaska. *The Nautilus* 120:30–33.

Oswald, W. W., L. B. Brubaker, and P. M. Anderson. 1999. Late Quaternary vegetational history of the Howard Pass area, northwestern Alaska. *Canadian Journal of Botany* 77:570–581.

Oswald, W. W., L. B. Brubaker, F. S. Hu, and D. G. Gavin. 2003a. Pollen-vegetation calibration for tundra communities in the arctic foothills, northern Alaska. *Journal of Ecology* 91:1022–1033.

Oswald, W. W., L. B. Brubaker, F. S. Hu, and G. W. Kling. 2003b. Holocene pollen records from the central arctic foothills, northern Alaska: Testing the role of substrate in the response of tundra to climate change. *Journal of Ecology* 91:1034–1048.

Sturm, M., C. Racine, and K. Tape. 2001. Increasing shrub abundance in the Arctic. *Nature* 411:546–547.

Walker, D. A. 2000. Hierarchical subdivision of arctic tundra based on vegetation response to climate, parent material and topography. *Global Change Biology* 6:19–34.

Walker, D. A., N. A. Auerbach, and M. M. Shippert. 1995. NDVI, biomass, and landscape evolution of glaciated terrain in northern Alaska. *Polar Record* 31:169–178.

Walker, D. A., N. A. Auerbach, J. G. Bockheim, F. S. Chapin, III, W. Eugster, J. Y. King, J. P. McFadden, G. J. Michaelson, F. E. Nelson, W. C. Oechel, C. L. Ping, W. S. Reeburg, S. Regli, N. I. Shiklomanov, and G. L. Vourlitis. 1998. Energy and trace-gas fluxes across a soil pH boundary in the Arctic. *Nature* 394:469–472.

5

Terrestrial Ecosystems at Toolik Lake, Alaska

Gaius R. Shaver
James A. Laundre
M. Syndonia Bret-Harte
F. Stuart Chapin, III
Joel A. Mercado-Díaz
Anne E. Giblin
Laura Gough
William A. Gould
Sarah E. Hobbie
George W. Kling
Michelle C. Mack
John C. Moore
Knute J. Nadelhoffer
Edward B. Rastetter
Joshua P. Schimel

Introduction

The broad aim of terrestrial research of the Arctic LTER is to develop a predictive understanding of the distribution of tundra ecosystems in the landscape; of the controls over their structure, functioning, and biogeochemical cycles; and of their interactions with each other and with the local and regional environment. Ultimately this leads to predictions of how they will respond to anthropogenic, development-related disturbances and to global environmental change. To meet this aim the research includes a combination of (1) comparative analyses of contrasting ecosystem types, (2) long-term monitoring of key system properties and processes in relation to local weather variation and secular climate change, and (3) long-term experimental manipulations of environmental factors related to global change. The conceptual framework for this research emphasizes interactions among multiple limiting factors in regulating biodiversity and overall carbon and other element budgets, with a particular emphasis on temperature, light, and soil nutrient availability as dominant controls. Effects of varying species composition on ecosystem properties and interactions of vegetation with herbivores are also important considerations (Shaver et al. 1992, 2000).

The following sections first describe and compare the overall structure and major functions of contrasting tundras near Toolik Lake. Key physical, chemical, and biological controls over distribution and function of these ecosystems are then described and compared. This leads to a discussion of timescales and trajectories of system-level change in response to changes in the controls, followed by a discussion of how system-level changes interact with and feed back on the controls. Finally, we predict how the expected changes in global and arctic climates may alter the landscape and terrestrial ecosystems near Toolik Lake, and suggest priorities for future research.

Biogeochemistry of Contrasting Tundras

A Mosaic of Ecosystems

The landscape near Toolik Lake consists of a diverse mosaic of tundra ecosystems. The dominant plants in the vegetation of these ecosystems may be evergreen, deciduous, or graminoid, woody or herbaceous, vascular or nonvascular. There are no trees except in a few isolated groves along rivers, and annual plants are extremely rare. The distribution of these ecosystems in the landscape is predictable and largely related to topographic position and landscape age. The fundamental environmental gradients in the landscape include a topographic-moisture gradient from dry uplands to wet lowlands (Figure 5.1) and a soil pH gradient related to time since deglaciation (chapter 3). Additional differences are due to local, fine-scale drainage at any point along the topographic gradient and to snow cover or winter exposure, which is also related to topography.

All of the tundra ecosystems near Toolik Lake are underlain by continuous permafrost ~200 m thick, so there is no deep drainage of soil water; unlike the freshwater springs and larger stream and lake systems in the area, there is little or no connection with groundwater (Hinzman et al. 1991). The depth of the annually thawed "active layer" of soil varies from ~30 cm to 1–2 m depending on topographic position, soil moisture and surface water flow paths, the thickness of the overlying organic and litter layers, and the structure and density of the vegetation canopy. In some locations the annual maximum depth of thaw does not reach below the surface organic mat, and in these locations the vegetation is essentially isolated from mineral soils. Soils are all gelisols (formed over permafrost), typically cold, wet, and with high organic content. The parent material within a few kilometers of Toolik Lake is fine-grained loess and glacial deposits of age ranging from ~10,000 to >300,000 yr BP; older glacial surfaces exist farther north (Hamilton 2003).

Vegetation Biomass and Productivity

The dramatic variation in species and functional-type composition of tundra vegetation (Figure 5.1 and chapter 3) is accompanied by an equally dramatic variation

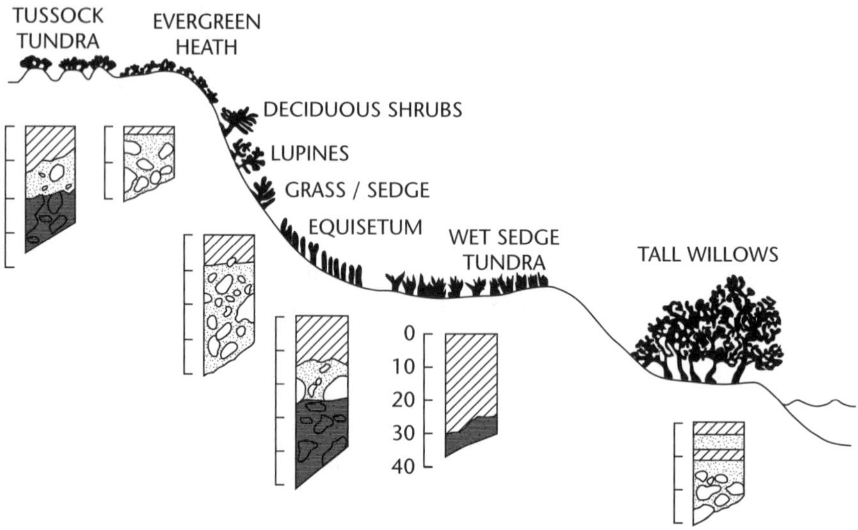

Figure 5.1 A typical toposequence of tundra vegetation and soils along a series of river terraces above the Sagavanirktok River. The horizontal distance is 50–250 m; vertical distance is 5–7 m. Soil profile depths are in centimeters. In the soil profiles, hatched areas indicate the upper organic mat with a mix of loess, till, and river gravel below. Dark shading at bottom indicates permafrost; where there is no dark shading, the profile does not extend to the bottom of the active (annually thawed) layer (from Giblin et al. 1991).

in standing biomass and its productivity. The live aboveground biomass of vegetation varies more than 15-fold among the major vegetation types in the Toolik Lake region, ranging from <90 to >1300 g m^{-2} (Figure 5.2). Although the majority of the total biomass (aboveground + belowground) is always composed of vascular plants, mosses account for >40% of the aboveground biomass in several vegetation types including some shrub tundras, some tussock tundras, and relatively well-drained wet sites (no standing water). Mosses are least abundant in extremely wet tundras (under standing water most or all of the summer) and in very dry heath tundras. Lichens are the largest component of aboveground biomass in some dry heath tundras and are important components of most other vegetation types except for tall shrub tundras and very wet tundras.

The vascular plant component of aboveground biomass varies at least 11-fold from ~80 to >900 g m^{-2}, including wide ranges in the relative abundance of major plant forms (Figure 5.2). In general the highest vascular biomass is found in tundras dominated by erect, woody shrubs, particularly deciduous *Betula* and *Salix* species; the lowest vascular biomass is found in very dry, evergreen lichen-heaths and very wet sedge tundras. Most of the live biomass of vascular plants lies below the soil-moss surface, including belowground stems and rhizomes (Figure 5.2) as well as roots. Belowground stem and rhizome mass alone is usually greater than aboveground vascular biomass, although the ratio of aboveground (A) biomass to belowground (B) stem mass varies from 0.3 to 1.3 among vegetation types, with

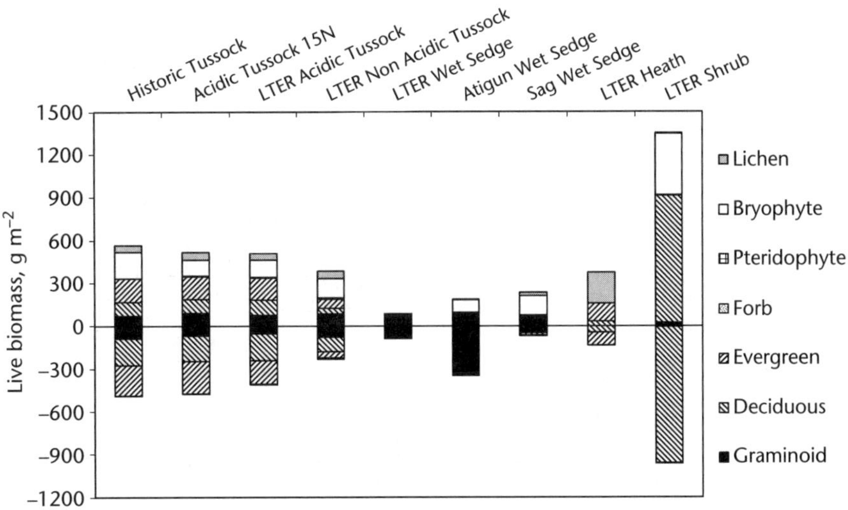

Figure 5.2 Peak-season (late July–early August) aboveground biomass and belowground stem mass in control plots of tundras near Toolik Lake. The first four sites are acidic and non-acidic tussock tundras; fifth through seventh sites are wet sedge tundras; eighth site is a dry heath; ninth site is a riparian shrub tundra. Aboveground biomass is shown as positive values on the vertical axis; belowground stem mass shown as negative values. Root biomass is not shown. Composition of the vegetation is indicated by separating biomass by plant functional type. Data are averages of harvests in 2–4 different years, except for the Atigun wet sedge and LTER shrub sites. (Original data in Shaver et al. 1998, 2001; Mack et al. 2004; Shaver and Chapin 1991; McKane et al. 1995; Gough et al. 1996, 2002; and Hobbie et al. 2005.)

belowground stems being relatively most important in wet sedge tundras. Root biomass has not been measured systematically in most of the tundras near Toolik Lake, but measurements in acidic tussock tundra range from 160 to 250 g m^{-2} (Mack et al. 2004; Sullivan et al. 2007; LTER unpublished data). Combining these root data with the belowground stem and rhizome data in Figure 5.2 gives a vascular A:B biomass ratio of 0.51:1 for the LTER moist acidic tundra site; including nonvascular plants, the A:B biomass ratio is 0.77:1.

Tussock tundra is the most widespread vegetation type in Alaskan upland tundra and is well represented by sites studied near Toolik Lake (chapter 3). Within tussock tundra, the most striking differences relate to soil pH. The three acidic tussock sites at Toolik Lake are all quite similar to each other in functional-type composition and in biomass (Figure 5.2), while the nonacidic tussock site is lower in biomass and has a similar functional-type composition but very different species composition (Hobbie et al. 2005). The nonacidic tussock site also has much higher species richness, including >40 vascular species (mostly forbs), while the acidic tussock sites contain only 9–14 vascular species (Gough et al. 2000). All four of these tussock sites are on relatively young (12,000–25,000 yr BP) soils. Tussock tundras on older soils are generally lower in vascular biomass with greater moss biomass, and have lower soil pH; examples include the tussock tundras at Imnavait

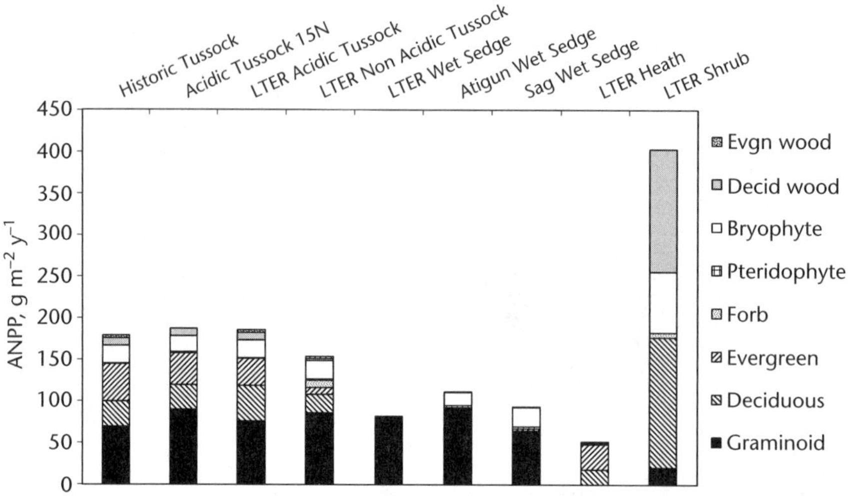

Figure 5.3. Aboveground net primary production at the same sites as Figure 5.2, determined from peak-season (late July–early August) harvests. The apical growth component of production was determined directly by separation of new leaves, twigs, and inflorescences in the harvests. Bryophyte-production and wood-production components were estimated by assuming constant percentage growth rates for bryophyte biomass and for evergreen and deciduous woody stems (Chapin et al. 1995). Lichen production was not measured or estimated at any site.

Creek and at the Sagavanirktok River, both within 50 km NE of Toolik Lake, and at Eagle Creek in central Alaska near Fairbanks (e.g., Shaver et al. 1991).

Annual aboveground net primary production of vascular plants also varies more than seven-fold, from <50 to >330 g m^{-2} (Figure 5.3). Most of this production is the result of growth from apical and intercalary meristems (i.e., growth of new leaves, twigs, and inflorescences). However, in sites dominated by woody deciduous shrubs, secondary growth from cambial meristems (i.e., wood production as annual rings) accounts for 30%–45% of aboveground production (Shaver 1986; Bret-Harte et al. 2002).

Belowground stem or rhizome production also has apical and secondary growth components. The apical component is small except in graminoids (Shaver et al. 1986a; Shaver and Chapin 1991) because most belowground stems of evergreen and deciduous species actually complete their apical growth above or on the surface of the ground and are later engulfed by upward growth of mosses, which buries them. Some evergreen species like *Vaccinium vitis-idaea* also produce apical growth from belowground stems, but this is small relative to apical growth aboveground. Secondary growth of belowground stems does not occur in graminoids and cannot be measured in woody belowground stems because stem diameters do not increase with age at a measureable rate once adventitious roots begin to appear after the stems are buried by mosses.

Estimates of root production at sites near Toolik Lake are available only for wet sedge tundra and moist acidic tussock tundra (Nadelhoffer et al. 2002; Sullivan et al. 2007). These estimates (75 g m^{-2} yr^{-1} in wet sedge tundra and 60–160 g m^{-2} yr^{-1} in tussock tundra) indicate that root production is about 70%–100% of aboveground vascular production, or 35%–50% of the total vascular production.

No direct estimates of moss or lichen production are available from the Toolik Lake region. However, if it is assumed that the relative growth rate of mosses is about 17% per year (Hobbie and Chapin 1998), moss productivity ranges from near zero to >120 g m^{-2} yr^{-1} (Figure 5.3). Moss production may account for 40%–60% of total aboveground production in old, relatively wet tussock tundras. Lichens probably account for more than half of total production in some dry heaths (Shaver and Chapin 1991), although this has not been measured.

Vignette 5.1. Plant Species Diversity in the Vicinity of Toolik Lake, Alaska, by Laura Gough

Plant species diversity in arctic tundra is limited overall by the low temperatures, short growing seasons, and other abiotic and biotic environmental conditions to which these species must be adapted in order to survive. There are no trees (except in a few, isolated groves) and annual plants are extremely rare. Most of the seed plants near Toolik Lake are long-lived perennials that produce new individuals clonally or vegetatively (without sexual reproduction). The graminoids, or grasslike species, survive the winter underground by storing large reserves of resources in belowground stems and roots. Dwarf shrubs, both deciduous and evergreen, maintain woody biomass aboveground through the winter and rely on snow cover during the coldest months to protect them from wind and frost damage. Perennial broad-leaved herbaceous plants, or forbs, also occur. Most arctic species produce flowers and thus are capable of reproducing sexually, although recruitment from seed is relatively rare. Mosses are a very important component of many tundra plant communities. In particular, *Sphagnum* mosses build peat and organic matter and maintain many tundra communities with highly acidic soil conditions. Lichens, consisting of a fungus and a photosynthetic organism (either a cyanobacterium or a green alga), are also important, and often serve as valuable food sources for mammals such as caribou.

Local abiotic conditions, often influenced by topographic position in the landscape, can restrict or facilitate occurrence of individual species and thus affect community diversity. Extremes of soil moisture, either very wet or quite dry, and rocky soils prevent certain species from germinating and surviving. Exposure during winter restricts some communities to very small plant species (less than 5 cm in height). There is a strong relationship between soil pH or acidity (influenced by soil age and dust inputs) and plant species diversity, so that less acidic soils support more species (Figure 5.V1). Several factors likely contribute to this relationship, including differences in element toxicity, nutrient availability, and occurrence of mosses that affect soil conditions. Soil nutrient availability has also been shown to affect plant species diversity experimentally, such that when nutrients are added, plant species diversity declines. This is likely occurring because nutrients allow one or two species to grow faster than the others, and they then competitively exclude the other species (see Vignette 5.4).

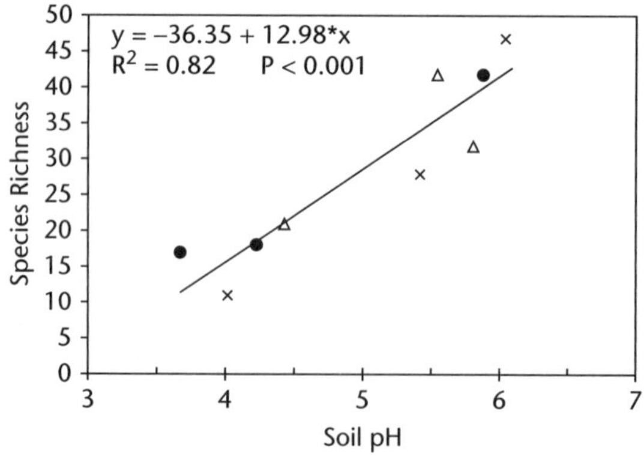

Figure 5.V1 Relationship between soil pH and vascular plant species richness for snowbed (triangle), watertrack (x), and tussock tundra (circles). From Gough et al. (2000).

Biotic factors also affect plant diversity. In the Toolik area, vascular plant species successfully partition resources such as light (by occupying different locations in the canopy) and soil nitrogen (by preferentially taking up different forms of nitrogen), suggesting niche differentiation that promotes coexistence and diversity. To date, insect herbivory has not been adequately measured, but observations suggest it is likely not very important for most plant species. Mammalian herbivory (by small mammals such as voles as well as caribou), however, may play an important role in particular tundra communities, and current research suggests it may become more important in the future with climate warming. Facilitation (positive interactions between plants) may also be important in arctic plant communities, although evidence from near Toolik is scarce to date.

Thus plant species diversity in the tundra communities around Toolik Lake is determined by the landscape position and particular characteristics of the area, such as soil moisture, nutrient availability, and exposure, as well as by interactions among plants and between plants and other organisms including mammals. As the region continues to warm, these factors are also changing, thus diversity of these communities will likely be affected as certain species benefit from these new conditions.

Consumers and Decomposers

Herbivores at Toolik Lake include small and large mammals, birds, insects, and other invertebrates. These animals affect the composition and biogeochemistry of the landscape through selective consumption of live and dead plant material, through their burrowing and trampling, and through their redistribution of elements in urine and feces. Although long-term, overall consumption of vegetation by arctic herbivores is relatively small in comparison with more temperate ecosystems (usually less than 5%–10% of annual aboveground primary production; Jefferies et al. 1994), herbivory in the Arctic is strongly influenced by spatial, temporal, and species-related variation in nutritional quality. This means that intense, short-term pulses of herbivory can have considerable long-term impacts on vegetation, particularly at fine scales (1–20 m).

The area around Toolik Lake supports a number of small and large mammalian herbivores. Five species of microtine rodents occur, the most abundant being tundra voles (*Microtus oeconomus*) and singing voles (*M. miurus*). These animals appear to be food limited as tundra vole abundance increases with food subsidies (Batzli and Lesieutre 1995), although voles show strong preferences for palatable plant species and are able to consume in quantity only some of the species present, in particular sedges and some dicots. Batzli and Hentonnen (1990) suggested that vole population dynamics—cycling every 3–5 years in areas close to waterways with more preferred food, and less frequently in upland areas (G. Batzli, personal communication)—might be correlated with predator abundance, particularly of foxes and weasels, as is true for lemmings in coastal Alaska (Batzli et al. 1980). The other common and often abundant small mammal is the arctic ground squirrel (*Spermophilus parryii*), which burrows extensively in dryer, deeply thawed soils. Small mammals serve as the primary food source for a number of bird and mammal predators including jaegers, owls, several species of raptors, foxes, wolves, and weasels. Caribou serve as a food source for larger predators such as wolves and grizzly bears.

Results of >10 years of limiting herbivore access to moist acidic tundra (MAT) and dry heath tundra at Toolik Lake indicate that small mammals have small but significant effects on plant community structure, primary production, and biomass under ambient nutrient conditions and at large spatial scales (Johnson 2008). At the level of the individual plant, however, and particularly under increased soil nutrient conditions, mammals clearly affect the vegetation (Gough et al. 2007, 2008). For example, *Eriophorum vaginatum*, the abundant tussock-forming sedge in MAT, does not easily recover from intense vole damage at the level of the individual tussock, even under increased nutrient conditions. One result of this may be competitive release of other co-occurring but less palatable species, such as *Betula nana*, which has greater growth in plots exposed to animal activity. In dry heath tundra, lichens are negatively affected by herbivory and by animal disturbances, and herbivory exacerbates the fertilizer-induced dominance of the grass *Heirochloe alpina*.

Caribou (*Rangifer tarandus*) are not major foragers in the Toolik Lake region in most years, as Toolik lies within the range of the central arctic herd but is not a calving ground (Griffith et al. 2002; Cameron et al. 2002; Lenhart 2002). Caribou are present, however, every year at Toolik Lake and every 5–8 years are abundant in late summer (August); their winter activities have not been documented, but their feces are found in abundance after snowmelt in some years, particularly in vegetation such as dry heaths, which has less snow cover than surrounding areas and also has abundant lichen cover.

Interannual variability in snowmelt and vegetation phenology propagates throughout the terrestrial food web. For example, several studies have shown that interannual variability in weather and plant growth exhibits significant control over the abundance and phenology of arthropods on the tundra (MacLean and Pitelka 1971; Pitelka 1973; Myers et al. 1979). Preliminary data collected at Toolik by

J. Wingfield (personal communication) show clear correlations between interannual variability in weather and arthropod abundance. Cold spring temperatures, for example, may result in late emergence and fewer flying insects than in other years.

Higher up on the trophic ladder, and thus dependent for food and shelter on climate, vegetation, and arthropod food sources, are arctic birds. Because the arrival time of migratory songbirds is strongly photoperiod dependent, breeding success of arctic avifauna has been shown to vary as a result of interannual climate variability, most notably with the timing of spring snowmelt (Hahn et al. 1995; Skinner et al. 1998; Martin and Wiebe 2004). Wingfield et al. (unpublished data) have shown that, near Toolik Lake, interannual variability in early season breeding activity (i.e., male call, song, and flight frequencies) of Lapland longspurs are strongly correlated with spring snowmelt date. Males sing more and show greater numbers of flight displays when the tundra is snow free. Stress hormone levels of the same species are elevated during storm events. Together, these results suggest that through a complex web of multi–trophic level interactions, interannual variability in climate dictates the breeding success of the tundra's migratory songbirds from year to year.

The most abundant consumer organisms in the Toolik Lake landscape are the decomposers, including bacteria, fungi, and other microbial and invertebrate members of the soil food web (Doles 2000). These soil organisms account for the majority of the heterotrophic respiration in the ecosystem (R_H) and for most of the recycling of essential elements like N and P from plant litter into forms that can be used by plants (discussed below). The abundance and composition of these communities is related to the productivity and composition of the vegetation aboveground (Gough et al. 2012). They also contain in their bodies a disproportionate amount of labile forms of these essential elements; for example, Schmidt et al. (2002) found that in moist acidic tundra and wet sedge tundra at Toolik Lake and in heath tundras at Abisko, Sweden, the amount of labile N and P in microbes typically equaled or exceeded the amounts in the vegetation.

Vignette 5.2. Belowground Food-Web Responses to Warming and Nutrient Additions, by John Moore

Soil ecological research at Toolik Lake has studied how current and anticipated changes in climate affect the dynamics of soils and soil biota in arctic tundra ecosystems. The soils of the dominant moist acidic tussock (MAT) ecosystem are characterized by a thick organic horizon overlying a poorly developed mineral layer. The microtopographic features include tussocks formed by the growth of sedges and shrubs and water-saturated inter-tussock depressions dominated by mosses. The organic layer that has formed is a result of the low pH of the plant debris (roots, stems, and leaves), soggy soils, and cold temperatures, which act to inhibit the activities of biota that decompose the materials. Over the past 30 years, the Arctic has experienced significant regional warming. A series of manipulations at Toolik Lake that included warming during the growing season with greenhouses (GH) and annual applications of nitrogen (10 g N m^{-2} yr^{-1} as NH_4NO_3) and phosphorus (and 5 g P m^{-2} yr^{-1} as P_2O_5) following snowmelt have driven shifts in plant species composition with a loss of mosses and increases in net primary production (NPP) and biomass, particularly among shrubs (Chapin et al. 1995; Gough et al. 2012). Nutrient additions dating back to 1982, have precipitated large net losses of soil organic C (40–100 g C m^{-2} yr^{-1}) in the organic and mineral horizons (Mack et al. 2004) and changes in the size distribution and quality of soil aggregates within the mineral horizon (Simpson 2010).

Nutrient additions and warming have initiated changes in the configuration of the belowground community as well (Figure 5.V2). The soil food web experienced structural shifts in terms of the diversity of functional groups, the vertical distribution of biomass with trophic position, and horizontal distribution of biomass within energy

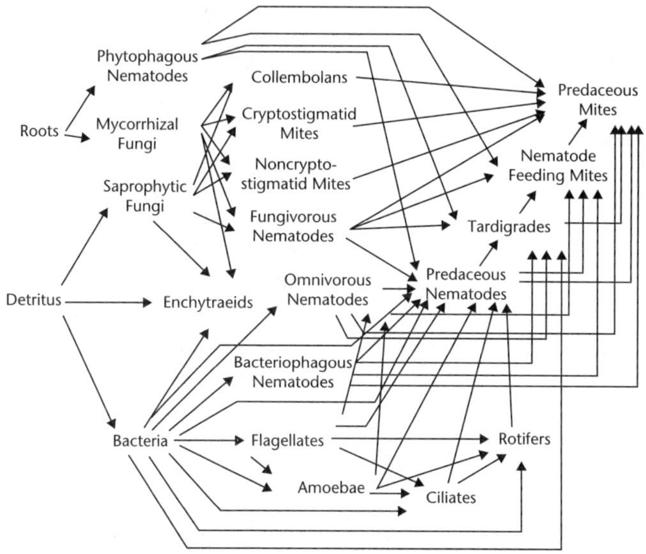

Functional Groups	Upper O-Horizon (0-5 cm)			Lower O-Horizon (5 cm-mineral)			Mineral (upper 5 cm)		
	Control	N+P	GH	Control	N+P	GH	Control	N+P	GH
Microbes g C m-2									
Bacteria	0.25	0.17	0.35	0.50ab	0.36b	0.97a	0.88	1.11	1.09
Fungi	41.47	66.63	78.79	287.91	99.16	148.02	111.12	52.15	92.42
Consumers mg C m-2									
Herbivores	0.41a	0.05b	0.08ab	0.09	0.04	0.02	--	0.02	0.02
Bacterivores	176.77	228.08	218.12	638.28a	334.41b	112.81b	288.35	565.33	218.84
Fungivores	21.37	11.77	14.73	15.37	5.14	10.89	1.88	0.63	1.97
Microbivores	--	1.35	1.26	--	4.23	16.69	--	--	--
Predators	45.16a	9.55b	53.55ab	31.63	4.31	21.59	1.68	1.82	2.43

Figure 5.V2 The belowground food web of the moist acidic tussock (MAT) tundra with biomass estimates of functional groups for three soil horizons from control, N + P, and greenhouse (GH) treatments. Sampling occurred during the summer of 2008. N + P treatments and GH treatments were initiated in 1996 and 1989, respectively. Fungi = saprophytic fungi + mycorrhizal fungi; Herbivores = phytophagous nematodes; Fungivores = collembolans, cryptostigmatid mites, non-cryptostigmatid mites, fungivorous nematodes; Bacterivores = bacteriophaous nemaodes + flagellates + amoebae + ciliates + rotifers; Microbivores = enchytraeids; Predators = omnivorous nematodes + predaceous nematodes + tardigrades + nematode feeding mites + predaceous mites. Letters that differ indicate significant differences among treatments within soil horizon.

channels. Phytophagous nematodes declined with nutrient additions and warming, but established themselves within the mineral layers of both treatments coincident with observed increases in root densities within the mineral layers. Enchytraeids, which were absent in controls, became established in treated plots. Fungi dominate the microbial biomass in all plots, but bacteriovores are the dominant consumers, indicating great activity within the bacterial energy channel. Activity within the fungal and bacterial energy channels appears less stratified in the treated plots. Predator biomass increased at all depths with warming and declined within the organic horizons following nutrient additions. These results and others point to an ecosystem that is vulnerable to changes in temperature and nutrient availability. The results also provide insights into the types of changes we can anticipate under continued climate warming.

Soil Organic Matter and Element Stocks

In tundra ecosystems, soils contain by far the largest pools of organic matter and essential elements such as C, N, and P. Near Toolik Lake the soils typically have an organic (O) horizon of varying thickness overlying a poorly developed mineral soil. The O horizon (often called the "peat layer") consists largely of poorly decomposed plant material with very little mineral content, while the lower, mostly mineral soil also contains significant organic matter and mineral elements. Because all of these soils are underlain by permafrost, the depth of the annually thawed "active layer" is a critical determinant of the volume of soil in which C and other elements actively cycle. Depending on the thickness of the O horizon and the depth of thaw, the active layer may consist entirely of organic soil, or it may include a thin (0–2 cm) O horizon above a meter or more of thawed, rocky mineral soil (Table 5.1). In most tundra soils there are also significant organic matter and element stocks below the active layer (within the top 1–2 m of permafrost), reflecting former conditions of deeper annual soil thaw and "cryoturbation" or mixing of the soil by freeze-thaw processes (Ping et al. 1997, 1998, 2008).

Both the thickness of the O horizon and the thickness of the active layer are highly variable (Table 5.1), reflecting a complex interaction among topography, soil moisture, drainage, the thermal properties of the peat and mineral soil, and the effects of surface litter and the plant canopy on heat flux into the soil. Near Toolik Lake the parent material is generally glacial till, outwash, or morainal material of various ages. The pH of the younger soils is generally close to neutral or slightly acidic (>5.5 or "nonacidic"), while the pH on older surfaces is often 3.5–4.5 or "acidic." Soil temperature and moisture also interact with the amounts and chemical composition of plant litter to determine the rate of decomposition and thus net organic matter accumulation or loss. The presence of permafrost is a key factor, because it means that soils are frozen solid in winter and thaw from the surface down only after snowmelt. Deep drainage of water does not occur through permafrost, meaning soil water must drain laterally over the permafrost surface and soils are always wet at depth. Although soil surface temperatures can exceed 30°C in the summer, the bottom of the thawed "active layer" is always at 0°C. The maximum thaw depth is not reached until the end of July or later, and when the soil refreezes

Table 5.1 Soil properties of LTER study sites near Toolik Lake.

Community	Site	Microsite	Horizon thickness, cm			pH		Bulk density, g cm^{-2}	
			Organic	Mineral	Active layer	Organic	Mineral	Organic	Mineral
Heath	hilltop heath		13	9	22	4.2	4.6	0.17	0.69
	LTER Heath		3	>1 m	>1 m	4.5		0.07	
Moist acidic tundra	LTER MAT	inter-tussock	25	15	40	3.7	4.6	0.07	0.67
	LTER MAT	tussock	17	17	34	3.7	4.3	0.27	1.01
	Sagwon acidic	inter-tussock	18	18	37	4.2	4.3	0.07	1.23
	Sagwon acidic	tussock	21	2	23	4.3		0.05	
	1981 acidic	tussock				3.8	4.2	0.07	0.76
Moist nonacidic tundra	LTER MNT	inter-tussock	11	9	20	5.9	5.9	0.13	1.05
	LTER MNT	tussock	27	8	35			0.14	1.56
	Sag tussock tundra	inter-tussock	22	0	22	5.6	5.3	0.13	
Shrub	LTER MAT	watertrack	24	13	36	4.0	4.3	0.07	0.47
	LTER MNT	watertrack	21	0	21	6.0		0.14	
	LTER Shrub		20	7	27			0.15	0.83
	riverside willow		7	7	14	6.8	6.7	0.27	0.80
	Sagwon acidic	watertrack	4	24	28	5.2	5.8	0.30	0.91
Snowbed	footslope		14	0	14	6.6		0.18	
	hillslope		22	0	22	6.2		0.27	
	LTER MAT	snowbed	3	6	9	5.6	5.5	0.23	0.48
	LTER MNT	snowbed	11	9	19	5.8	6.3	0.34	0.84
	Sagwon acidic	snowbed	8	20	28	4.4	4.3	0.34	0.95
Wet sedge	LTER wet sedge		51	0	51	5.8		0.12	
	Sag wet sedge		29	0	29	6.1		0.12	
Spring sites	Echooka	poplar	6	15	21	7.0	7.4	0.29	0.93
	Echooka	source	8	0	8	6.8		0.12	

in autumn it does so both from the top down and from the bottom up. This autumn refreezing may take eight to ten weeks to complete after the mean air temperatures go below freezing in September, depending mainly on snow cover, and even then microbial activity may continue as soil temperatures descend to below –5°C to –10°C (Grogan and Chapin 1999; Schimel et al. 2006).

In the O horizon, C stocks vary 40-fold, from <500 to almost 20,000 g m^{-2} (Figure 5.4; Giblin et al. 1991). Much of this variation is due simply to differences

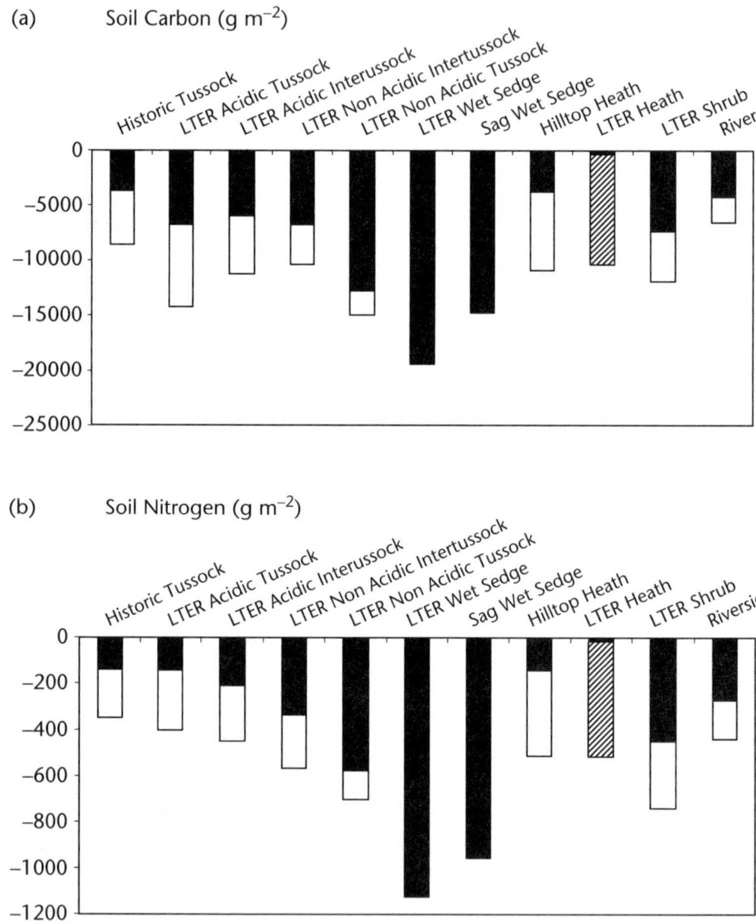

Figure 5.4. Soil C stocks (top) and N stocks (bottom) in tussock, wet sedge, heath, and shrub tundras. Shown are the total amounts of C and N within the seasonally thawed "active layer." The upper (black) segment indicates amounts within the organic horizon, and the lower (clear) segment indicates amounts within the mineral soil. Where there is no clear segment, the soil did not thaw beneath the organic mat. In the "LTER heath" site the depth of thaw is unknown so the C and N stocks in the upper 50 cm of soil are estimated (hatched segment). Inclusion of C and N stocks frozen within the upper meter of permafrost would increase the total stocks by at least 100% (Ping et al. 1997, 2008).

in the thickness of the O horizon, but variation in bulk density (usually ~0.06 g cm^{-3} but varying from 0.04 to 0.14) and C concentration (usually >35% but as low as 10%–15% depending on how much mineral material is incorporated) also contribute. N stocks of the O horizon are smaller but show similar patterns of variation (from <20 to >1100 g m^{-2}).

In the mineral soil, C and N stocks also vary greatly but generally inversely with the O horizon, so overall C and N stocks within the active layer are less variable than in the individual organic or mineral layers (Figure 5.4). In the mineral soil, C and N concentrations are much lower but bulk density is much higher, so the mineral soil per unit area often contains more C and N than the O horizon. The mineral soil horizons, however, are also colder than the O horizons, they spend a larger portion of the year frozen, and the organic matter they contain is less readily usable by soil microbes (Weintraub and Schimel 2003, 2005; Shaver et al. 2006). For these reasons the highest rates of element turnover occur in the O horizon.

Inputs, Outputs, and Turnover of Carbon, Nitrogen, and Phosphorus

Carbon

As in virtually all terrestrial ecosystems, the main input of C to tundras is by photosynthesis (gross primary production, GPP) and the main losses of C are as respiration by plants (R_A, autotrophic respiration) and by heterotrophs (R_H), largely microorganisms; respiration of the whole ecosystem (R_E) is the sum of R_A and R_H. In tundras most of the R_H occurs in the soil in association with decomposition processes, and soil respiration (not including root respiration) is often considered a measure of R_H. Soil carbon emissions in tundras also often include a significant methane component, particularly in wet sites with anaerobic soils (Walter et al. 2006). In addition to these gaseous fluxes, C may enter and leave patches of tundra as dissolved organic and inorganic C moving in soil water across the surface of the permafrost (Kling et al. 1991; Judd and Kling 2002), and as leaf litter redistributed by wind and snow (Fahnestock et al. 2000). Lateral losses of dissolved inorganic plus organic C range from ~2–5 g C m^{-2} yr^{-1} (see chapter 6).

In the landscape near Toolik Lake, GPP varies with vegetation composition, topographic location, time of year, and recent and current weather conditions including radiation, rainfall, and temperature (Figure 5.5). Despite considerable variation in photosynthetic rates at the leaf level among species (Starr and Oberbauer 2003), there is a strong convergence in canopy leaf area:leaf N relationships suggesting an "optimal" canopy structure in which GPP increases similarly with both leaf area and canopy N content (Williams and Rastetter 1999). About 75% of the variation in GPP is explained by leaf area and light (photosynthetic photon flux density, PPFD) alone (Shaver et al. 2007). In the middle of the growing season and at a constant PPFD of 600 μmol photons m^{-2} s^{-1}, GPP ranges from ~2 to 8 μmol CO_2 m^{-2} s^{-1} in wet, moist, and dry tundras at Toolik Lake and is closely correlated with canopy leaf area. Light-saturated (mid-day) GPP can be much higher (up to 15–20 μmol CO_2 m^{-2} s^{-1}), but the daily average rates are in the range of 1–3 μmol CO_2 m^{-2} s^{-1} (Oberbauer et al. 2007).

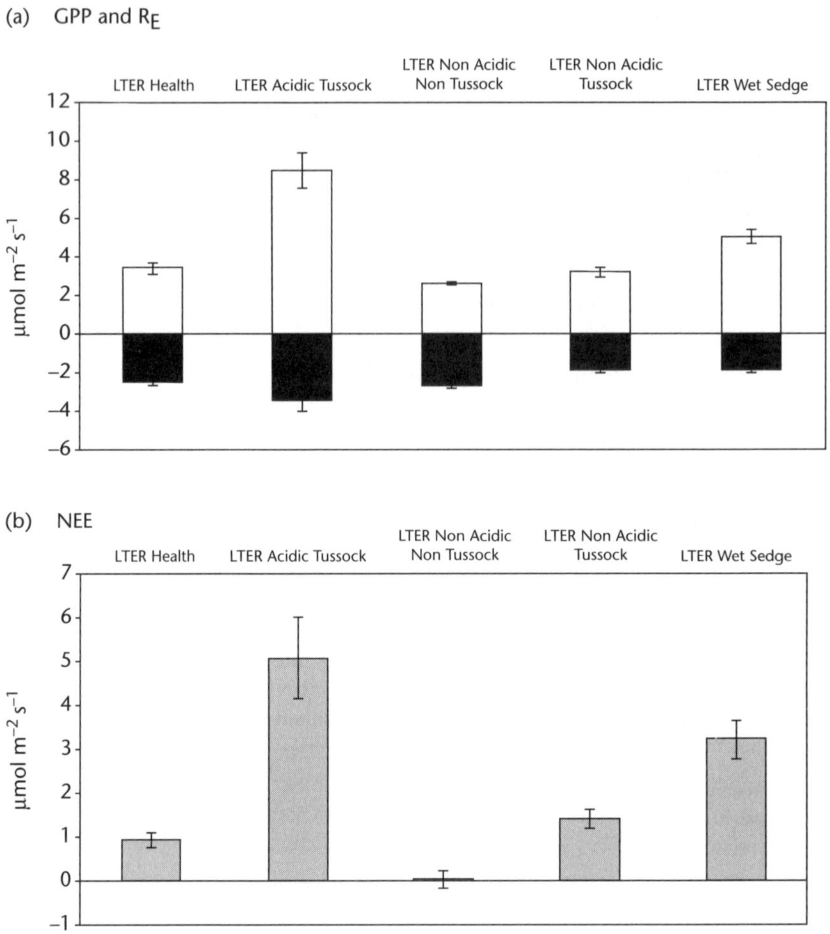

Figure 5.5 Midseason summer values for (A) GPP (open) and R_E (closed) and (B) NEP (same as NEE) at core LTER sites. GPP and NEP are predicted values at 600 μmol m^{-2} s^{-1} PPFD (photosynthetically active photon flux density), from light response curves in Street et al. (2007).

Most studies of respiration in the Toolik Lake region have focused on whole-ecosystem respiration (R_E; e.g., Hobbie and Chapin 1998) or have used bulk soil respiration (without roots) as an estimate of R_H (e.g., Hobbie et al. 2002; Schimel et al. 2006). Overall, R_E is correlated with GPP; in the long term, GPP is the ultimate source of all C lost in respiration, so R_E cannot exceed GPP indefinitely. Nonetheless, extrapolations of short-term measurements of R_E to seasonal and annual estimates consistently exceed estimates of GPP in the Toolik Lake region even during midsummer (Oberbauer et al. 2007).

Winter respiration probably accounts for at least 20% of annual R_E although there is great variability among estimates (3%–50%; Grogan and Chapin 1999).

One reason for the variability in "winter" respiration estimates is variation in the definition of when the "winter" starts and ends; most of the cold-season respiration actually occurs during the autumn (late September through December) and a smaller portion during the spring (May and early June; Schimel et al. 2006). Respiration rates of soil microorganisms (R_H) during the coldest parts of the winter are extremely low (e.g., Fahnestock et al. 1999). Such respiration is possible because even down to temperatures as low as –10ºC, liquid water films remain on soil particles and allow microbes to remain physiologically active. Although R_A has not been measured through the arctic winter, there is little reason to expect that R_A is significant except in the same fall and spring periods as with R_H; in fact it is likely that the aboveground component of R_A is very low through most of the autumn as air temperatures are colder than soil temperatures in autumn, winter, and spring.

Carbon losses from methane are rarely more than 5% of the total C loss from arctic landscapes; however, because methane has a higher greenhouse warming potential than CO_2, there may be a net greenhouse warming effect even when GPP exceeds R_E. Methane losses are greatest in wet, anaerobic soils and carbon-rich sediments such as the "yedoma" (loess) soils of northeast Siberia (Walter et al. 2006). Near Toolik Lake, these wet anaerobic soils include wet sedge tundras and moist tussock tundras where water is close to or above the soil surface and methane can be lost by diffusion and ebullition without passing through a surface horizon, where it may be consumed by methane oxidizers (Christensen et al. 2003). Methane losses may also be facilitated by passage through internal plant air spaces (Schimel 1995; Torn and Chapin 1993).

Net ecosystem exchange (NEE) is the balance of GPP and R_E. Midday, midsummer measurements of NEE near Toolik Lake range from a small net loss of C to the atmosphere (0–6 μmol m^{-2} s^{-1}), especially in wetter sites, to net gains of ~10 μmol m^{-2} s^{-1} to the tundra (Figure 5.5B). Nighttime values of NEE show consistent small C losses to the atmosphere despite the lack of full darkness in summer months. The summer value of NEE is most strongly correlated with leaf area and PPFD, reflecting the strong influence of these variables on the GPP component, although there is also a significant effect of daily variation in temperature on R_E (Williams et al. 2006). Integrated over a full day, midsummer NEE of tundra usually shows a net C gain of 0.5–1.5 g C m^{-2} d^{-1} (0.04–0.13 mol C m^{-2} d^{-1}) and is also correlated with leaf area (McFadden et al. 2003). Over the three summer months when GPP exceeds R_E on most days, cumulative NEE at Imnavait Creek near Toolik Lake varies from 50 to 95 g C m^{-2} stored in the tundra (Euskirchen et al. 2012). Over longer periods, NEE may be positive or negative and is related to summer weather (especially summer rainfall) and to multiyear changes in temperature and evapotranspiration.

Nitrogen

The main N inputs are by bacterial fixation and by deposition in rainfall and snow (Figure 5.6). The highest rates of N fixation (per g tissue) occur in the roots of

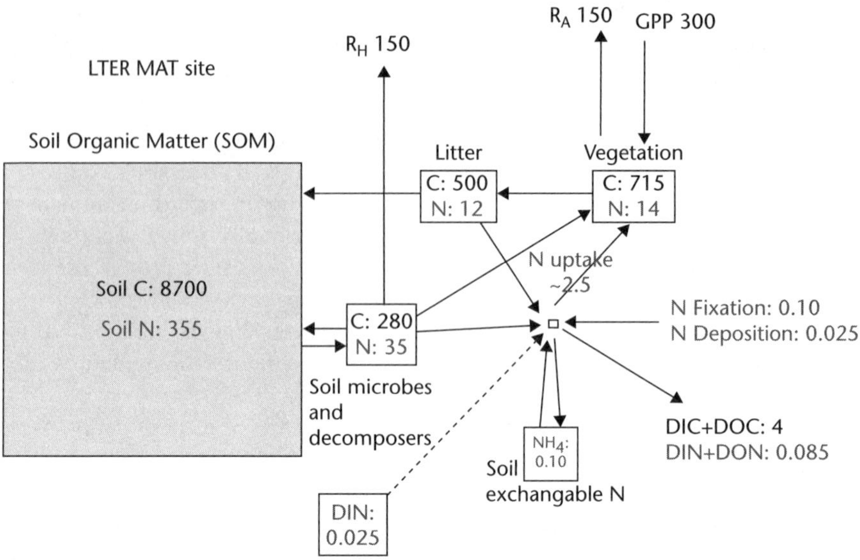

Figure 5.6 Summary of C (black) and N (grey) budgets in moist acidic tundra at Toolik Lake. Pool sizes (boxes) in g m^{-2}; fluxes (arrows) in g m^{-2} yr^{-1}; the dashed box and arrow indicate the DIN pool size (very small). Values are estimates, chosen for consistency among the numerous individual studies described in this chapter.

legumes such as *Lupinus* species and actinorrhizal shrubs such as *Alnus* species, but the majority of the N fixation (per m^2 ground area) occurs in free-living algae or in moss-algal associations and in several lichen species. The lichen *Peltigera apthosa* in particular is widely distributed especially in moist tundra and is capable of very high rates of fixation (0.77 μmol N g lichen^{-1} h^{-1}; Weiss et al. 2005). Although these lichens constitute important hotspots of N fixation, the algal-associated inputs are probably more important over the whole landscape (Hobara et al. 2006). At Imnavait Creek near Toolik Lake, the overall average rate of N fixation is 80–131 mg N m^{-2} yr^{-1}. These inputs by fixation are five- to ten-fold larger than the long-term average summer (unfrozen) deposition rate of ~14 mg N m^{-2} yr^{-1}. Winter deposition in snowfall is similar on average to the summer deposition but much more variable due to redistribution of snow, from near zero in windblown, snow-free hilltops to >50 mg N m^{-2} yr^{-1} in snow accumulation areas (Shaver et al. 1991).

The gross inputs of N by fixation and deposition (0.1–0.2 g m^{-2} yr^{-1}) are considerably smaller than the annual vegetation uptake requirement for N used in NPP (0.4–4.5 g m^{-2} yr^{-1}), indicating that most of the vegetation N supply must come from recycling of N already in the ecosystem. This recycling occurs as the mineralization of soil organic N to ammonium or nitrate, followed by plant uptake, and by short-circuits of this process including uptake via mycorrhizae (Hobbie and Hobbie 2006) and direct plant uptake of organic N (Kielland 1994; Schimel and Chapin 1996).

Table 5.2 Annual net N mineralization rates (g N m^{-2} y^{-1}) using buried bags (Giblin et al. 1991; J. Laundre, unpublished data). In this case the year runs from early August to early August; values are means among years

	NH_4	NO_3	Total N	Years
LTER acidic tussock	0.40	0.02	0.42	1990–2006
LTER nonacidic tussock	0.03	0.00	0.03	1998–2006
Sag river tussock	0.15	0.03	0.18	1987–2005
Sag river riparian shrub	0.16	0.23	0.40	1985–1989
Sag river wet sedge	0.41	0.02	0.43	1985–1989
Sag river equisetum	0.06	0.01	0.08	1985–1989
Sag river shrub/lupine	0.08	0.04	0.12	1985–1989
Sag river heath	0.64	0.04	0.67	1985–1989

Net N mineralization rates have been estimated in several studies near Toolik Lake, under the assumption that the inorganic N supply to plants is essentially the remainder of the gross N mineralization after microbial N uptake (Table 5.2). Net N mineralization, however, is insufficient to account for total plant N uptake, usually amounting to only about one-third of the amount needed. Typically during the growing season, net N immobilization occurs, with net N mineralization occurring only during the winter. Thus, by this measure, plants appear to compete directly with microbes for uptake of N produced by gross mineralization processes, in which case net mineralization underestimates the mineral N available to plants (Schmidt et al. 2002). Alternatively, plants may use organic N in addition to mineral N (McKane et al. 2002). Both of these forms appear to be used at least for abundant, nonmycorrhizal graminoid species such as *Eriophorum vaginatum*. In the case of the majority of species that are mycorrhizal, including dominant shrubs such as *Betula nana*, N uptake via mycorrhizae probably accounts for more than half of N uptake (Hobbie and Hobbie 2006). According to these authors, the plants with low values (^{15}N depleted) of ^{15}N in the leaves in Figure 5.7 are receiving large amount of N from mycorrhizal fungi.

The main losses of N from tundra ecosystems are by leaching as dissolved organic and inorganic N and by denitrification. Leaching losses, determined as the amounts of N leaving tundra watersheds in stream flow, are ~50–100 mg N m^{-2} yr^{-1} (Peterson et al. 1992; McNamara et al. 2008), about the same magnitude and perhaps smaller than the sum of the N inputs by fixation and deposition. Losses by denitrification are still uncertain and have not been measured directly; the only study to date (Alexander-Ozinskas 2007) indicates that under optimum conditions N loss by denitrification may be as high as 1–2 g m^{-2} yr^{-1} in wet and moist tundras where soils are frequently anaerobic, and near zero in drier, well-aerated heath soils. These high rates of potential denitrification are unlikely to occur under field conditions.

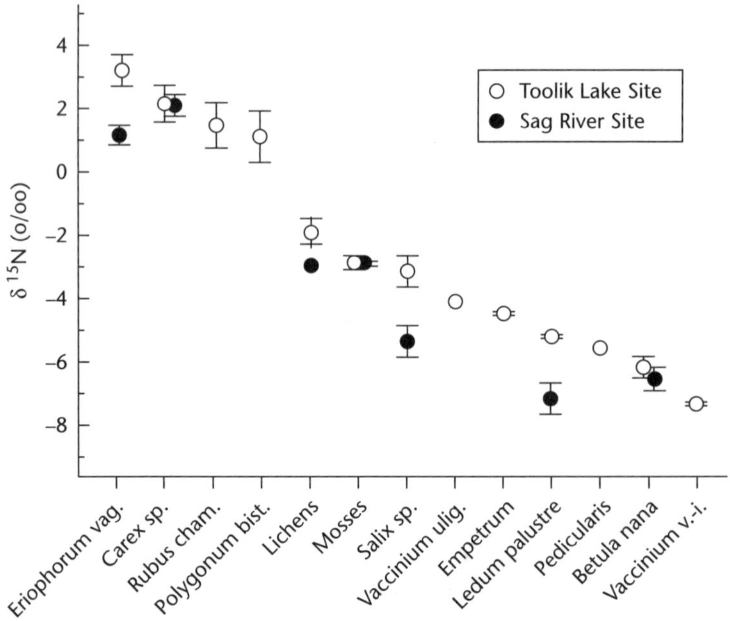

Figure 5.7. Natural abundance N isotope ratios of leaves of common species of moist acidic tundra (from Nadelhoffer et al. 1996). Leftmost species are graminoids, Rightmost species are mostly woody evergreen Ericaceae or deciduous shrubs.

Vignette 5.3. Mycorrhizal Fungi Provide Nitrogen to Plants: A Vital Symbiosis, by John Hobbie

Most of the plants in the Toolik region (in fact, 90% of all plants) are symbiotic with fungi; plants provide sugars to the fungi, and the fungi provide several kinds of nutrients and probably water to the plants. Threads of fungal hyphae, each 5–15 μm in diameter, attach to plant roots in several ways and extend out as far as a meter to mine the soil for carbon, nitrogen, and phosphorus. At Toolik, the ectomycorrhizal (ECM) fungi are symbiotic with birch and willow roots, while the ericoid mycorrhizae are symbiotic with ericaceous plants such as *Ledum*, blueberry, and cranberry. Exactly how much carbon and nutrients are transferred is difficult to quantify but the benefits to the plants must be large or they would not give up an estimated 20% of their net photosynthetic sugars to the fungi (Hobbie and Hobbie 2008). The functional relationship of ectomycorrhizal and ericoid mycorrhizal fungi to plants is just beginning to be understood; pure cultures show that many fungi have the enzymes to decompose proteins.

The low stature of the vegetation, the abundance of fruiting bodies of mycorrhizal fungi, and the nitrogen limitation of plant growth made the tussock tundra at Toolik an ideal location for quantification of the fungal nitrogen cycle. This is accomplished through analysis of the ^{15}N content of soil nitrogen, of the ectomycorrhizal fruiting bodies (mushrooms) that are composed of hyphae, and of the ectomycorrhizal plant stems and foliage; it has the great advantage over other ways of studying a soil process in that the natural isotope abundance is measured and the belowground system remains undisturbed.

The process that makes the analysis possible occurs during the transfer of nitrogen from the hyphae to the plant. At Toolik (Hobbie and Hobbie 2006), the soil organic matter has a $\delta^{15}N$ of +1–2‰, the plant values average ~–5‰ for ectomycorrhizal and ericoid mycorrhizal plants, and the ectomycorrhizal fungi average +7‰. Note that in Figure 5.7 this analysis applies only to the plants with negative $\delta^{15}N$ values, the ectomycorrhizal and ericoid mycorrhizal plants. At the tip of the hyphae, enzymes are produced that break down soil proteins into amino acids. The hyphae take up the amino acids and transport them back to the roots. Before transfer, all the amino acids are transformed to the amino acid glutamine; this transamination process favors the light isotope, ^{14}N, and results in glutamine with less ^{15}N being transferred to the plant. The amino acids and chitin with more ^{15}N remain in the hyphae. From a mass balance of the ^{15}N, 61%–86% of the nitrogen in plants came from the fungal pathway, while 8%–17% of the net photosynthetic carbon was transferred to fungi. A slightly different analysis (Yano et al. 2010b), using the $\delta^{15}N$ of hydrolysable organic nitrogen, estimated that 30%–60% of the plant nitrogen came from fungi. The warming of arctic plants, already known commonly to increase growth, also leads to a shift in their allied ECM fungal community; moreover, the new community of fungi has a distinctly different ecological function (Deslippe et al. 2011). A warming experiment took place over 18 years in acidic tussock tundra at Toolik where greenhouses raised the temperature of plants and soil ~2°C causing a dramatic increase in height of the *Betula nana*. An analysis of the internal transcribed spacer sequences of ECM fungi in 1,000 root tips from *Betula* found that the ECM fungi in control, fertilized, and fertilized-plus-warming treatments were dominated by a *Russula*-associated community, while the fungi in the warming treatment changed to a *Cortinarius*-dominated community (Deslippe et al. 2011). Based on data on fungal exploration types, the warming enhanced growth of *Betula,* which responded to an increasing nutrient limitation (likely nitrogen) by increasing organic carbon transport to roots. Growth of roots and the *Russula*-associated fungal community depleted the more labile organic and inorganic nitrogen compounds in the near vicinity of the roots. As a result, the community of ECM fungi shifted to *Cortinarius* species, which can mine nitrogen from widely dispersed and recalcitrant organic matter and transport solutes back to fruiting bodies and plants in hydrophobic rhizomorphs (a rootlike mass of hyphae associated with transport). The hyphae and rhizomorphs of *Cortinarius* species can be up to a meter long. The ECM community shifts in the experiments described here not only help in understanding how ecological function differs among mycorrhizal taxa, but may also presage changes in the diversity of ECM communities as the arctic tundra continues to become warmer and shrubbier.

Phosphorus

Much less is known about the P cycle near Toolik Lake, although fertilizer experiments have indicated P-limitation to productivity of the vegetation in most wet sedge tundras and some moist tussock tundras (Shaver and Chapin 1986, 1995). Dissolved inorganic P in soil solution was low (0.3–0.4 μmol L^{-1}) and frequently below detection limits along a sequence of six tundras on a hillslope above the Sagavanirktok River, but KCl-extractable PO_4-P varied more than 20-fold (Giblin et al. 1991). The greatest amounts of KCl-extractable and resin-exchangeable P occurred in deeply thawed soils such as the hilltop heath, *Equisetum* tundra on a steep hillslope, and riparian shrub tundra (Figure 5.1). In younger soils most of the P occurred as primary mineral P (HCl-extractable and "residual" P), while the amounts of Fe- and Al-bound P increased in older soils.

The largest pool of actively cycling P is held in the top 30 cm of soil organic matter, about 19.5 and 22 g P m^{-2}, respectively, in moist tussock tundra and wet sedge tundra at Toolik Lake (Schmidt et al. 2002). Soil microbial P is the second largest organic P pool, accounting for about 3 g m^{-2} in moist tundra and 1.3 g m^{-2} in wet sedge. The P content of live vegetation is the smallest of these pools, less than 2 g m^{-2} in moist tussock tundra and less than 1.2 g m^{-2} in wet sedge (Shaver and Chapin 1991).

Overall Element Budgets and Stoichiometry

An overall conceptual model of element cycling and element interactions is needed to integrate and compare the above information. We do this in the context of a generalized organic matter budget for a terrestrial ecosystem, accounting for the fact that all organic matter is composed of multiple, essential, nonsubstitutable elements and that the relative abundance of those elements changes among different kinds of organic matter. When the data on C and N cycles are combined in this way (Figure 5.6; Shaver et al. 1992), several key facts emerge:

First, the budget components include organic matter pools that differ widely in their magnitudes, turnover rates, and C:N ratios, yet they are all linked through the exchanges of these elements among pools, either as organic matter containing both C and N or as inorganic N.

Second, the "external" inputs and outputs of N (fixation, deposition, denitrification, and leaching losses) are all small relative to the "internal" exchanges of N among organic matter pools (plant uptake, litter fall, mineralization). In other words, the N cycle of this model tundra is relatively closed and depends strongly on internal recycling of N.

Third, in contrast to the situation for N, the "external" inputs and outputs of C (GPP, R_A, R_H, leaching losses) are mostly large relative to "internal" exchanges of C (litter fall, plant uptake of organic molecules, consumption by microbes, leaching losses, and [not shown in figure 5.6] herbivores); in other words, the C cycle is relatively "open" in comparison with the N cycle.

These facts indicate that controls on short- and long-term change in either the C cycle or the N cycle depend on controls over *both* elements. In particular, in the short term the C balance of such a system can be thought of as C gains ($NPP = GPP–R_A$) associated with plant N uptake balanced against C losses (R_H) associated with the breakdown of litter and soil organic matter into forms of N that can be taken up by plants. In the long term there are three general kinds of controls that should mediate change in overall organic matter stocks of this model ecosystem:

(1) *Controls on variability of C:N ratios in both pools and fluxes.* As the range in possible C:N ratios changes, the greater the possibilities for changes in the C budget without change in the N budget, and vice versa. The C:N ratios of pools and fluxes may change for many reasons, including changes in species composition (and thus the chemical composition of plant tissues), changes in processes like N resorption at leaf litterfall (and thus the C:N ratio of the litter), and changes in microbial growth efficiencies (and thus the amount of CO_2 lost per unit of microbial biomass produced).

(2) *Controls over the distribution of organic matter among plants, soils, and microbes.* For example, because soil organic matter always has a lower C:N ratio than plant biomass, the transfer of N from soil organic matter to plant biomass will result in a higher carbon storage in the ecosystem as a whole, even if there is no change in total N in the ecosystem. Microbes, in contrast, have very low C:N ratios, so consumption of soil organic matter by microbes, associated with increases in microbial biomass, results in loss of C while N is retained.

(3) *Changes in the balance of element inputs and outputs and thus total element stocks.* In the case of this model tundra, the C cycle is relatively "open" and the N cycle is relatively "closed." This means that short-term increases in C inputs (such as the expected photosynthetic response to warming) cannot be sustained without a corresponding increase in N inputs, changes in the distribution of N, or changes in C:N ratios as described above. In general it means that long-term changes in organic matter stocks and turnover in this system should be more sensitive to changes in factors controlling N balance, N turnover, and N distribution than to changes in factors controlling C balance and distribution.

By incorporating these controls and relationships between C and N, the "simple arctic model" (SAM; Shaver et al. 1992) outlined in Figure 5.6 provides an overall, integrated framework for interpretation of experimental and monitoring studies of the terrestrial Arctic LTER. This framework has proven particularly useful in developing long-term predictions (decades to centuries) of change in response to climate change and human impacts on the tundra near Toolik Lake.

Controls on Ecosystem Structure and Function

Climatic and Other Environmental Controls

The major environmental controls on tundra ecosystems near Toolik Lake include temperature, light, soil moisture, snow cover, and length of the growing season. Effects of changes in these controls have been studied in a wide range of short-term and long-term experiments, often involving factorial combinations of two or more controls. Because past research showed the productivity of these systems is consistently limited by N or by N and P availability to plants, N and P fertilization experiments were also included in all of the major terrestrial ecosystems near Toolik Lake. The major overall lessons from these experiments include:

(1) On time-scales of 1–20 years, the greatest changes in primary production, plant biomass, species composition, and soil processes are caused by N and P fertilizers, indicating strong control of the C cycle by the availability and turnover of these elements (Figure 5.8).
(2) Essentially all ecosystem processes in the tundra are temperature-limited in the sense that increases in temperature result in immediate, short-term increases in process rates. However, long-term changes in element budgets and stocks take place relatively slowly and less dramatically in response to increases in air temperature (Figure 5.8). Changes in air temperature also lead to much smaller changes in soil temperature, leading to smaller and slower changes in temperature-limited processes in the soils, including mineralization of N and P.
(3) Changes in photosynthetically active radiation have an immediate impact on photosynthesis at both leaf and canopy levels, but longer-term changes in C cycling are smaller and depend more on regulation of leaf area, which is also limited by nutrient availability and plant allocation patterns. In multiyear experiments a 50% reduction in photosynthetically active photon flux density (PPFD) during the summer results in a 25%–50% reduction in both production and aboveground biomass (Figure 5.8).

Results from tundra at Toolik Lake (Figure 5.8) are consistent with similar experiments throughout the Arctic (Shaver and Jonasson 1999; van Wijk et al. 2003; Dormann and Woodin 2002), which generally show a greater whole-system responsiveness to changes in N or P availability than to changes in temperature or light.

At present it is not possible to say which kinds of tundra (wet, moist, dry, or shrubby) are more or less responsive to these manipulations of growing conditions. Overall they all respond similarly to the changes in nutrients, temperature, and light. In experiments where N and P fertilizers have been added factorially, the response to N generally dominates especially in moist and dry tundras, although some moist tundras are P-limited and co-limitation has also been observed (Shaver et al. 1986a; Shaver and Chapin 1995). The productivity of wet sedge tundras near Toolik Lake is typically (but not always) P-limited or co-limited by P and N (Shaver et al. 1998). The greater importance of P limitation in wetter sites is probably related to the chemical immobility of P at low pH and low soil oxygen status, as well as to the common isolation of these tundras from mineral soils (as a source of P) due to their thick O horizons and shallow thaw depths (Table 5.1).

Species composition of all tundras responds strongly to changes in nutrients, temperature, light, and other manipulations such as changes in herbivory and snow cover. Here again, the greatest changes occur in response to fertilizer addition, and changes in species composition in response to other treatments often appear to be linked to secondary changes in N and P cycling. In moist acidic tundra, grasses and sedges often dominate in fertilized plots in the first 1–6 years of treatment; these graminoids may retain their dominance where *Betula* is absent (Bret-Harte et al. 2008). The long-term response (6–20 years) to nutrient addition in moist acidic

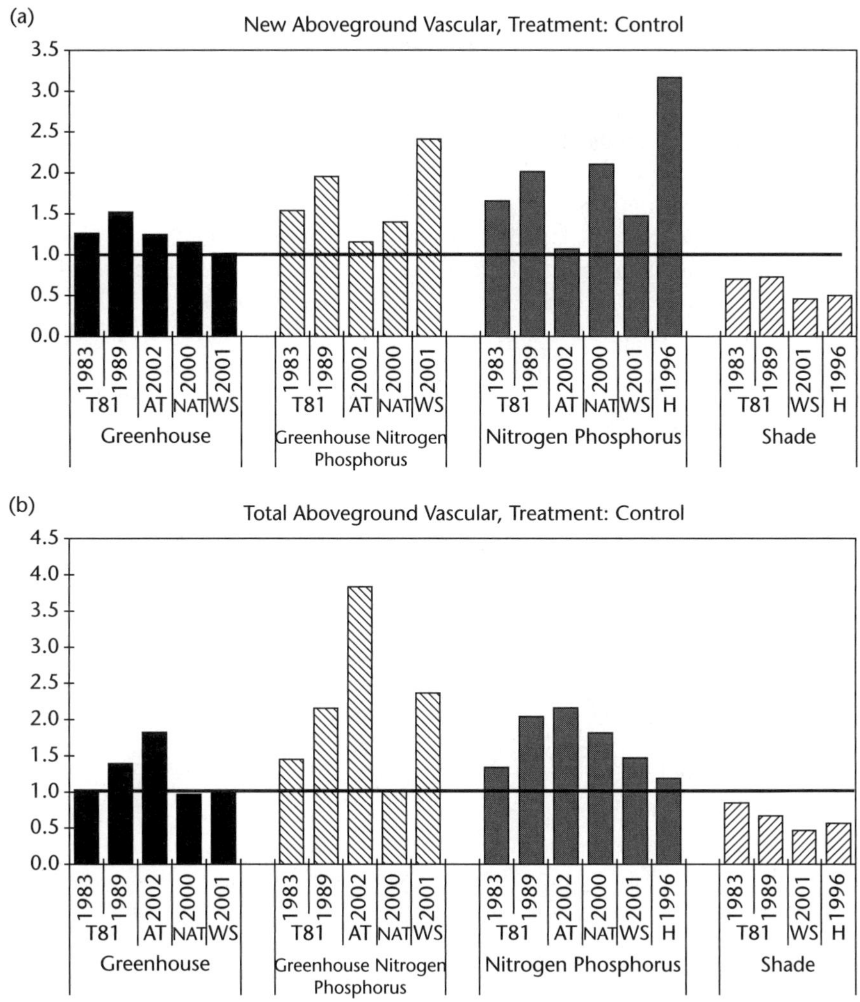

Figure 5.8 Response index (treatment value : control value) of (A) new aboveground vascular plant growth and (B) total aboveground vascular biomass in greenhouse, fertilizer, and shade treatments in acidic and nonacidic tussock, wet sedge, and heath tundras. (N.B.: Results of 1994 harvest of LTER wet sedge tundra not shown due to anomalously low production and biomass in control plots, leading to anomalously high response indices in all treatments in 1994.) The horizontal axis indicates treatment (greenhouse, greenhouse nitrogen phosphorus, nitrogen phosphorus, or shade), site (T81 = 1981 historic tussock, AT = LTER acidic tussock, NAT = LTER nonacidic tussock, WS = LTER wet sedge, H = LTER heath), and year of sampling.

tundra, however, is a dramatic increase in the rapidly growing woody shrub *Betula nana* (Shaver et al. 2001). In moist nonacidic tundra where *Betula* is rare or absent, there is a general increase in abundance of all plant functional types, resulting in an overall increase in biomass and productivity similar in magnitude to that of moist acidic tundra (Hobbie et al. 2005). In other systems, such as dry heath tundra initially dominated by evergreen shrubs and lichens, the dominant species of fertilized plots is the grass *Hierochloe alpina* (Gough et al. 2002). Warming also changes the species composition although more slowly than fertilizer addition, probably because the species changes are more limited by the temperature effect on nutrient availability than by the direct effect of warming on growth. As for the nutrient response, deciduous shrubs and forbs are typically the most responsive plant functional types to warming (Walker at al. 2006).

Mosses and lichens are also strongly affected by these treatments. Although the direct response is often similar to the vascular plants (i.e., increased growth with higher nutrients or warming), in the long term mosses and lichens generally decline or disappear in treatments where taller shrubs or other species form a closed, dense canopy (Shaver et al. 2001; Cornelissen et al. 2001). Lichen abundance is reduced by herbivores and increases in response to herbivore exclusion in dry heath and moist acidic tundras (Gough et al. 2008).

Effects of Species Composition on Biogeochemistry

Variability and change in species composition are important because different species have very different responses to environmental change, and species composition has important impacts on biogeochemical cycles. A multitude of effects of species on biogeochemistry derive from differences in their growth rates, morphology, and allocation patterns. For example, because leaves of grasses and sedges grow from basal, intercalary meristems, grass and sedge species are less constrained in terms of the size and number of leaves that can be produced each year (Shaver and Laundre 1997). For this reason, grasses and sedges respond very quickly to changes in environment and dominate the initial changes in LTER fertilized plots. Other species, in particular the dominant evergreens like *Ledum palustre*, are limited in their ability to respond within the same growing season to change in the environment because the current season's growth is limited by the characteristics of the buds produced at the end of the previous season; the growth of individual shoots cannot respond fully until the following season, with growth from a new set of buds (Shaver 1981, 1983).

Vegetative demography is thus a key aspect of controls over the multiyear response to environmental change in the tundra near Toolik Lake. One of the most important differences among species in this regard is the ability to branch rapidly and to grow taller as biomass accumulates with more favorable growing conditions. Although graminoids can produce new tillers quickly (Fetcher and Shaver 1982), they cannot grow tall because they do not produce woody stems. The deciduous shrub *Betula nana*, on the other hand, is typically the dominant species in nutrient-rich moist and wet tundras because it (1) can rapidly produce new, upward-growing "long shoot" branches from the store of meristems it maintains as "short shoots" under less favorable environmental conditions and (2) can rapidly increase stem wood production by secondary growth from cambial meristems (Bret-Harte et al. 2001, 2002).

A second important effect of species composition is related to differences in their nutrient uptake, particularly N uptake in these typically N-limited systems.

Experiments using isotopically labeled N sources have shown that tundra plant species are physiologically able to acquire N in a wide range of different chemical forms, at different depths, and at different times of year (Kielland 1994; Schimel and Chapin 1996; McKane et al. 2002). The common tundra plant species are also known to differ consistently in $\delta^{15}N$, a measure of the relative abundance of N isotopes, in their tissues (Figure 5.7; Nadelhoffer et al. 1996). Because changes in $\delta^{15}N$ within plants are assumed to be negligible, the differences in $\delta^{15}N$ among species must be due to differences in sources of N taken up by those species (i.e., NO_3^-, NH_4^+, amino acids, and other forms of organic N taken up directly by roots, and N acquired via mycorrhizal symbionts). Surveys of $\delta^{15}N$ in leaves (e.g., Figure 5.7) consistently show that evergreen and deciduous shrub species have low $\delta^{15}N$ values (in the range –4 to –8‰), while graminoids and most forb species have much higher $\delta^{15}N$ values (in the range +4 to 0‰). The shrub species all have ectomycorrhizae or ericoid mycorrhizae associated with their roots, and these mycorrhizae are known to transfer N to their plant hosts at low $\delta^{15}N$. Much of the N supply to evergreen and deciduous shrubs appears to be via mycorrhizae. The graminoid and forb species, on the other hand, are mostly nonmycorrhizal or endomycorrhizal and appear to meet most of their N requirement by direct uptake of a wide range of N forms (*Pedicularis*, a forb in the Scrophulariaceae, is a root hemiparasite of *B. nana* and may obtain much of its N in this way). These differences in N uptake source among species may promote diversity in the tundra vegetation and may increase the community-total N uptake in this N-limited system.

In addition to these species effects, there is also considerable within-species variation in responses to environment that must be considered in extrapolating these responses over large areas. Populations ("ecotypes") of *Eriophorum vaginatum* from south of the Brooks Range, for example, are significantly more responsive to environmental variation in reciprocal transplant experiments than are northern populations (Shaver et al. 1986b; Fetcher and Shaver 1990). A recent resurvey of these now 30-year-old reciprocal transplant experiments shows that individuals of the same species from local populations are more fit than foreign populations in common gardens (Bennington et al. 2012).

One of the best-documented effects of species composition on biogeochemistry is that of litter "quality" (relative decomposability) on decomposition and nutrient mineralization. In a comparison of decomposition rates using litter from 18 sites around the Arctic (Cornelissen et al. 2007), differences among plant functional types (deciduous, evergreen, graminoid, moss) accounted for about 30% of the explained variance in decomposition rates while "life zone" (overall differences in environment between the two experimental sites where the litter decomposed) accounted for about half of the explained variance (Figure 5.9). Among the species and ecosystems at Toolik Lake, the different kinds of litter also decompose at very different rates and immobilize and release N differently; at constant temperature the range of decomposition rates among litter types from Toolik Lake is greater than the effect of a 6°C change in temperature (Hobbie 1996). However, when weighted-average decomposition rates are used to estimate overall community litter decomposition at different sites, the site effects (moist acidic versus

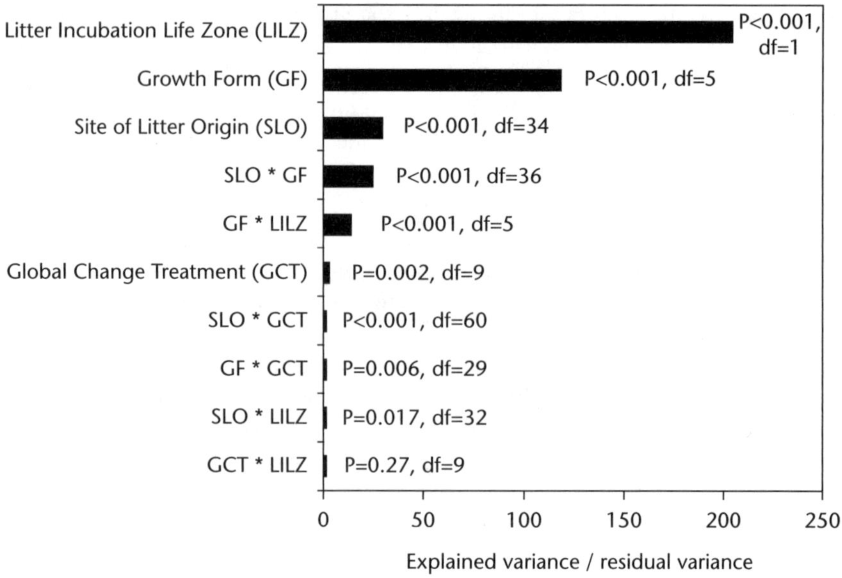

Figure 5.9 Results of a four-way analysis of variance unraveling the key biotic and abiotic effects on percentage of mass loss of leaf litters, based on the predominant plant species in 33 experiments in arctic and alpine sites in the Northern Hemisphere (see Figure 5.1). Mass loss percentage data were arcsine(square-root(100/x)) transformed prior to analysis. Total df = 1824 (from Cornellisen et al. 2007.)

nonacidic tundra) on decomposition become more important than species composition effects (Hobbie and Gough 2004). The combination of species effects on initial litter "quality" and site effects on decomposition and mineralization is ultimately related to the chemical composition and nutrient mineralization of soil organic matter (Shaver et al. 2006), and likely to the microbial community composition (Zak and Kling 2006).

Finally, differences in growth and allocation patterns among species can converge on similar overall ecosystem responses to environment despite very different component responses. For example, the concentration of N in plant tissues varies widely both among species and among tissues (leaves, stems, roots, inflorescences) as well as among sites and experimental treatments. However, when the overall N requirements of production and biomass accumulation by whole tundra vegetation are calculated, there is very little variation in overall N concentration across a ten-fold range in production and biomass in vegetation ranging from dry heath to tall shrub tundra (Shaver and Chapin 1991). In unmanipulated vegetation, this convergence in overall N use results more from changes in the mix of species and tissues with different N concentrations than from variation in N concentration within species or tissues. However, even in long-term fertilized plots where species composition has changed dramatically and N concentrations in all tissues have increased, because increases in woody stem mass (with low N concentrations) are much greater than increases in leaf mass (with high N concentrations), the overall

N concentration in the whole vegetation has not changed despite a doubling of biomass, production, and N content (Shaver et al. 2001). This convergence in N use appeared even when dominant species had been removed from the vegetation for six years, and regrowth by remaining species resulted in no change in overall N concentration (Bret-Harte et al. 2008). Together, these studies indicate strong constraints on N-allocation and N-use efficiency in these typically N-limited tundras, resulting in similar overall N use in all tundras.

The convergence among species and vegetation types in overall response to environment is particularly well illustrated by controls over canopy-level carbon exchange. In a comparison of 14 different tundras in the Kuparuk River basin (including Toolik Lake), Williams and Rastetter (1999) found a constant, linear relationship between canopy leaf area and canopy N content, irrespective of species or plant functional-type composition, and showed that this relationship was optimal for GPP at any leaf area or N content. A comparison of long-term harvest data from LTER study sites indicates that there is a very close correlation between aboveground net primary production (ANPP) and leaf area among years in control plots despite some year-to-year variation in species composition (Figure 5.10A–C) and that this relationship is continuous

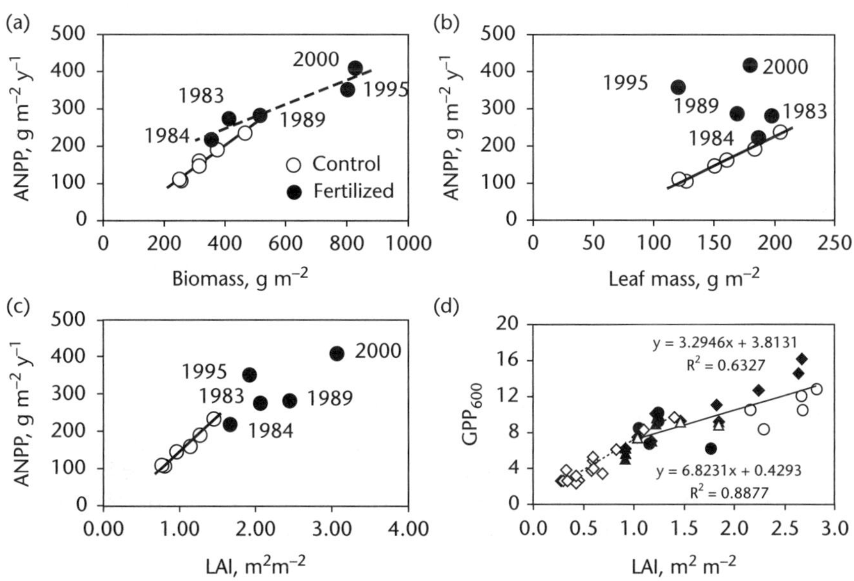

Figure 5.10 Gross primary production (GPP) and aboveground net primary production (ANPP) are closely correlated with leaf area even as species composition changes among years or with fertilizer addition. Data in panels A, B, C from control and fertilized moist tussock tundra at Toolik Lake. Data in panel D from control and fertilized plots in five different tundras at Toolik Lake. In panel D, GPP_{600} is GPP at 600 μmol m^{-2} s^{-1} photosynthetically active photon flux. Open and filled symbols represent data from control and fertilized plots, respectively. Circles = LTER acidic tussock tundra; diamonds = nonacidic nontussock tundra; triangles = heath tundra. (Original data in Chapin et al. 1995; Shaver et al. 2001; and Mack et al 2004.)

with the relationship between ANPP and leaf area in fertilized plots despite major changes in species composition and total N mass in these plots. A similar relationship exists between canopy photosynthesis (GPP) and leaf area across contrasting tundra types both with and without fertilization (Figure 5.10D). Here again, these studies taken together indicate very strong constraints on how canopy-level CO_2 exchange is regulated, with ~80% of the variation in CO_2 exchange explained knowing only leaf area, light intensity, and temperature, irrespective of species composition (Street et al. 2007; Shaver et al. 2007). Despite a wide range of leaf morphology, leaf longevity, and leaf chemistry among the common species and tundra types, CO_2 exchange at the canopy level is regulated mainly by leaf area, by leaf aspect, and by leaf distribution through the canopy.

Vignette 5.4. Species Effects of Terrestrial Plants, by M. Syndonia Bret-Harte

It has been proposed that plant species with similar effects on ecosystem function, such as species with highly decomposable leaf litter that stimulates decomposition, can be grouped into a single "functional type." Chapin et al. (1996) identified six functional types in tussock tundra, corresponding to the recognized physiognomic growth forms (deciduous shrubs, evergreen shrubs, graminoids, forbs, mosses, and lichens). Species in a given functional type are expected to be better at replacing each other than species from other functional types, because they share similar patterns of resource acquisition and use (Symstad 2003). However, there have been few tests of this idea.

To understand how species affect ecosystem functioning and the trajectory of ecosystem response to increased nutrient availability, we variously removed (a) the dominant evergreen shrub (*Ledum palustre*), (b) the dominant deciduous shrub (*Betula nana*), (c) all mosses, and (d) the combination of mosses, *B. nana*, and *L. palustre* in the presence and absence of fertilization. We did not remove graminoids because the dominant graminoid is a tussock-forming sedge upon which many other species grow; removing it would have changed microtopography and drainage.

After six years, the remaining plants had grown enough to restore total biomass to control levels (i.e., biomass compensation had occurred) in most unfertilized removal treatments (Bret-Harte et al. 2008). Net primary productivity was not different from controls in unfertilized removal treatments. Compensation was remarkably rapid, considering how slow plant growth is in arctic ecosystems and how much biomass was removed. Contrary to prediction, the species that provided most of the compensatory growth were not from the same functional type as the removed species, but were instead the most abundant species in other functional types.

Fertilization did not increase total plant community biomass, because some growth forms and species benefited at the expense of others (Bret-Harte et al. 2008). However, under fertilization, deciduous shrub biomass increased more than any other growth form when *Betula nana* was present, but graminoid biomass increased the most when *B. nana* had been removed. *Betula nana* and many graminoid species can rapidly increase their numbers of active meristems under fertilization, which makes them good competitors under increased nutrient availability.

Changes in plant species composition might alter ecosystem N-use efficiency, because different plant species have different uptake rates and use different forms of N (McKane et al. 2002). However, N-use efficiency of plant biomass (total live plant biomass divided by total N content) was not significantly different among most unfertilized removal treatments (Figure 5.V4A), although removal changed the relative abundance of species. Only removing the combination of moss, *Betula*, and *Ledum* decreased the N-use efficiency of biomass relative to control plots; this treatment was dominated by graminoids and had not completely compensated in biomass (Bret-Harte et al. 2008). Removal did not alter N-use efficiency in net primary production, either (Figure 5.V4B). These similar N-use efficiencies probably result from the strong N limitation of plant productivity in unfertilized tundra (Shaver and Chapin 1980, 1986; Chapin and Shaver 1985; Chapin et al. 1995; Shaver et al. 2001).

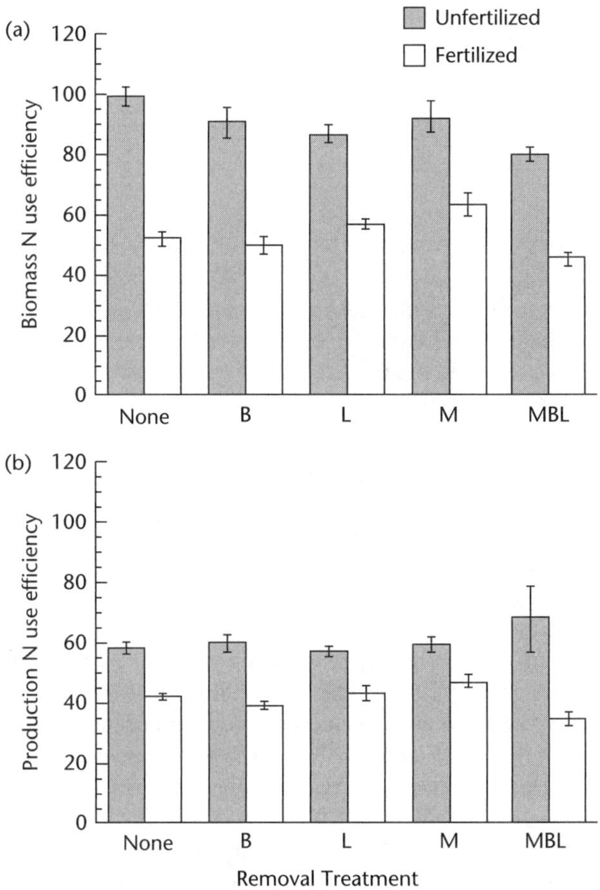

Figure 5.V4 Ecosystem N-use efficiencies of (A) total plant biomass and (B) vascular production after six years of removal and fertilization. Removal treatments were as follows: None = no removal (controls); B = removal of *Betula nana*; L = removal of *Ledum palustre*; M = removal of all mosses; MBL = combined removal of mosses, *B. nana*, and *L. palustre*.

In contrast, fertilization decreased N-use efficiency of biomass (Figure 5.V4A), increased the total N pool in live biomass, and increased N concentrations in plant tissues. Fertilization decreased the N-use efficiencies of both biomass and production most where *B. nana* biomass was the least, where plots were dominated by graminoids. Fertilized *B. nana* produces a large amount of wood of low N content, which is not produced by graminoids.

Our results suggest that plant species composition does not affect capture of N by vegetation in unfertilized tundra, but does so when nutrient limitation is released by fertilizer addition. Species composition changes the trajectory of ecosystem response to fertilizer, and can affect other aspects of ecosystem functioning. Large shrubs decrease albedo in spring, and thus alter tundra energy balance (Chapin et al. 2005). *Betula nana* is unpalatable to most mammalian herbivores, and a transition to dominance by shrubs would negatively impact caribou if it occurs over a large area. In a future shaped by climate warming and increased N deposition, shrub dominance and graminoid dominance may be alternative outcomes, depending on the local abundance of these growth forms now.

Time-Scales and Trajectories of Change

In the absence of major disturbances like fire or permafrost thawing, change at the ecosystem level takes time because multiple, linked processes are involved and not all processes change at the same rate (Shaver et al. 2000). For this reason, annual variation in production and biomass of the tundras near Toolik Lake is poorly correlated, if it is correlated at all, with annual variation in weather variables like temperature and radiation. Productivity of moist acidic tundra, for example, varied more than two-fold between 1982 and 2000 (Figure 5.11A), but there was little indication of a long-term trend in productivity despite persistent, strong summer warming throughout the late 1980s and 1990s. Although the highest production and biomass were both measured in the final year, 2000, neither was significantly different from the long-term mean production or biomass. During this time there was also little change in species or functional-type composition of moist acidic tundra. Although the biomass of *E. vaginatum* was 30% lower in 2000 versus 1982, most of that decline took place during the early 1980s (before climate warming) and graminoid biomass was, if anything, recovering during the 1990s (Figure 5.12A). There were no clear trends in biomass of the dominant deciduous and evergreen shrubs (*Betula nana, Ledum palustre, Vaccinium vitis-idaea*), with the exception that between the final two harvests, in 1995 and 2000, *Betula* biomass doubled (Figure 5.12A).

Even the responses to strong, artificial manipulation of the environment may take many years to develop. In fertilized plots in moist acidic tundra, aboveground productivity increased 2- to 2.5-fold relative to control plots in every year they were harvested (Figure 5.11B). Aboveground biomass, on the other hand, took much longer to change because initially strong increases in production by graminoid species declined after the mid-1980s while the long-term dominant species in the fertilized plots, *Betula nana*, took about five years to emerge (Figure 5.12B). Thus the species

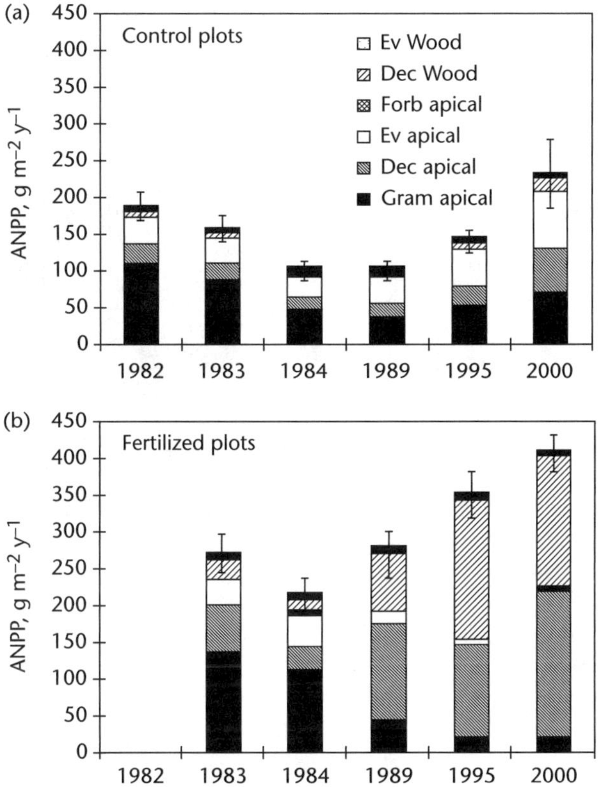

Figure 5.11 (A) Aboveground net primary production (ANPP) in unfertilized control plots of moist acidic tundra at Toolik Lake, Alaska, over six harvests between 1982 and 2000. Although the 2000 harvest occurred after 20 years of climate warming, we cannot say for sure whether the greater total ANPP and the greater productivity of deciduous shrubs in 2000 is the result of warming or is within the "normal" range of ANPP. (B) ANPP in fertilized plots at the same site, where fertilizer addition (N + P) began in 1981. The initial fertilizer response was dominated by graminoids, with a long-term dominance by deciduous shrubs and a large increase in wood production apparent only after 1989. Ev wood = wood production in evergreen shrubs; Dec wood = wood production in deciduous shrubs; Ev apical = leaf and twig production in evergreens; Dec apical = leaf and twig production in deciduous shrubs; Gram apical = leaf and culm production by grasses and sedges. Details are published in Chapin et al. (1995); Shaver et al. (2001); and Mack et al. (2004).

replacement process delayed the increase in community biomass in fertilized plots despite an immediate and sustained increase in production.

Almost all the changes in species composition that were observed in experimental plots within the first 10–15 years of treatment took place as a result of differential growth of the same individual plants that were present at the start of the experiment, with little or no establishment of new individuals by seed and thus little or no addition of new species (Shaver et al. 2001). Exceptions occurred in the dry heath

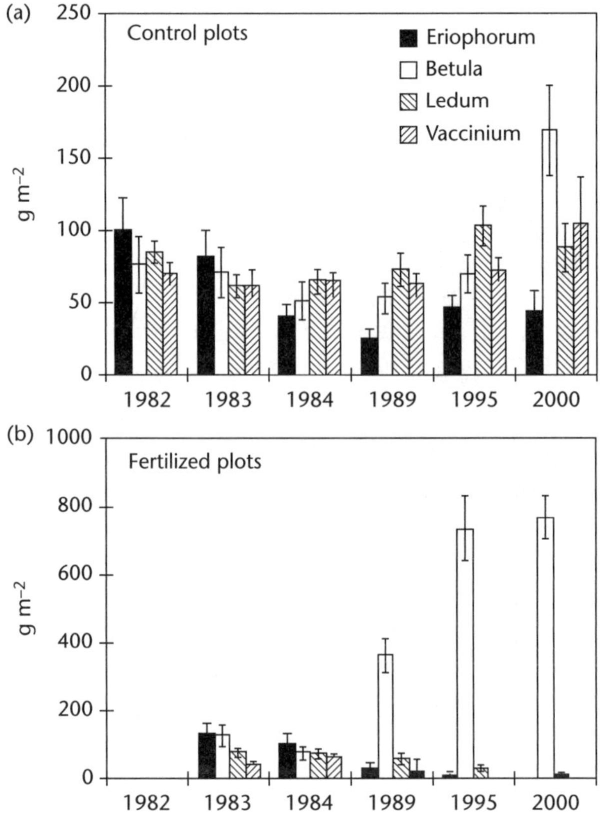

Figure 5.12 Biomass of the four most abundant species in moist acidic tundra at Toolik Lake, in (A) control and (B) fertilized plots. Here biomass includes aboveground biomass and belowground stems but not roots. Original data in Chapin et al. (1995); Shaver et al. (2001); and Mack et al. (2004).

tundra and in some tussock tundras with higher-pH soils where several grass species increased several-fold in abundance in fertilized plots, by both seed establishment and vegetative growth (Gough et al. 2002). In a few individual plots in wet sedge and heath tundras, weedy species like *Epilobium angustifolium* invaded and became very well established (Shaver et al. 1998). Nonetheless, decade-scale changes in community composition are possible by addition of new species and have been observed, for example, in the establishment of *Betula nana* in fertilized wet sedge tundra after 12 years (Boelman et al. 2003) and the invasion of fertilized, acidic tussock tundra by *Calamagrostis lapponica* and *Stellaria* species after ~15 years. These slow additions of species to experimental plots indicate potential for continued change over many decades, although their effects, if any, on overall productivity, organic matter accumulation, or biogeochemistry are both unknown and unpredictable at present.

Even processes that vary dramatically from year to year are unrelated to annual variation in weather. For example, flowering of *Eriophorum vaginatum* at Toolik

Lake varies more than 100-fold among years (Figure 5.13, top) and is correlated with the annual variation in flowering along a 300-km transect running from the Yukon River to the northern coastal plain (Figure 5.13, bottom; Shaver et al. 1986b). The most likely explanation for this observation is that flowering is a process that is controlled over several years of weather and plant response. Only after 30 years of observation is it becoming clear that years of high flowering appear to follow at least two warm summers with good growing conditions and favorable conditions for soil N mineralization and N uptake by plants (J. Laundre unpublished; LTER

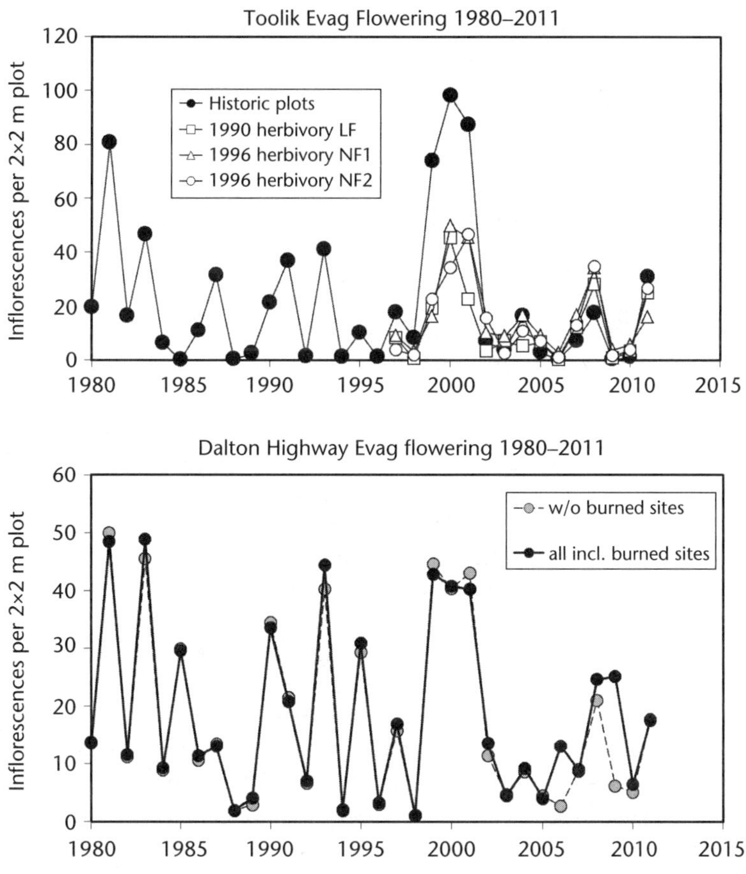

Figure 5.13 Mean number of inflorescences of *Eriophorum vaginatum* in 2 × 2 m plots at Toolik Lake (top) and in similar plots along the Dalton Highway (bottom; in most years n = 21–23 sites, 1979–2011). In the upper panel, different symbols represent different sets of unmanipulated control plots in moist acidic tundra at Toolik Lake. In this panel, "Historic plots" are the same as the "Historic tussock" plots in Figures 5.2–5.4, while "1990 herbivory" and "1996 herbivory" plots are separate, unmanipulated plots at the "LTER acidic tussock" site, also shown in Figures 5.2–5.4. In the lower panel, black symbols are means including all sites while grey symbols are means not including 3 sites that were burned in 2004-2005. Data for 1979-1982 were published in Shaver et al. (1986).

database); the same high flowering also occurs in fertilized plots in the second (but not the first) year of treatment, and secondary peaks in flowering occur in older, abandoned fertilized plots in years *following* warm years.

Overall, these long-term observations of variability and change in individual species and whole vegetation suggest that changes in production and biomass are well buffered against short-term variation in weather. The major components of this "buffering" include:

(1) *Physiological* characteristics such as the internal storage and recycling of C and nutrient resources used in growth, which make each year's production a function of resources acquired over more than one year;
(2) *Morphological* characteristics such as (for some but not all species) the need to form new buds from which each year's growth is produced, and constraints on the rate and pattern of new meristem production;
(3) *Species differences* in physiology and morphology, which means that growth of each species has a different relationship to short-term variation in weather or sustained changes in climate; and
(4) *Population dynamics* of the different species in the community, which tend to reduce overall variation in vegetation biomass because increases in some species are compensated by simultaneous decreases in others.

Soil-plant interactions also play a key role in regulating long-term change in tundra ecosystems, in particular because production and biomass accumulation are so strongly linked to soil nutrient supply. For this reason, the limited responses to greenhouse warming that were observed in tussock and wet sedge tundras (Figure 5.6) are interpreted as resulting from a relatively slow increase in soil nutrient availability in response to modest temperature increases, in contrast to the larger and more rapid responses to fertilizer addition. Similarly, in open-topped chamber warming experiments completed at more than 20 arctic and alpine sites (including Toolik Lake) as part of the International Tundra Experiment (ITEX), initial increases in individual plant growth in the first two years of treatment were frequently not sustained over longer periods, suggesting some secondary limitation, perhaps related to nutrients, that restricted the potential for sustained increases in growth (Arft et al. 1999; Walker et al. 2006; Elmendorf et al. 2011).

In addition to revealing the long-term trajectories of change in composition and processes in response to manipulation, the Arctic LTER experiments allow us to document overall changes in standing stocks of elements and organic matter. In fertilized moist tussock tundra, a total of 200 g m^{-2} N and 100 g m^{-2} P were added over the 20 years from 1981 through 2000 (10 g m^{-2} yr^{-1} N and 5 g m^{-2} yr^{-1} P). Increased N and P incorporation into plant biomass (Figures 5.10A–C, 5.11, 5.12) can account for only a small proportion of these amounts. This leads to the related questions: "Where did the added fertilizer nutrients go, and can we track their accumulation in some other part of the ecosystem?" When complete C and N budgets were calculated for these plots at the 2000 harvest (Mack et al. 2004), it was discovered that the total amount of N in fertilized plots was, if anything, *less than* the amount in control plots (Figure 5.14C), whereas if all of the added N had remained on the fertilized plots the total amount should have been at least 50% greater. All of the losses of N were from (a) mineral soils and (b) portions of the organic mat below 10 cm; other components of the soil-plant system gained N (Figure 5.14D). Because the N losses were from deeper soil pools, this suggests that much of the N that was lost was not from fertilizer N, but from N in older organic matter that was mineralized and either leached or denitrified. The net losses of N in fertilized plots were matched by a similar pattern of net C loss (Figure 5.14A, B), despite a doubling of productivity that should have increased total C inputs by at least 1500–2000 g m^{-2} over 20 years, relative to control plots. All of the net C losses were from deeper levels of organic soil, especially from the mineral soil (Figure 5.14B), again indicating that the C lost was old-soil organic C and not recently fixed C or surface litter C.

The net long-term losses of *both* C and N in this "N-limited" moist acidic tussock tundra were not expected but still are consistent with the strong control of organic matter cycling by interactions among the C and N cycles as discussed in relation to Figure 5.6. Overall, C and N were lost or gained in similar proportions in each of the pools described in Figure 5.14, with little change in C:N ratios overall or

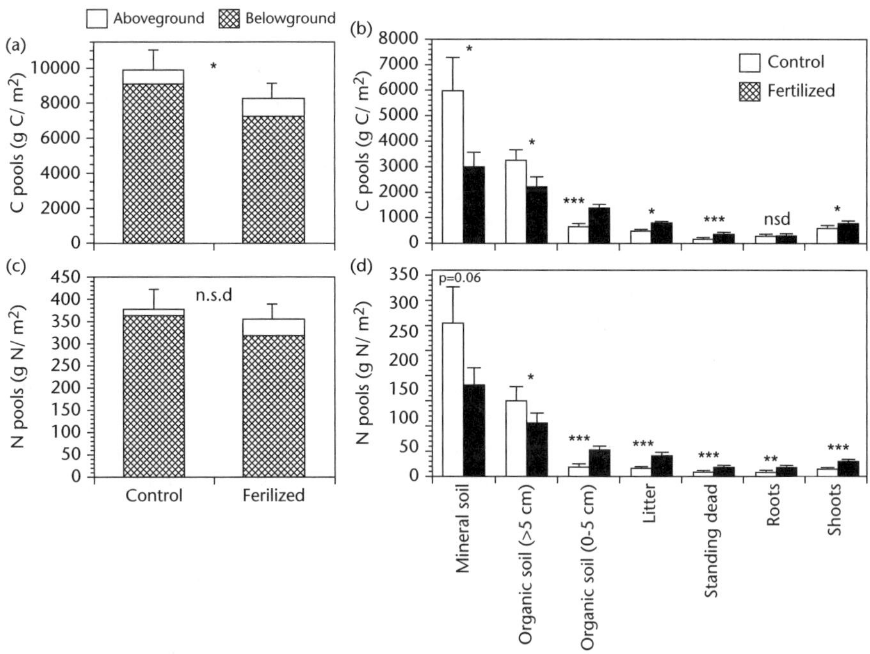

Figure 5.14 Effects of fertilizer on tundra carbon and nitrogen pools after 20 years of fertilization. Mean (±1 standard error) above- and belowground carbon (A) and nitrogen (C) pools in unmanipulated control and fertilized treatments of moist acidic tundra near Toolik Lake, Alaska. Aboveground pools include shoots, standing dead plant material, and rhizomes. Belowground pools include surface litter, roots, and organic and mineral soil. Mean (±1 standard error) carbon (B) and nitrogen (D) pools in plant and soil compartments (from Mack et al. 2004).

in individual pools. The large losses of deep, presumably older soil organic matter after 20 years of fertilization support the hypothesis that soil-decomposition processes, as well as primary production, are N-limited in these ecosystems (Schimel and Weintraub 2003). Addition of readily available N (as NO_3^- and NH_4^+) in the fertilizer appears to have stimulated *both* C fixation by plants and also respiratory C loss by soil microbes (Nowinski et al. 2008). Much of the increased C loss must have come in association with increased denitrification and gaseous N losses (Alexander-Ozinskas 2007).

Increased downslope leaching losses of C and N are unlikely to have contributed greatly to the overall losses from fertilized plots. In the case of N, for example, the background rate of leaching loss (~0.1 g N m^{-2} yr^{-1}, or ~2 g m^{-2} over 20 years) would have had to increase ten-fold to account for only ~20 g, or ~8%, of the "missing" 240 g m^{-2} of N in the fertilized plots (including 200 g N added as fertilizer plus a reduction in standing stock of ~40 g m^{-2}). Because the amount of water leaching from the fertilized plots should, if anything, have decreased due to higher leaf area and presumably higher evapotranspiration in fertilized plots, any large increase in downslope leaching could only have occurred as a result of large increases (ten-fold or greater) in dissolved C and N concentrations in the water moving downslope. However, the concentrations of dissolved N that were actually measured within fertilized plots were only about 3–5 times control values (e.g., Chapin et al. 1995), and any excess N in soil water is likely depleted rapidly to control levels within 1–5 m downslope (Yano et al. 2010). The maximum contribution of increased leaching to the total N loss from fertilized moist acidic tundra is thus <4%–5% of the total loss over 20 years. Using the same argument for C, increased C leaching from fertilized plots cannot account for more than about 4% of the "missing" C.

Feedbacks and Interactions: Changes in Surface Energy Balance, C and N Balance, and Disturbance Regime

Each patch of tundra interacts with its neighbors, with the atmosphere, and with downslope aquatic ecosystems through exchanges of energy, elements, and water. These spatial interactions play an important role in regulating long-term change in tundras, and constitute an important system of feedbacks on the external drivers of change, including climate change, disturbances such as fire and thawing of permafrost, and direct human impacts. As terrestrial ecosystems respond to these drivers, changes in their surface energy balance, overall C and N balance, and leaching losses to downslope tundras and aquatic systems will create both positive and negative feedbacks.

One example of this system of feedbacks is related to the expected general increase in abundance of shrubs, especially deciduous shrubs, in response to climatic warming. An increase in shrub abundance has been predicted for over 25 years based on observed species-level responses to weather, climate, human and natural disturbance, and experimental manipulations (e.g., Chapin and Shaver 1985). The principal basis for this prediction is the higher abundance of shrubs in warmer, better-drained, and higher fertility sites, and the consistent increases in the abundance of deciduous shrubs in disturbed, warmed, and (especially) fertilized tundras throughout the Arctic (e.g., Shaver and Chapin 1986; Walker et al. 2006). Analyses of pairs of photographs taken 30–40 years apart appear to support this prediction for many alder-dominated sites in northern Alaska (Sturm et al. 2001), although isolating long-term increases in shrub abundance from annual variation and from concurrent changes in other functional types gives more equivocal results for the area around Toolik Lake (e.g., Figure 5.12A; Wahren et al. 2005).

The expected general increase in "shrubbiness" has very substantial implications for both surface energy balance and C and N cycling (Figure 5.15). Because shrub canopies absorb more energy from incoming solar radiation, particularly in spring when shrub branches emerge from the snow, they reflect less of that energy and warm up more, ultimately transferring more energy back to the air as latent and sensible heat rather than as reflected shortwave radiation. The change in the atmospheric heating is quite large, potentially as large as 8.9 W m^{-2} if the tundra landscape is completely converted to shrubland (Chapin et al. 2005); this amount is *double* the global average 4.4 W m^{-2} heating experienced since the last glacial maximum (or predicted due to greenhouse gas accumulations by the end of this century), suggesting that the regional feedback on climate resulting from a complete change to shrub dominance could have a greater effect on regional climate than the original driver, global warming.

Changes in shrub dominance will have major impacts on soil thermal regime and on the cycling of C and N in soils (Figures 5.15, 5.16), in addition to the direct impacts of warmer temperatures and decreases in the period of snow cover (Euskirchen et al. 2006). Because shrub vegetation accumulates deeper snow in

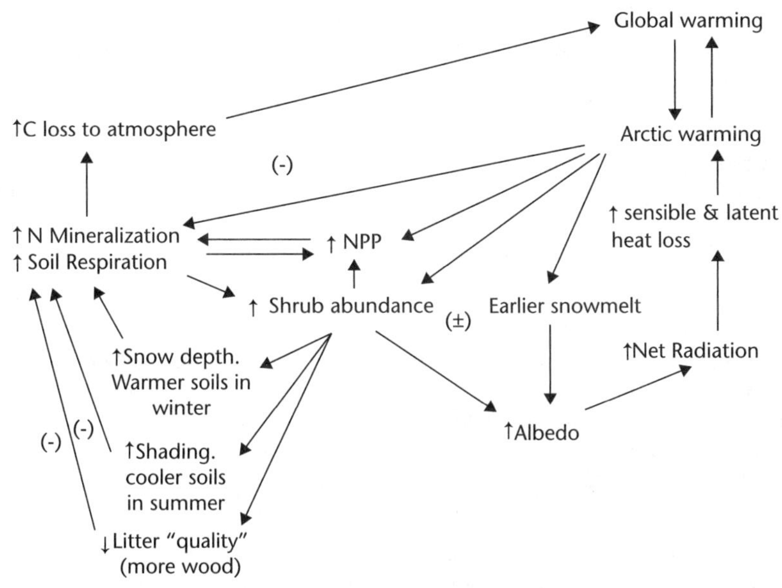

Figure 5.15 Feedbacks on drivers of change, showing the dominant connections (arrows) and whether there will be a positive or negative feedback.

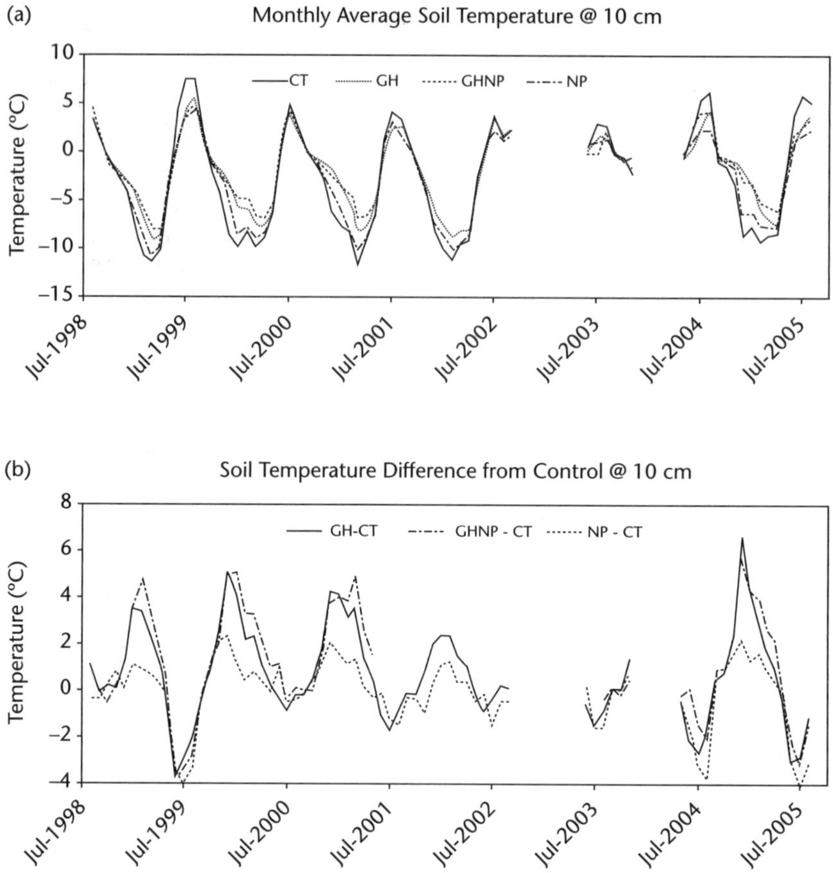

Figure 5.16 Soil temperatures beneath shrub-dominated experimental plots are warmer in winter and colder in summer than in controls. (A) Monthly average soil temperature at 10-cm depth in LTER experimental plots in moist acidic tundra (MAT) at Toolik Lake, 1993–2005. (B) Difference from control (treatment-control) in monthly average soil temperature at the same plots. CT = control plots; GH = greenhouse plots; NP = N + P fertilized plots; GHNP = greenhouse plus fertilizer plots. Source: Arctic LTER database.

winter (Sturm et al. 2001), the soils beneath shrub canopies are better insulated from very cold air temperatures, taking longer to freeze and not falling below the extremely low temperatures that would stop microbial activity (Schimel et al. 2004). Thus it is likely that mineralization of C and N and respiratory loss of C will continue for longer periods into the middle of winter and at higher rates than in tundras with less shrub cover (Borner et al. 2008). In contrast to this winter warming beneath shrub canopies, during the summer the denser shrub canopy intercepts incoming solar radiation above the soil surface and less energy reaches and warms the soil, leading to cooler soils in the summer (Figure 5.16) and shallower depth of thaw. This *negative feedback* may act to limit the magnitude of the change in community composition in the long term. However, at present it is not known which of these two opposing feedbacks will predominate in regulating the response to climate warming.

Exchange of CO_2 with the atmosphere will also be affected by changes in species composition, including increased shrub abundance. Here again, one of the main factors determining the net effect will be the effect of species composition on soil respiration via the effect of the canopy on soil temperatures (Figure 5.16). Another key factor is the effect of plant litter quality (decomposability) on soil and litter respiration. Shrub vegetation is not only more productive than other tundras (Figure 5.3), it also produces more wood than other tundras (Figure 5.11; Weintraub and Schimel 2005). Although higher productivity means higher litter production and thus more respiration from litter, woody litter decomposes much more slowly than most leaf litters (Hobbie 1996). Shrub tundras do have higher soil and litter respiration rates than other tundras, at least during the summer (Grogan and Chapin 1999), but they do not have above-average soil C accumulation (Figure 5.4). Thus it is still not clear whether a shift to shrub tundras should be accompanied by a net increase or a net decrease in ecosystem C stocks, although the overall rate of C turnover should be increased.

In addition to the direct effects of climate change on species composition, surface energy balance, and biogeochemical processes, changes in the regional disturbance regime have the potential to affect the entire arctic region even though the area of disturbance is relatively small. For example, in 2007 a single wildfire north of Toolik Lake released >2 Tg of carbon to the atmosphere, mostly due to combustion of the upper few centimeters of soil organic matter over an area of 1039 km^2, about one-half of 1% of the area of the North Slope of Alaska (Mack et al. 2011). This was the largest wildfire known to have occurred in arctic tundra, and the amount of C it released was equal to about half the annual net C sequestration of the entire arctic region (McGuire et al. 2009). In the years following this wildfire, changes in C balance in the burned tundra were also large as the burned tundra slowly regrew, initially losing C to the atmosphere as R_E exceeded GPP, and then sequestering C as plant canopies recovered and GPP increased (Rocha and Shaver 2011a and unpublished data). These changes in C balance in regrowing, burned tundra were large enough to cancel any predicted increases in C sequestration due to warming climate over much larger areas, suggesting that future increases in wildfire frequency, severity, and area burned have the potential to dominate the regional changes in C balance as the climate warms. Although wildfires in northern Alaska have been rare or absent for at least 5,000 years (Hu et al. 2010), clearly the tundra plants are well adapted to at least occasional burning, as shown by the spectacular flowering of plants in 2010 and 2011 that survived the burning (M. S. Bret-Harte, A. Rocha and others, personal observation).

Disturbances such as wildfire also interact with changes in permafrost that are already occurring as the climate warms. Because wildfire reduces surface albedo and increases net radiation in burned sites, soils are warmer and depth of thaw is greater (Rocha and Shaver 2011b), leading to abundant thermokarst features on burned land. Several recent studies have found increases in thermokarst activity in northern Alaska (e.g., Bowden et al. 2008; Gooseff et al. 2009), often leading to loss of much of the organic matter in thermokarst-impacted sites. These "hot spots" of C and other element turnover have the potential to affect regional element balances because the rates of change in element stocks per unit area can be several orders of magnitude greater than slower, climate-driven rates of change over much larger areas.

Vignette 5.5. Decadal-Scale Changes of Vegetation from Long-Term Plots in Alaskan Tundra, by William A. Gould and Joel A. Mercado-Díaz

Climate-related changes in vegetation structure and composition may be both subtle and slow. These changes have been studied in the tundra near Toolik Lake using both experimental manipulations and long term monitoring. As part of the International Tundra Experiment (ITEX; Walker et al. 1999; Elmendorf et al. 2012), manipulations of summer temperatures have been carried out using open-topped chambers, and snow depths have been manipulated using snow fences; long-term response of plant communities to ambient climate has been monitored in two 1-km^2 grids located at Toolik Lake and nearby Imnavait Creek (Walker et al. 1989). Monitoring in the two grids involves resampling a set of permanent control plots at five- to seven-year intervals, beginning in 1989, using the point-frame method. In this method an aluminum frame with a paired-grid of wires is placed over vegetation plots and aligned to a set of permanent markers. By sighting downward and aligning the paired crosshairs (100 sampling points per frame), the same point can be resampled over many years. The sampling is designed to measure changes in species abundance and vegetation structure over time to understand plant community dynamics in the Alaskan tundra.

Prediction and experimental observation of tundra vegetation changes in recent decades has indicated a likely increase in the relative abundance of shrubs in response to climate warming (Sturm et al. 2001). The ITEX manipulation of summer temperature and snow depths began in 1994 and continues to show effects on vegetation. Changes in snow regimes have the largest impact. Both warming and snow effects are strongest in the moist tussock tundra relative to dry heath tundra. The most common changes associated with increased snow depths included increases in canopy height and shrub abundance and decreases in species diversity and lichen cover. Experimental warming was associated with a relative increase in shrub abundance in the moist tussock tundra. In contrast, analysis of our 20-year record of tundra vegetation structure and composition from 155 permanent monitoring plots at Toolik Lake and Imnavait Creek indicates a general increase in vascular plants. Over the last two decades the relative abundance of vascular vegetation increased by 18.6% whereas abundance of the nonvascular component of vegetation has decreased significantly: lichens by 9.3%, non-*Sphagnum* mosses by 20%, and *Sphagnum* by 28%. Graminoids, herbaceous dicots, and shrubs all increased significantly in abundance: graminoids by 25.5%, herbaceous dicots by 24%, and shrubs by 13% ($p < 0.05$) (Figure 5.V5). The canopy height, as well as the horizontal extent of the canopy, increased over time with the amount of horizontal surface having multiple strata increasing from about 60% to 80% (Mercado-Díaz 2011). The increase of canopy overstory in these plots represents a structural response to the shift in the relative abundance of vascular plants vs. bryophytes. The increase in

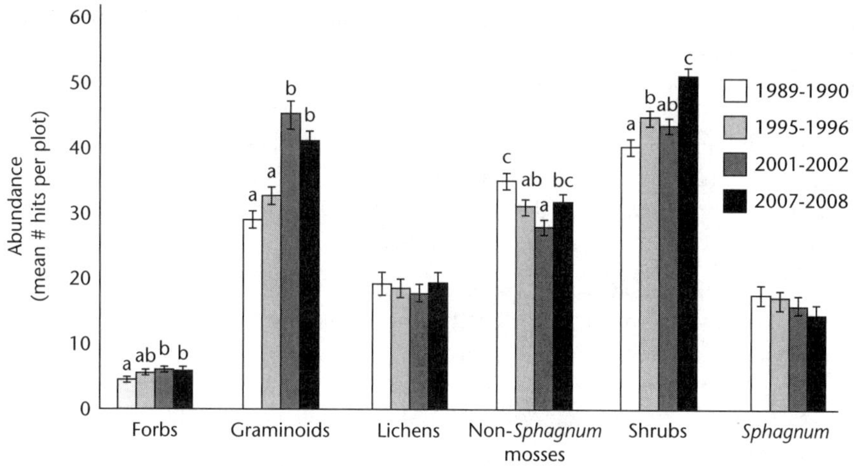

Figure 5.V5. The mean number of total hits per plot of different vegetation growth forms located at the Imnavait Creek and Toolik grids from 1989–1990 to 2007–2008 (letters above the bars indicate significant differences, error bars represent standard errors).

the abundance of vascular vegetation and in canopy height and complexity will likely affect snow redeposition and winter biological processes and will have feedbacks to nutrient allocation and cycling (Sturm et al. 2005).

Predictions of Future State and Key Uncertainties

As global warming and other climate changes continue, tundra ecosystems will continue to respond through many decades and centuries. Over the long term (a century or more) it is reasonable to expect that tundra ecosystems will look more like the warmer ecosystems farther south, with taller, woodier vegetation including trees, higher overall species richness, higher productivity, higher organic matter accumulations in plants and soils, and higher overall rates of element cycling. Short-term changes (a decade or less) are also relatively predictable—in the short term one would again expect higher productivity and element turnover (unless there is a dramatic decline in precipitation, not currently predicted). Most of the short-term changes in species composition will result from changes in relative abundance within the existing flora and fauna, and rates of species loss will initially exceed gains so species richness may decline slightly. Most of the losses will be among subdominant species. In the vegetation, grass and sedge species will be the most responsive to change in the first 1–5 years, while it will take 5–10 years for others (like shrubs) to fully respond. The overall C balance is more likely to show a net C loss in the short term, as increases in soil respiration accompanied by increased N mineralization and plant N uptake must precede an expected increase in NPP.

Prediction of ecosystem changes between 20 and 100 years in the future is much more difficult than over shorter or longer periods. The longest-running experimental manipulation at Toolik Lake was in its 30th year in 2010 and shows no sign of reaching an endpoint in its long-term responses to fertilizer addition. Because we are generally unable to observe directly the changes in tundra ecosystems at a time scale of 20–100 years, model predictions are not well corroborated at this scale (Rastetter 1996). We can say, though, what we expect the major drivers of ecosystem change to be at this time scale, including:

- Changes in permafrost, soil drainage, and surface energy balance
- N retention and overall balance of N inputs and losses
- Changes in coupling or synchrony between processes like N mineralization and N uptake, or between plant processes and soil microbial processes
- Spatial linkages in C, nutrient, and water cycling among landscape units, especially along hillslopes (see chapter 6)
- Plant species invasions, including especially invasions of trees and shrubs
- Changes in herbivory
- Changes in disturbance regime, especially thermokarst failures and fire

Modeling studies consistently show that long-term changes are constrained by C-N interactions (Rastetter et al. 1997; McKane et al. 1997) as well as by the characteristics of competing plant species (Herbert et al. 1999). Models of the physiology of tundra canopies have also proven very successful and are well corroborated (Williams et al. 2000, 2001). However, it is clear that models lack mechanisms to explain large losses of N (and hence C) from soils in fertilized plots (Mack et al. 2004). This failure of the models is particularly surprising given how well the models predict C and N losses in litter decomposition studies (Moorhead et al. 1999). The missing mechanisms in these models might be associated with the forms of N lost from tundra ecosystems and the availability of these N forms to plants and microbes (Rastetter et al. 2005). Nevertheless, models are vital in scaling the results of plot-scale experiments to regions and the pan-Arctic (Le Dizès et al. 2003; Rastetter et al. 2003). However, as with the intermediate scale in time (20–100 yr), intermediate scales in space present particular problems for model applications and corroboration. Most prominent among these intermediate scale problems are the spatial interactions among ecosystem patches on hillslopes (Rastetter et al. 2004), the understanding of which requires the linkage of ecosystem and hydrological models (see chapter 6, Vignette 6.3).

Conclusions

Understanding of tundra ecosystems has advanced greatly since the first studies began at Toolik Lake in the mid-1970s, including much research done at other arctic sites and not discussed in this chapter. Before the 1970s, most research on tundra was focused on specific, species-level adaptations to a cold, short growing season, on the distribution of species in relation to fine-scale environmental variation, and on broad regional patterns of distribution and abundance. Interactions among species, feedbacks on ecosystem properties in relation to species composition, soil processes in general, and nutrient limitation of plant growth received relatively little attention because the Arctic was viewed as a "cold-dominated" landscape where survival was primary and extremes of the physical environment limited the importance of both species interactions and biogeochemical limitations (e.g., Billings and Mooney 1968). There were just a few notable early studies that took an ecosystem perspective (e.g., Summerhayes and Elton 1923; Pitelka 1973), leading the way to later work on trophic interactions and whole-system biogeochemistry and feedbacks.

The focus of research on tundra ecosystems began to change rapidly in the early 1970s, in large part due to the International Biological Program's tundra biome study, which produced the first reasonably complete descriptions of tundra biogeochemical cycles (Brown et al. 1980; Bliss et al. 1981); these descriptions indicated that slow inputs and turnover of elements other than C, particularly N and P, were likely important limiting factors in tundra biogeochemistry. This knowledge was a major factor influencing the design and initial hypotheses of the long-term fertilizer, shade, greenhouse, and other experiments at Toolik Lake and other arctic sites (Shaver et al. 1992).

By the end of the 1980s it was clear that the species that live in tundra ecosystems were generally very well adapted to the cold environment and short summers. Although essentially all species and all processes within the tundra ecosystem are responsive to temperature in the short term (within a year), the primary limitation and regulator of long-term response to temperature (years to decades) is the slow input and turnover of elements like N and P, and the long time it takes for temperature change to affect this input and turnover. At the same time it was becoming clear that species interactions in tundra plant communities were much more important in determining community composition than previously believed, and that the species composition of tundra ecosystems has a major impact on biogeochemistry and on vegetation, soil, and even air temperatures. Together, these species effects and the slow change in element cycles make the tundra ecosystem much more resistant and resilient to climate warming than was expected when the research at Toolik Lake began.

Since the 1990s, the importance of physical disturbance in effecting major, long-term change has become particularly clear. Fine-scale disturbances like frost boils, ice-wedge polygon formation, and other periglacial phenomena have long been viewed as important sites of change in tundras (Walker et al. 2008); however, as the Arctic has warmed, the increasing frequency, severity, and area affected by major disturbances like thermokarst and fire has shown that these disturbances are not only where rapid change is possible, but also that extremely rapid change in a relatively small proportion of the tundra landscape can dominate the carbon or energy balance of a large region.

In sum, arctic tundras have been viewed as model systems for ecological research for decades, but the focus of that research has changed greatly since the 1970s. Before the 1970s the tundra was viewed as a useful model system because it was dominated by the physical environment, and because specific adaptations to that environment could be studied in a system where species interactions were relatively unimportant and low temperatures and short growing seasons were the primary limiting factors. After another four decades of research, at Toolik Lake and elsewhere in the Arctic, tundra is still viewed as a model system but one in which the low stature of the vegetation, low species richness, and relatively fine-grained environmental heterogeneity make tundra ecosystems particularly amenable to ecosystem-level experimentation and to studies of individual species effects on ecosystem properties. Overall, this is a major change in how ecologists think about tundra ecosystems, and it opens the way for a rich new era of whole-system experimentation, observation, and understanding.

Acknowledgments

Funding for this research was provided by the National Science Foundation (NSF). Toolik Field Station, VECO Polar Services, and CH2MHill Polar Services provided outstanding logistics support (also funded by the NSF). Helpful reviews of the draft manuscript were provided by Theresa Hollingsworth, Philip Wookey, and two anonymous reviewers.

References

Alexander-Ozinskas, M. O. 2007. Denitrification contributes to nitrogen loss in fertilized arctic tundra sites. M.S. thesis, Brown University, RI.

Arft, A. M., M. D. Walker, J. Gurevitch, J. M. Alatalo, M. S. Bret-Harte, M. Dale, M. Diemer, F. Gugerli, G. H. R. Henry, M. H. Jones, R. Hollister, I. S. Jónsdóttir, K. Laine, E. Lévesque, G. M. Marion, U. Molau, P. Mølgaard, U. Nordenhäll, V. Raszhivin, C. H. Robinson, G. Starr, A. Stenström, M. Stenström, Ø. Totland, L. Turner, L. Walker, P. Webber, J. M. Welker, and P. A. Wookey. 1999. Response patterns of tundra plant species to experimental warming: A meta-analysis of the International Tundra Experiment. *Ecological Monographs* 69:491–512.

Batzli, G. O., and H. Hentonnen. 1990. Demography and resource use by microtine rodents near Toolik Lake, Alaska, U.S.A. *Arctic and Alpine Research* 22:51–64.

Batzli, G. O., and C. Lesieutre. 1995. Community organization of arvicoline rodents in northern Alaska. *Oikos* 72:88–98.

Batzli, G. O., R. G. White, S. F. MacLean, Jr., F. A. Pitelka, and B. D. Collier. 1980. The herbivore-based trophic structure. Pages 335–410 in J. Brown, P. C. Miller, L. L. Tieszen, and B. L. Bunnell, editors, *An Arctic Ecosystem: The Coastal Tundra at Barrow, Alaska.* Stroudsburg, PA: Dowden, Hutchinson, and Ross.

Bennington, C., N. Fetcher, M. C. Vavrek, G. R. Shaver, K. Cummings, and J. B. McGraw. 2012. Home site advantage in two long-lived arctic plant species: Results from two thirty-year reciprocal transplant studies. *Journal of Ecology* 100:841–851.

Billings, W. D., and H. A. Mooney. 1968. The ecology of arctic and alpine plants. *Biological Reviews* 43:481–529.

Bliss, L. C., O. W. Heal, and J. J. Moore, editors. 1981. Tundra ecosystems: A comparative analysis. International Biological Program, vol. 25. Cambridge: Cambridge University Press.

Boelman, N. T., M. Steiglitz, H. Rueth, M. Sommerkorn, K. L. Griffin, G. R. Shaver, and J. Gamon. 2003. Establishing relationships between optical measurements and ecosystem level processes across long term warming and fertilization plots in wet sedge tundra. *Oecologia* 135:414–421.

Borner, A. P., K. Kielland, and M. D. Walker. 2008. Effects of simulated climate change on plant phenology and nitrogen mineralization in Alaskan arctic tundra. *Arctic, Antarctic, and Alpine Research* 40:27–38.

Bowden, W. B., M. N. Gooseff, A. Balser, A. Green, B. J. Peterson, and J. Bradford. 2008. Sediment and nutrient delivery from thermokarst features in the foothills of the North Slope, Alaska: Potential impacts on headwater stream ecosystems. *Journal Geophysical Research* 113:G02026. doi:10.1029/2007JG000470.

Bret-Harte, M. S., G. R. Shaver, J. P. Zoerner, J. F. Johnstone, J. L. Wagner, A. S. Chavez, R. F. Gunkelman, S. C. Lippert, and J. A. Laundre. 2001. Developmental plasticity allows *Betula nana* to dominate tundra subjected to an altered environment. *Ecology* 82:18–32.

Bret-Harte, M. S., G. R. Shaver, and F. S. Chapin, III. 2002. Primary and secondary stem growth in arctic shrubs: Implications for community response to environmental change. *Journal of Ecology* 90:251–267.

Bret-Harte, M. S., M. C. Mack, G. R. Goldsmith, D. B. Sloan, J. DeMarco, G. R. Shaver, P. M. Ray, Z. Biesinger, and F. S. Chapin, III. 2008. Plant functional types do not predict biomass responses to removal and fertilization in Alaskan tussock tundra. *Journal of Ecology* 96:713–726.

Brown, J., P. C. Miller, L. L. Tieszen, and F. L. Bunnell, editors. 1980. *An arctic ecosystem: The coastal tundra at Barrow, Alaska*. US/IBP Synthesis Series 12. Stroudsburg, PA: Dowden, Hutchinson, and Ross.

Cameron, R. D., W. T. Smith, R. G. White, and B. Griffith. 2002. Pages 38–45 in D. C. Douglas, P. E. Reynolds, and E. B. Rhode, editors, The central arctic Caribou herd. *Arctic Refuge Coastal Plain Terrestrial Wildlife Research Summaries.* Biological Science Report USGS/BRD BSR-2002-0001. U.S. Geological Survey, Biological Resources Division.

Chapin, F. S., III, and G. R. Shaver. 1985. Arctic. Pages 16–40 in B. Chabot and H. A. Mooney, editors, *Physiological Ecology of North American Plant Communities.* London: Chapman and Hall.

Chapin, F. S., III, and G. R. Shaver. 1996. Physiological and growth responses of arctic plants to a field experiment simulating climatic change. *Ecology* 77:822–840.

Chapin, F. S., III, G. R. Shaver, A. E. Giblin, K. J. Nadelhoffer, and J. A. Laundre. 1995. Responses of arctic tundra to experimental and observed changes in climate. *Ecology* 76:694–711.

Chapin, F. S., III, M. Sturm, M. C. Serreze, J. P. McFadden, J. R. Key, A. H. Lloyd, A. D. McGuire, T. S. Rupp, A. H. Lynch, J. P. Schimel, J. Beringer, W. L. Chapman, H. E. Epstein, E. S. Euskirchen, L. D. Hinzman, G. Jia, C.-L. Ping, K. D. Tape, C. D. C. Thompson, D. A. Walker, and J. M. Welker. 2005. Role of land-surface changes in arctic summer warming. *Science* 310:657–660.

Christensen, T. R., A. Ekberg, L. Ström, M. Mastepanov, N. Panikov, M. Öquist, B. H. Svensson, H. Nykänen, P. J. Martikainen, and H. Oskarsson. 2003. Factors controlling large scale variations in methane emissions from wetlands. *Geophysical Research Letters* 30:1414. doi:10.1029/2002GL016848.

Cornelissen, J. H. C., T. V. Callaghan, J. M. Alatalo, A. Michelsen, E. Graglia, A. E. Hartley, D. S. Hik, S. E. Hobbie, M. C. Press, C. H. Robinson, G. H. R. Henry, G. R. Shaver, G. K. Phoenix, D. Gwynn Jones, S. Jonasson, F. S. Chapin, III, U. Molau, C. Neill, J. A. Lee, J. M. Melillo, B. Sveinbjörnsson, and R. Aerts. 2001. Global change and arctic ecosystems: Is lichen decline a function of increases in vascular plant biomass? *Journal of Ecology* 89:984–994.

Cornelissen J. H. C., P. M. van Bodegom, R. Aerts, T. V. Callaghan, R. S. P. van Logtestijn, J. Alatalo, F. S. Chapin, III, R. Gerdol, J. Gudmundsson, D. Gwynn-Jones, A. E. Hartley, D. S. Hik, A. Hofgaard, I. S. Jónsdóttir, S. Karlsson, J. A. Klein, J. Laundre, B. Magnusson, A. Michelsen, U. Molau, V. G. Onipchenko, H. M. Quested, S. M. Sandvik, I. K. Schmidt, G. R. Shaver, B. Solheim, N. A. Soudzilovskaia, A. Stenström, A. Tolvanen, Ø. Totland, N. Wada, J. M. Welker, X. Zhao, and M. O. L. Team. 2007. Global negative vegetation feedback to climate warming responses of leaf litter decomposition rates in cold biomes. *Ecology Letters* 10:619–627.

Deslippe, J. R., M. Hartmann, W. W. Mohn, and S. W. Simard. 2011. Long-term experimental manipulation of climate alters the ectomycorrhizal community of *Betula nana in* Arctic tundra. *Global Change Biology* 17:1625–1636.

Doles, J. 2000. A survey of soil biota in the arctic tundra and their role in mediating terrestrial nutrient cycling. M.S. thesis, University of Northern Colorado, Greeley.

Dormann, C. F., and S. J. Woodin. 2002. Climate change in the Arctic: Using plant functional types in a meta-analysis of field experiments. *Functional Ecology* 16:4–17.

Elmendorf, S. C., G. H. R. Henry, R. D. Hollister, R. G. Björk, A. J. Bjorkman, T. V. Callaghan, E. J. Cooper, J. H. C. Cornelissen, T. A. Day, A. M. Fosaa, W. A.Gould, J. Grétarsdóttir, J. Harte, L. Hermanutz, D. A. Hik, A. Hofgaard, F. Jarrad, I. S. Jónsdóttir, F. Keuper, K. Klanderud, J. A. Klein, S. Koh, G. Kudo, S. Lang, V. Lowen, J. L. May, J. Mercado, A. Michelsen, U. Molau, S. Pieper, C. H. Robinson, L. Siegart, I. Myers-Smith, S. F. Oberbauer, E. Post, C. Rixen, N. M. Schmidt, G. R. Shaver, A. Tolvanen, O. Totland, T. Troxler, C.-H. Wahren, P. J. Webber, J. M. Welker, and P. A. Wookey. 2011. Global assessment of simulated climate warming on tundra vegetation: Heterogeneity over space and time. *Ecology Letters* 15:164–175.

Euskirchen, E. S., A. D. McGuire, D. W. Kicklighter, Q. Zhuang, J. S. Clein, R. J. Dargavillle, D. G. Dyeki, J. S. Kimball, K. C. McDonald, J. M. Melillo, V. E. Romanovsky, and N. V. Smith. 2006. Importance of recent shifts in soil thermal dynamics on growing season length, productivity, and carbon sequestration in terrestrial high-latitude ecosystems. *Global Change Biology* 12:731–750.

Euskirchen, E. S., M. S. Bret-Harte, G. J. Scott, C. Edgar, and G. R. Shaver. 2012. Seasonal patterns of carbon and water fluxes in three representative ecosystems in the northern foothills of the Brooks Range, Alaska. *Ecosphere* 3:4.

Fahnestock, J. T., M. H. Jones, and J. M. Welker. 1999. Wintertime CO_2 efflux from arctic soils: Implications for annual carbon budgets. *Global Biogeochemical Cycles* 13:775–780.

Fahnestock, J. T., K. L. Povirk, and J. M. Welker. 2000. Ecological significance of litter redistribution by wind and snow in arctic landscapes. *Ecography* 23:623–631.

Fetcher, N., and G. R. Shaver. 1982. Growth and tillering patterns within tussocks of *Eriophorum vaginatum. Holarctic Ecology* 5:180–186.

Fetcher, N., and G. R. Shaver. 1990. Environmental sensitivity of ecotypes as a potential influence on primary productivity. *American Naturalist* 136:126–131.

Giblin, A. E., K. J. Nadelhoffer, G. R. Shaver, J. A. Laundre, and A. J. McKerrow. 1991. Biogeochemical diversity along a riverside toposequence in arctic Alaska. *Ecological Monographs* 61:415–436.

Gooseff, M. N., A. Balser, W. B. Bowden, and J. B. Jones. 2009. Effects of hillslope thermokarst in Northern Alaska. *Eos Transactions: American Geophysical Union* 90:29–36.

Gough, L., G. R. Shaver, J. Carroll, D. Royer, and J. A. Laundre. 2000. Vascular plant species richness in Alaskan arctic tundra: The importance of soil pH. *Journal of Ecology* 88:54–66.

Gough, L., P. A. Wookey, and G. R. Shaver. 2002. Dry heath arctic tundra responses to long-term nutrient and light manipulation. *Arctic, Antarctic, and Alpine Research* 34:211–218.

Gough, L., E. A. Ramsey, and D. R. Johnson. 2007. Plant-herbivore interactions in Alaskan arctic tundra change with soil nutrient availability. *Oikos* 116:407–418.

Gough, L., K. Shrestha, D. R. Johnson, and B. Moon. 2008. Long-term mammalian herbivory and nutrient addition alter lichen community structure in Alaskan dry heath tundra. *Arctic, Antarctic, and Alpine Research* 40:65–73.

Gough, L., J. C. Moore, G. R. Shaver, R. T. Simpson, and D. R. Johnson. 2012. Above- and belowground responses of Arctic tundra ecosystems to altered soil nutrients and mammalian herbivory. *Ecology* 93:1683–1694.

Griffith, B., D. C. Douglas, N. E. Walsh, D. D. Young, T. R. McCabe, D. E. Russell, R. G. White, R. D. Cameron, and K. R. Whitten. 2002. The Porcupine caribou herd. Pages 8–37 in D. C. Douglas, P. E. Reynolds, and E. B. Rhode, editors, *Arctic Refuge Coastal Plain Terrestrial Wildlife Research Summaries*. Biological Science Report USGS/BRD BSR-2002-0001. U.S. Geological Survey, Biological Resources Division.

Grogan, P., and F. S. Chapin, III. 1999. Arctic soil respiration: Effects of climate and vegetation depend on season. *Ecosystems* 2:451–459.

Hahn, T. P., J. C. Wingfield, P. Deviche, and R. Mullen. 1995. Spatial and temporal opportunism in Arctic birds. *American Zoologist* 35:259–273.

Hamilton, T. D. 2003. *Glacial geology of Toolik Lake and the Upper Kuparuk River regions*. Biological papers of the University of Alaska, no. 26, scale 1:100,000. Fairbanks, AK: Institute of Arctic Biology.

Herbert, D. A., E. B. Rastetter, G. R. Shaver, and G. Ågren. 1999. Effects of plant growth characteristics on biogeochemistry and community composition in a changing climate. *Ecosystems* 2:367–382.

Hinzman, L. D., D. L. Kane, R. E. Gieck, and K. R. Everett. 1991. Hydrologic and thermal properties of the active layer in the Alaskan arctic. *Cold Regions Science and Technology* 19:95–110.

Hobara, S., C. McCalley, K. Koba, A. E. Giblin, M. S. Weiss, G. M. Gettel, and G. R. Shaver. 2006. Nitrogen fixation in an arctic tundra watershed: A key atmospheric N source. *Arctic, Antarctic, and Alpine Research* 38:363–372.

Hobbie, J. E., and E. A. Hobbie. 2006. ^{15}N in symbiotic fungi and plants estimates nitrogen and carbon flux rates in arctic tundra. *Ecology* 87:816–822.

Hobbie, E. A., and J. E. Hobbie. 2008. Natural abundance of ^{15}N in nitrogen-limited forests and tundra can estimate nitrogen cycling through mycorrhizal fungi: A review. *Ecosystems* 11:815–830.

Hobbie, S. E. 1996. Temperature and plant species control over carbon and nitrogen cycling through litter and soil in Alaskan tundra. *Ecological Monographs* 66:503–522.

Hobbie, S. E., and F. S. Chapin, III. 1998. An experimental test of limits to tree establishment in arctic tundra. *Journal of Ecology* 86:449–461.

Hobbie, S. E., and L. Gough. 2004. Litter decomposition in moist acidic and nonacidic tundra with different glacial histories. *Oecologia* 140:113–124.

Hobbie, S. E., L. Gough, and G. R. Shaver. 2005. Species compositional differences on different-aged glacial landscapes drive contrasting responses of tundra to nutrient addition. *Journal of Ecology* 93:770–782.

Hobbie, S. E., T. A. Miley, and M. S. Weiss. 2002. Carbon and nitrogen cycling in soils from different glacial surfaces in northern Alaska. *Ecosystems* 5:761–774.

Hu, F. S., P. E. Higuera, J. E. Walsh, W. L. Chapman, P. A. Duffy, L. B. Brubaker, and M. L. Chipman. 2010. Tundra burning in Alaska: Linkages to climatic change and sea ice retreat. *Journal of Geophysical Research-Biogeosciences* 115:G04002.

Jefferies, R. L., D. R. Klein, and G. R. Shaver. 1994. Herbivores and northern plant communities: Reciprocal influences and responses. *Oikos* 71:193–206.

Johnson, D. R. 2008. How herbivores affect plant growth, community structure and Litter decomposition in Alaskan tundra: Implications for responses to climate change. Ph.D. dissertation, University of Texas at Arlington.

Johnson, L., G. R. Shaver, D. Cades, E. Rastetter, K. J. Nadelhoffer, A. E. Giblin, J. Laundre, and A. Stanley. 2000. Carbon-nutrient interactions control CO_2 exchange in Alaskan wet sedge ecosystems. *Ecology* 81:453–469.

Judd, K. E., and G. W. Kling. 2002. Production and export of dissolved C in arctic tundra mesocosms: The roles of vegetation and water flow. *Biogeochemistry* 60:213–234.

Kielland, K. 1994. Amino acid absorption by arctic plants: Implications for plant nutrition and nitrogen cycling. *Ecology* 75:2373–2383.

Kling, G. W., G. W. Kipphut, and M. C. Miller. 1991. Arctic lakes and streams as gas conduits to the atmosphere: Implications for tundra carbon budgets. *Science* 251:298–301.

Le Dizès, S., B. L. Kwiatkowski, E. B. Rastetter, A. Hope, J. E. Hobbie, D. Stow, and S. Daeschner. 2003. Modeling biogeochemical responses of tundra ecosystems to temporal and spatial variations in climate in the Kuparuk River basin (Alaska). *Journal of Geophysical Research-Atmospheres* 108(D2):8165. doi:10.1029/2001JD000960.

Lenhart, E. A. 2002. Central arctic herd. Pages 296–292 in C. Brown, editor, *Caribou management report of survey of inventory activities.* Alaska Department of Fish and Game, Division of Wildlife Conservation.

Mack, M. C., E. A. G. Schuur, M. S. Bret-Harte, G. R. Shaver, and F. S. Chapin, III. 2004. Ecosystem carbon storage in arctic tundra reduced by long-term nutrient fertilization. *Nature* 431:440–443.

Mack, M. C., M. S. Bret-Harte, T. N. Hollingsworth, R. Jandt, E. A. G. Schuur, G. R. Shaver, and D. L. Verbyla. 2011. Carbon loss from an unprecedented arctic tundra wildfire. *Nature* 475:489–492.

MacLean, S. F., and F. A. Pitelka. 1971. Seasonal patterns of abundance of tundra arthropods near Barrow. *Arctic* 24:19–40.

Martin, K., and K. L. Wiebe. 2004. Coping mechanisms of alpine and arctic breeding birds: extreme weather and limitations to reproductive resilience. *Integrative and Comparative Biology* 44:177–185.

McFadden, J. P., W. Eugster, and F. S. Chapin, III. 2003. A regional study of the controls on water vapor and CO_2 fluxes in arctic tundra. *Ecology* 84:2762–2776.

McGuire, A. D., L. G. Anderson, T. R. Christensen, S. Dallimore, L. Guo, D. J. Hayes, M. Heimann, T. D. Lorenson, R. W. MacDonald, and N. Roulet. 2009. Sensitivity of the carbon cycle in the Arctic to climate change. *Ecological Monographs* 79:523–555.

McKane, R. B., L. C. Johnson, G. R. Shaver, K. J. Nadelhoffer, E. B. Rastetter, B. Fry, A. E. Giblin, K. Kielland, B. L. Kwiatkowski, J. A. Laundre, and G. Murray. 2002. Resource-based niches provide a basis for plant species diversity and dominance in arctic tundra. *Nature* 415:68–71.

McKane, R. B., E. B. Rastetter, G. R. Shaver, K. J. Nadelhoffer, A. E. Giblin, and J. A. Laundre. 1997. Climatic effects on tundra carbon storage inferred from experimental data and a model. *Ecology* 78:1170–1187.

McNamara, J. P., D. L. Kane, J. E. Hobbie, and G. W. Kling. 2008. Hydrologic and biogeochemical controls on the spatial and temporal patterns of nitrogen and phosphorus in the Kuparuk River, arctic Alaska. *Hydrological Processes* 22:3294–3309.

Mercado-Díaz, J. A. 2011. Plant community responses of the Alaskan arctic tundra to environmental and experimental changes in climate. M.S. thesis, University of Puerto Rico, Río Piedras.

Moorhead, D. L., W. S. Currie, E. B. Rastetter, W. J. Parton, and M. E. Harmon. 1999. Climate and litter quality controls on decomposition: An analysis of modeling approaches. *Global Biogeochemical Cycles* 13:575–589.

Myers, J. P., P. G. Connors, and F. A. Pitelka. 1979. Territory size in wintering sanderlings: The effects of prey abundance and intruder density. *Auk* 96:551–561.

Nadelhoffer, K. J., G. Shaver, B. Fry, L. Johnson, and R. McKane. 1996. ^{15}N natural abundances and N use by tundra plants. *Oecologia* 107:386–394.

Nadelhoffer, K. J., L. Johnson, J. Laundre, A. E. Giblin, and G. R. Shaver. 2002. Fine root production and nutrient use in wet and moist arctic tundras as influenced by chronic fertilization. *Plant and Soil* 242:107–113.

Nowinski, N., S. Trumbore, E. Schuur, M. Mack, and G. Shaver. 2008. Nutrient addition prompts rapid destabilization of organic matter in an arctic tundra ecosystem. *Ecosystems* 11:16–25.

Oberbauer, S. F., C. E. Tweedie, J. M. Welker, J. T. Fahnestock, G. H. R. Henry, P. J. Webber, R. D. Hollister, M. D. Walker, A. Kuchy, E. Elmore, and G. Starr. 2007. Tundra CO_2 fluxes in response to experimental warming across latitudinal and moisture gradients. *Ecological Monographs* 77:221–238.

Peterson, B. J., T. Corliss, K. Kriet, and J. E. Hobbie. 1992. Nitrogen and phosphorus concentrations and export for the upper Kuparuk River on the North Slope of Alaska in 1980. *Hydrobiologia* 240:61–71.

Ping, C.-L., G. J. Michaelson, W. M. Loya, R. J. Candler, and R. L. Malcolm. 1997. Characteristics of soil organic matter in Arctic ecosystems of Alaska. Pages 157–167 in R. Lal, J.M. Kimble, and B.A. Stewart, editors, *Carbon sequestration in soils*. Chelsea, MI: CRC Lewis Publishers.

Ping, C.-L., J. G. Bockheim, J. M. Kimble, G. J. Michaelson, and D. A. Walker. 1998. Characteristics of cryogenic soils along a latitudinal transect in arctic Alaska. *Journal of Geophysical Research* 103:28,917–928,928.

Ping, C.-L., G. J. Michaelson, M. T. Jorgensen, J. M. Kimble, H. Epstein, V. E. Romanovsky, and D. W. Walker. 2008. High stocks of soil organic carbon in the North American arctic region. *Nature Geoscience* 1:615–619.

Pitelka, F.A. 1973. Cyclic patterns in lemming populations near Barrow, Alaska. *Arctic Institute of North America Technical Papers* 25:199–215.

Rastetter, E. B. 1996. Validating models of ecosystem response to global change. *BioScience* 46:190–198.

Rastetter, E. B., R. B. McKane, G. R. Shaver, K. J. Nadelhoffer, and A. E. Giblin. 1997. Analysis of CO_2, temperature, and moisture effects on carbon storage in Alaskan arctic tundra using a general ecosystem model. Pages 437–451 in W. C. Oechel, T. Callaghan, T. Gilmanov, J. I. Holten, B. Maxwell, U. Molau, and B. Sveinbjörnsson, editors, *Global Change and Arctic Terrestrial Ecosystems*, New York: Springer-Verlag.

Rastetter, E. B., J. D. Aber, D. P. C. Peters, D. S. Ojima, and I. Burke. 2003. Using mechanistic models to scale ecological processes across space and time. *BioScience* 53:68–76.

Rastetter, E. B., B. L. Kwiatkowski, S. Le Dizès, and J. E. Hobbie. 2004. The role of down-slope water and nutrient fluxes in the response of arctic hill slopes to climate change. *Biogeochemistry* 69:37–62.

Rastetter, E. B., S. S. Perakis, G. R. Shaver, and G. I. Ågren. 2005. Carbon sequestration in terrestrial ecosystems under elevated CO2 and temperature: Role of dissolved organic N loss. *Ecological Applications* 15:71–86.

Rocha, A., and G. R. Shaver. 2011a. Burn severity influences post-fire CO2 exchange in arctic tundra. *Ecological Applications* 21:477–489.

Rocha, A., and G. R. Shaver. 2011b. Post-fire energy exchange in arctic tundra: the importance and climatic implications of burn severity. *Global Change Biology* 17:2831–2841.

Schimel, J. P. 1995. Plant transport and methane production as controls on methane flux from arctic wet meadow tundra. *Biogeochemistry* 28:183–200.

Schimel, J. P., and M. N. Weintraub. 2003. The implications of exoenzyme activity on microbial carbon and nitrogen limitation in soil: A theoretical model. *Soil Biology and Biochemistry* 35:549–563.

Schimel, J. P., and F. S. Chapin, III. 1996. Tundra plants compete effectively with soil microbes for amino-acid nitrogen. *Ecology* 77:2142–2147.

Schimel, J. P., C. Bilbrough, and J. M. Welker. 2004. Increased snow depth affects microbial activity and nitrogen mineralization in two arctic tundra communities. *Soil Biology and Biochemistry* 36:217–227.

Schimel, J. P., J. Fahnestock, G. Michaelson, C. Mikan, C.-L. Ping, V. E. Romanovsky, and J. M. Welker. 2006. Cold-season production of CO_2 in arctic soils: Can laboratory and field estimates be reconciled through a simple modeling approach? *Arctic, Antarctic, and Alpine Research* 38:249–256.

Schmidt, I. K., S. Jonasson, G. R. Shaver, A. Michelsen, and A. Nordin. 2002. Mineralization and allocation of nutrients by plants and microbes in four arctic ecosystems: Responses to warming. *Plant and Soil* 242:93–106.

Shaver, G. R. 1981. Mineral nutrition and leaf longevity in an evergreen shrub, Ledum palustre ssp. *decumbens*. *Oecologia* 49:362–365.

Shaver, G. R. 1983. Mineral nutrition and leaf longevity in *Ledum palustre*: The role of individual nutrients and the timing of leaf mortality. *Oecologia* 56:160–165.

Shaver, G. R. 1986. Woody stem production in Alaskan tundra shrubs. *Ecology* 67:660–669.

Shaver, G. R., and F. S. Chapin, III. 1980. Response to fertilization by various plant growth forms in an Alaskan tundra: Nutrient accumulation and growth. *Ecology* 61:662–675.

Shaver, G. R., and F. S. Chapin, III. 1986. Effect of fertilizer on production and biomass of tussock tundra, Alaska, U.S.A. *Arctic and Alpine Research* 18:261–268.

Shaver, G. R., and F. S. Chapin, III. 1991. Production/biomass relationships and element cycling in contrasting arctic vegetation types. *Ecological Monographs* 61:1–31.

Shaver, G. R., and F. S. Chapin, III. 1995. Long-term responses to factorial NPK fertilizer treatment by Alaskan wet and moist tundra sedge species. *Ecography* 18:259–275.

Shaver, G. R., and S. Jonasson. 1999. Response of arctic ecosystems to climate change: Results of long-term field experiments in Sweden and Alaska. *Polar Biology* 18:245–252.

Shaver, G. R., and J. A. Laundre. 1997. Leaf exsertion, leaf elongation, and leaf senescence in *Eriophorum vaginatum* and *Carex bigelowii* in northern Alaska. *Global Change Biology* 3, supplement 1:146–157.

Shaver, G. R., F. S. Chapin, III, and B. L. Gartner. 1986a. Factors limiting growth and biomass accumulation in *Eriophorum vaginatum* L. in Alaskan tussock tundra. *Journal of Ecology* 74:257–278.

Shaver, G. R., N. Fetcher, and F. S. Chapin, III. 1986b. Growth and flowering in *Eriophorum vaginatum*: Annual and latitudinal variation. *Ecology* 67:1524–1525.

Shaver, G. R., K. J. Nadelhoffer, and A. E. Giblin. 1991. Biogeochemical diversity and element transport in a heterogeneous landscape, the North Slope of Alaska. Pages 105–126 in M. G. Turner and R. H. Gardner, editors, *Quantitative methods in landscapes.* New York: Springer-Verlag.

Shaver, G. R., W. D. Billings, F. S. Chapin, III, A. E. Giblin, K. J. Nadelhoffer, W. C. Oechel, and E. B. Rastetter. 1992. Global change and the carbon balance of arctic ecosystems. *BioScience* 42:433–441.

Shaver, G. R., L. C. Johnson, D. H. Cades, G. Murray, J. A. Laundre, E. B. Rastetter, K. J. Nadelhoffer, and A. E. Giblin. 1998. Biomass accumulation and CO_2 flux in three Alaskan wet sedge tundras: Responses to nutrients, temperature, and light. *Ecological Monographs* 68:75–99.

Shaver, G. R., J. Canadell, F. S. Chapin, III, J. Gurevitch, J. Harte, G. Henry, P. Ineson, S. Jonasson, J. Melillo, L. Pitelka, and L. Rustad. 2000. Global warming and terrestrial ecosystems: A conceptual framework for analysis. *BioScience* 50:871–882.

Shaver, G. R., M. S. Bret-Harte, M. H. Jones, J. Johnstone, L. Gough, J. Laundre, and F. S. Chapin, III. 2001. Species composition interacts with fertilizer to control long-term change in tundra productivity. *Ecology* 82:3163–3181.

Shaver, G. R., A. E. Giblin, K. J. Nadelhoffer, K. K. Thieler, M. R. Downs, J. A. Laundre, and E. B. Rastetter. 2006. Carbon turnover in Alaskan tundra soils: Effects of organic matter quality, temperature, moisture, and fertilizer. *Journal of Ecology* 94:740–753.

Shaver, G. R., L. E. Street, E. B. Rastetter, M. T. van Wijk, and M. Williams. 2007. Functional convergence in regulation of net CO_2 flux in heterogeneous tundra landscapes in Alaska and Sweden. *Journal of Ecology* 95:802–817.

Simpson, R. 2010. Soil organic matter and aggregate dynamics in an arctic ecosystem. Ph.D. dissertation, Colorado State University, Fort Collins.

Skinner, W. R., R. L. Jefferies, T. J. Carleton, R. F. Rockwell, and K. F. Abraham. 1998. Prediction of reproductive success and failure in lesser snow geese based on early season climactic variables. *Global Change Biology* 4:3–16.

Starr, G., and S. F. Oberbauer. 2003. Photosynthesis of arctic evergreens under snow: Implications for tundra ecosystem carbon balance. *Ecology* 84:1415–1420.

Street L. E., G. R. Shaver, M. Williams, and M. T. van Wijk. 2007. What is the relationship between changes in leaf area and changes in photosynthetic CO_2 flux in arctic ecosystems? *Journal of Ecology* 95:139–150.

Sturm, M., C. R. Racine, and K. Tape. 2001. Increasing shrub abundance in the Arctic. *Nature* 411:546–547.

Sturm, M., J. Schimel, G. Michaelson, J. M. Welker, S. F. Oberbauer, G. E. Liston, J. Fahnestock, and V. E. Romanovsky. 2005. Winter biological processes could help convert arctic tundra to shrubland. *Bioscience* 55:17–26.

Sullivan, P. F., M. Sommerkorn, H. M. Rueth, K. J. Nadelhoffer, G. R. Shaver, and J. M. Welker. 2007. Climate and species affect fine root production with long-term fertilization in acidic tussock tundra near Toolik Lake, Alaska. *Oecologia* 153:643–652.

Summerhayes, V. S., and C. S. Elton. 1923. Contributions to the ecology of Spitzbergen and Bear Island. *Journal of Ecology* 11:214–286.

Symstad, A. J. 2003. An overview of ecological plant classification systems: Linking functional response and functional effect groups. Pages 13–51 in R. S. Ambasht and N. K. Ambasht, editors, *Modern trends in applied terrestrial ecology*. Boston: Kluwer Academic Press.

Torn, M. S., and F. S. Chapin, III. 1993. Environmental and biotic controls over methane flux from arctic tundra. *Chemosphere* 26:357–368.

van Wijk, M. T., K. Clemmensen, G. R. Shaver, M. Williams, T. V. Callaghan, F. S. Chapin, III, J. H. C. Cornelissen, L. Gough, S. E. Hobbie, S. Jonasson, J. A. Lee, A. Michelsen, M. C.

Press, S. J. Richardson, and H. Rueth. 2003. Long-term ecosystem level experiments in Toolik Lake, Alaska, and Abisko, Northern Sweden: Generalizations and differences in ecosystem and plant type responses to global change. *Global Change Biology* 10:105–123.

Wahren, C.-H. A., M. D. Walker, and M. S. Bret-Harte. 2005. Vegetation responses in Alaskan arctic tundra after eight years of a summer warming and winter snow manipulation experiment. *Global Change Biology* 11:537–552.

Walker, D. A., E. Binnian, B. M. Evans, N. D. Lederer, E. Nordstrand, and P. J. Webber. 1989. Terrain, vegetation and landscape evolution of the R4D research site, Brooks Range foothills, Alaska. *Holarctic Ecology* 12:238–261.

Walker, D. A., H. E. Epstein, and J. M. Welker. 2008. Introduction to special section on biocomplexity of tundra ecosystems. *Journal of Geophysical Research* 113:G03000. doi:10.1029/2008JG000740.

Walker, M. D., D. A. Walker, J. M. Welker, A. M. Arft, T. Bardsley, P. D. Brooks, J. T. Fahnestock, M. H. Jones, M. Losleben, A. N. Parsons, T. R. Seastedt, and P. L. Turner. 1999. Long-term experimental manipulation of winter snow regime and summer temperature in arctic and alpine tundra. *Hydrological Processes* 13:2315–2330.

Walker. M. D., C.-H. Wahren, R. D. Hollister, G. H. R. Henry, L. E. Ahlquist, J. M. Alatalo, M. S. Bret-Harte, M. P. Calef, T. V. Callaghan, A. B. Carroll, H. E. Epstein, I. S. Jonsdottir, J. A. Klein, B. Magnusson, U. Molau, S. F. Oberbauer, S. P. Rewa, C. H. Robinson, G. R. Shaver, K. N. Suding, C. C. Thompson, A. Tolvanen, Ø. Totland, P. L. Turner, C. E. Tweedie, P. J. Webber, and P. A. Wookey. 2006. Plant community responses to experimental warming across the tundra biome. *Proceedings National Academy of Sciences* 103:1342–1346.

Walter, K. M., F. S. Chapin, III, S. A. Zimov, and J. P. Chanton. 2006. Methane bubbling from Siberian thaw lakes: A positive feedback to climate change. *Nature* 443:71–75.

Weintraub, M. N., and J. P. Schimel. 2003. Interactions between carbon and nitrogen mineralization and soil organic matter chemistry in arctic tundra soils. *Ecosystems* 6:129–143.

Weintraub, M. N., and J. P. Schimel. 2005. Seasonal dynamics of amino acids and other nutrients in arctic tundra soils. *Biogeochemistry* 73:359–380.

Weiss, M. S., S. E. Hobbie, and G. M. Gettel. 2005. Contrasting responses of nitrogen-fixation in arctic lichens to experimental and ambient nitrogen and phosphorus availability. *Arctic, Antarctic and Alpine Research* 37:396–401.

Williams, M., and E. B. Rastetter. 1999. Vegetation characteristics and primary productivity along an arctic transect: Implications for scaling-up. *Journal of Ecology* 87:885–898.

Williams, M., W. Eugster, E. B. Rastetter, J. P. McFadden, and F. S. Chapin, III. 2000. The controls on net ecosystem productivity along an arctic transect: A model comparison with flux measurements. *Global Change Biology* 6, supplement 1:116–126.

Williams, M., E. B. Rastetter, G. R. Shaver, J. E. Hobbie, E. Carpino, and B. L. Kwiatkowski. 2001. Primary production in an arctic watershed: An uncertainty analysis. *Ecological Applications* 11:1800–1816.

Williams, M., L. E. Street, M. van Wijk, and G. R. Shaver. 2006. Identifying differences in rates of carbon exchange among vegetation types along an arctic toposequence. *Ecosystems* 9:288–304.

Yano, Y., G. R. Shaver, A. E. Giblin, E. B. Rastetter, and K. J. Nadelhoffer. 2010a. Nitrogen dynamics in a small arctic watershed: Retention and downhill movement of ^{15}N. *Ecological Monographs* 80:331–351.

Yano, Y., G. R. Shaver, A. E. Giblin, and E. B. Rastetter. 2010b. Depleted N-15 in hydrolysable-N of arctic soils and its implication for mycorrhizal fungi-plant interaction. *Biogeochemistry* 97:183–194.

Zak, D. R., and G. W. Kling. 2006. Microbial community composition and function across an arctic tundra landscape. *Ecology* 87:1659–1670.

6

Land–Water Interactions

George W. Kling
Heather E. Adams
Neil D. Bettez
William B. Bowden
Byron C. Crump
Anne E. Giblin
Kristin E. Judd
Katy Keller
George W. Kipphut
Edward R. Rastetter
Gaius R. Shaver
Marc Stieglitz

Introduction

At the scale of whole catchments, ecosystems function through the interactions among land, groundwater, surface water, and the atmosphere. The dominant forms of interaction within and among catchments are the fluxes of materials and organisms and the fluxes of energy through climate forcing. For example, carbon moves from land to water and from water to the atmosphere, while the land and atmosphere exchange carbon in both directions. The amounts, chemical forms, and timing of delivery of water and materials from land to water are critical to the functioning of aquatic ecosystems. Thus understanding how the production and chemical transformations of material in terrestrial environments respond to changes in climate and land use are critical for developing a predictive understanding of the response of surface waters. In our research on land–water interactions, we consider the controls on production, transformation, and export of materials from land, and study how those materials are processed within lakes and streams.

Organizing Concepts

The simple idea of a hydrological catchment as a study ecosystem has provided a clear framework of biogeochemical cycling within and between ecosystems for several decades (e.g., Likens et al. 1967; Bormann et al. 1974). Yet it has proved an extraordinary challenge to measure the outputs of energy and biochemical elements and relate them back to the underlying processes controlling the structure and function of the terrestrial ecosystem. We still understand little about the complex dynamics and rates of production of dissolved materials on land and their delivery to surface waters. For example, a review of 42 studies of dissolved organic carbon (DOC) and nitrogen (DON) concentrations and fluxes in temperate forests found that lab and field studies differed greatly in their results, and that site-specific controls such as temperature or C:N ratios were rarely evident at regional scales (Michalzik et al. 2001). To study these complex dynamics we follow a general conceptual model (Figure 6.1) that illustrates that in any biome there are four major controls on the dynamics: (1) climate and vegetation influence the production of dissolved materials, especially C, N, and P moving from land to water, and (2) the effects of hydrology and biological processing exert a strong control on the export of materials from land to aquatic systems. In addition, the integration of hydrological and biological processes appears to depend on (3) hillslope processes and interactions between the hillslope and broader landscape scales including (4) the response of aquatic ecosystems. These four major control steps, shown in Figure 6.1, combine to determine what is exported from land to water, and thus in part govern the response of the receiving aquatic ecosystems. In this chapter we describe and synthesize our research in these four major areas within the framework of land–water interactions.

History of Land–Water Interactions Research

In the 1980s terrestrial researchers at Toolik Lake began studying the transfers of nutrients across the landscape and focused on a toposequence of different ecosystems from the upland terrace down a slope to the Sagavanirktok River. This work showed dramatic variation in soil-nutrient cycling among adjacent ecosystems

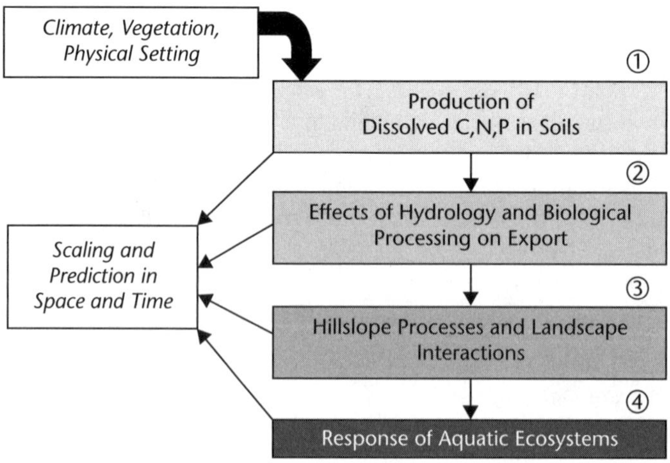

Figure 6.1 Conceptual model for organizing the Arctic LTER land-water interactions research. Each of the numbered boxes on the right represents major research efforts.

along the toposequence (Giblin et al. 1991), and suggested that transport of N and P in soil water moving through the unfrozen soil above the permafrost was potentially an important component of landscape nutrient budgets (Shaver et al. 1991). These ideas were extended to carbon cycling with the discovery that C transport in soil water was a major landscape-level C flux from land to streams and lakes. Some of this C from land was then transported downstream to oceans, and some was released from surface waters to the atmosphere as CO_2 (Kling et al. 1991). Since the early 1990s our LTER research has focused on the controls of these land–water–atmosphere transfers in terms of specific processes within ecosystems as well as the interactions between ecosystems, including terrestrial hillslopes and aquatic networks of lakes and streams.

Production and Fate of Dissolved Materials on Land

The general topic of land–water linkages, and specifically the movement of C and nutrients from arctic terrestrial ecosystems to the oceans, is of critical importance to understanding how the arctic system functions (McGuire et al. 2009). Although the amounts, forms, and timing of terrestrial exports have been studied extensively (e.g., Lewis and Saunders 1989), we understand little about the specific plant and soil origins of these materials or the controls on their production and delivery. Previous work in the Arctic along a toposequence of tundra soil showed that different vegetation types and locations differed strongly in the inorganic nutrient chemistry of their soil waters. Beyond the inorganic nutrients, what little information exists for temperate systems on dissolved organic N, P, and C in soil waters (e.g., McDowell and Likens 1988; Binkley et al. 1992) suggests that vegetation type and soil composition exert a major control on concentrations as well as dissolved organic matter (DOM) composition and processing (e.g., Cory and Kaplan 2012). However, information on the rates of production is rare.

In our LTER we approached this problem by making measurements of soil water chemistry under different vegetation types and at different locations within catchments. In addition we have added ^{14}C-CO_2 for photosynthesis by plants to act as a tracer to determine the origins and rates of production of C in soil waters, and have manipulated plant–soil mesocosms by adding water to determine the impact of hydrological flushing (Judd and Kling 2002). We also developed coupled hydrological-biogeochemical models that are capable of simulating the flow of water from the hillslope to the river network and that incorporate biological processes to predict the fixation and respiration of CO_2 and its flux between land and the atmosphere (Stieglitz et al. 1999, 2000). These models also predict the uptake and release of nutrients, including carbon, by terrestrial vegetation (Rastetter et al. 2004). As one of the first steps in determining the magnitude of material movements, we investigated the flux of carbon gases from land to water and water to the atmosphere as it relates to overall terrestrial carbon budgets.

Lateral Carbon Transfers and Atmospheric Evasion from Surface Water

The movement of C gases from land and water to the atmosphere and back (e.g., respiration and photosynthesis) is well understood, but the sources of C to surface waters and the interactions between the terrestrial, aquatic, and atmospheric systems are less clear. In the Arctic, the cycle begins by fixation of atmospheric CO_2 by tundra vegetation, and the subsequent respiration of plant organic matter in the soil to produce CO_2 and CH_4. These gases then dissolve in groundwaters to the point of extreme excess of atmospheric partial pressures (from ~3 to 200 times atmospheric), and are transported to lakes and streams where they are subsequently released to the atmosphere to complete the cycle (Figure 6.2).

Net positive gas fluxes from surface waters to the atmosphere were unexpected because lakes were generally thought to be in equilibrium with the atmosphere on an annual time-step, especially for dissolved CO_2 and CH_4. In addition, the lateral transport of C (especially C gases) from land to water was assumed to be minor in the overall landscape C budget. These general views of lake and stream C cycles began to change with the research done near Toolik Lake in the early 1990s and with the realization that lateral C fluxes from land to water are large in tundra ecosystems—on the order of 4 g C per m^2 of land surface per year (Table 6.1). This flux is also consistent enough over time to help explain why short-term measurements of C accumulation in soils are much higher than long-term accumulation rates assessed from peat cores—the short-term measurements ignore the lateral losses, whereas this loss is included in the time-averaged net storage rates from cores (Kling et al. 1991). The net, areal rates of fluxes of CO_2 to the atmosphere per square meter of land are similar in magnitude to the aquatic fluxes from a square meter of lake or stream (~20 g C m^{-2} yr^{-1}). But because the water surface area is much smaller than the land surface area on the landscape overall, when the total flux from surface waters is expressed per square meter of all land area in the catchment, the aquatic loss is reduced to ~4 g C m^{-2} (of land) yr^{-1}. However, this 4 g C m^{-2} is still substantial, and when these lateral C losses are accounted for in the total C loss it reduces the net, global terrestrial sink of C in the Arctic by ~20%. Models indicate that the 20-year running mean of arctic terrestrial net ecosystem productivity (NEP, which is the net land–atmosphere CO_2 flux) varies from ~–15 to +10 g C m^{-2} yr^{-1} (McKane et al. 1997), and when the aquatic loss term is accounted for in this estimate it can represent up to 100% of the NEP in years when terrestrial NEP is between –4 to +4 g C m^{-2}. Once recognized, this land–water–atmosphere flux was also found to be important in most ecosystems worldwide (Cole et al. 1994), and has led to a large body of research showing that on a global scale the movement of C from land to inland waters and the evasion of gases to the atmosphere from lakes and streams are both greater than the total global C flux from land to the oceans through rivers (Figure 6.2; Battin et al. 2009; Aufdenkampe et al. 2011).

One of the surprising aspects of this pattern of landscape C cycling is that it appears to be scale invariant; that is, the amount of C lost from aquatic systems per square meter of land was similar regardless of the size of the catchment. We constructed C budgets for three different-sized catchments in arctic Alaska, including

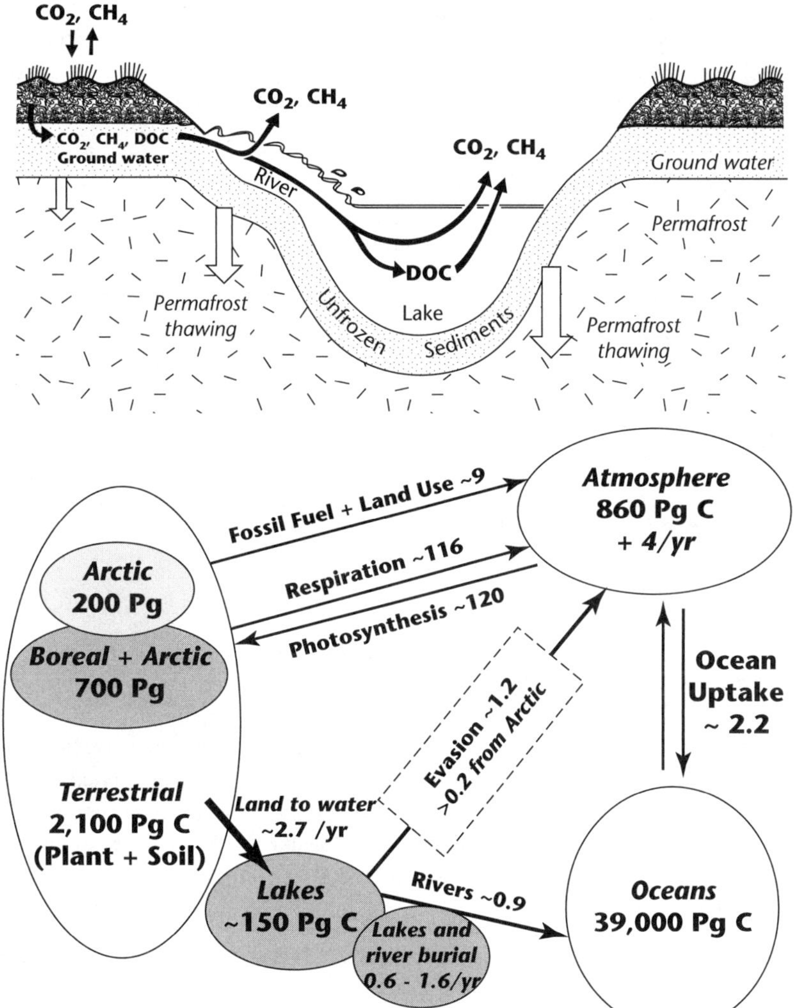

Figure 6.2 Schematic showing land–water transfer paths of C gases and how hydrological flowpaths may deepen with future permafrost thawing (large arrows in soil), especially beneath streams and lakes (top). The importance of these water-to-atmosphere gas transfers in the global C cycle (bottom). (Data from Kling et al. 1991, 1992; Kling 1995; Canadell et al. 2007; Battin et al. 2009; Aufdenkampe et al. 2011.)

dissolved losses in rivers and gaseous losses to the atmosphere from lakes and streams (Table 6.1). Catchment size ranged over five orders of magnitude, and the surface water extent ranged from ~4% in the largest catchment, which contained both lakes and streams, to <0.1% in the smallest catchment, which contained only a first-order (temporary) stream. The evasion of CO_2 was the largest component of aquatic flux in the basin with the highest surface water area, and it was the lowest

Table 6.1 Aquatic C flux per area of land, including the evasion of CO_2 directly from lakes and rivers to the atmosphere (methods in Kling et al. 1992, 2000). Sites include the entire Kuparuk basin (river mouth at the Arctic Ocean), the catchment of Toolik Lake, and the Tussock watershed, a small subcatchment immediately adjacent to and draining into Toolik Lake

Basin	Area	Total Export	DOC	DIC	POC	CO_2 evasion
	(km^2)	(g C m^{-2} yr^{-1})	(g C m^{-2} yr^{-1})	(g C m^{-2} yr^{-1})	(g C m^{-2} yr^{-1})	(g C m^{-2} yr^{-1})
Kuparuk River	9200	4.1 ± 0.24	1.0	1.3	0.1	1.7
Toolik Lake	65	2.3 ± 0.26	0.7	0.5	0.04	1.1
Tussock Watershed	0.02	4.7 ± 1.9	3.7	0.6	0.05	0.3

Values are means of export rates (± standard error, 3 years of study 1994–1996) expressed as C per m^2 of land surface per year (assuming 100-day open-water season).

component in the smallest basin with no lakes and the least surface water area. However, despite this variation in the proportion of dissolved versus gaseous export in the three basins, the total C export was quite similar among basins at ~2–5 g C per m^2 (land surface) yr^{-1} (Table 6.1). The lower gas evasion in the smallest catchment with the least surface water area was compensated for by an increase in DOC flux. We interpret this higher DOC flux as a consequence of (a) a closer connection to soil water in the small stream and (b) shorter water residence times in the catchment leading to less bacterial or photochemical decomposition of DOC to CO_2 during transport.

Soil Water Production of Dissolved Carbon

On the basis of these findings for whole catchments, we examined the production rates of dissolved C in soils, and applied a ^{14}C tracer to intact tundra ecosystems (Figure 6.3, Table 6.2). The tracer was added as $^{14}CO_2$ and the tundra plants were allowed to take up this CO_2 from a 1-m^2 chamber placed over the ground surface in upland moist acidic tundra (tussock tundra) and in lowland wet sedge tundra (see chapter 5 for more detail on these vegetation types). The ^{14}C tracer was taken up by the plants in photosynthesis and then followed into plant pools and dissolved pools of inorganic and organic carbon in the soils. The results from these studies indicate an unexpectedly fast transition from CO_2 taken up by leaves and converted into methane (King et al. 2002) and into DOC (minutes), and an unexpectedly large exudation of dissolved C by roots (see also Judd and Kling 2002). For example, the DOC exudation from plant roots alone was extremely high, from 1 to 4 g C m^{-2} per day as estimated by the ^{14}C tracer experiments. This amount is comparable to the *annual* DOC export from nearby catchments, which is only ~2–5 g C m^{-2} of

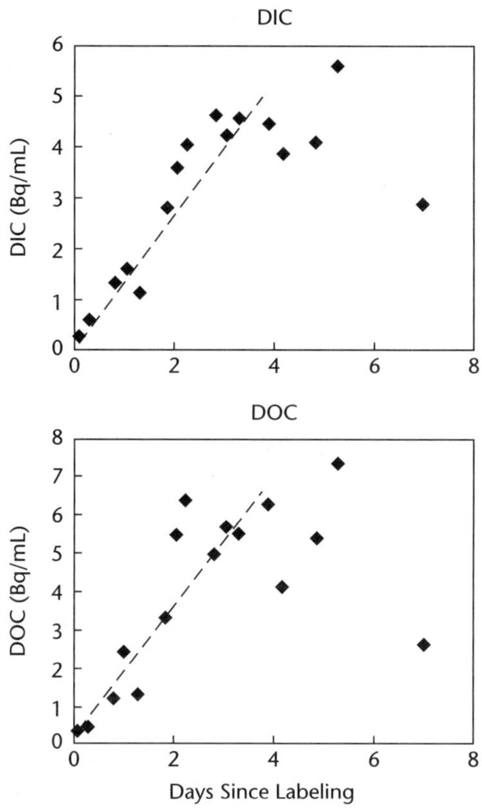

Figure 6.3 Activity of ^{14}C in dissolved inorganic C (DIC) and dissolved organic C (DOC) in soil waters of tussock tundra vegetation versus days since labeling plants with $^{14}CO_2$. Dashed line represents the production rate of ^{14}C compounds originating in roots and released into the soil waters. (Data from LTER; G. Kling, K. Nadelhoffer, M. Sommerkorn, K. Judd, unpublished.)

catchment per year (Table 6.1). By dividing the standing stock of DOC measured in soil waters by this DOC supply rate, we calculate that the turnover time for DOC is <10 days in both ecosystem types (Table 6.2). We similarly calculated that the dissolved CO_2 and CH_4 pools in the soil turned over within ~5–18 days at the times of measurement in this experiment. Overall, from 5% to 20% of the DOC, CO_2, and CH_4 dissolved pools were added to the soil water from roots each day (Table 6.2).

Our conclusion, given that soil C production rates are about two orders of magnitude higher than net C catchment export, is that microbial processing of this C must be substantial during transport through the landscape. This importance of microbial processing as DOM moves from land to oceans is not restricted to the Arctic; consider that in most terrestrial systems net primary production (NPP) is very large, hundreds of g C m^{-2} per year, and yet the net ecosystem production

Table 6.2 Dynamics of belowground dissolved C production in two ecosystem types, upland tussock tundra (acidic) and lowland wet sedge tundra

Vegetation	C pool	Mass of C	C added from roots	Turnover time
		(g C m^{-2})	(% per day)	(days)
Tussock tundra	DOC	31	11	9
	Dissolved CO_2	14	6	17
	Dissolved CH_4	0.2	20	5
Wet sedge tundra	DOC	6	9	9
	Dissolved CO_2	17	17	18
	Dissolved CH_4	1	5	18

Soils were sampled with a steel needle and syringe that withdraws gravity water (not tension water) from the soils, and the volume of soil water per m^2 of land surface was estimated by measuring wet weight and dry weight differences in soil cores. (LTER data; Kling, Nadelhoffer, Sommerkorn, Rastetter, unpublished.)

(NEP) is usually near zero and dissolved export averages only ~6 g C m^{-2} yr^{-1} worldwide (Hope et al. 1994). Thus the huge difference in terrestrial C production and aquatic export must be due mainly to microbial processing in soils, and perhaps to a lesser extent to DOC adsorption or photochemical reactions. Given this strong biogeochemical influence across landscape scales, we can ask questions about the spatial pattern of microbial processing: Where are the control points, and who is responsible?

Microbial Processing across the Landscape

The microbial transformations of nutrients and organic matter are a dominant aspect of aquatic ecology, and microbial transformations of these materials underlie primary production, respiration, and atmospheric gas-exchange in ecosystems. But consideration of how these processes are linked and interact across the landscape is relatively new, and requires an integration of concepts in microbial and landscape ecology. For example, we must first consider the spatial boundaries of ecotones and ecosystems with respect to the rates of microbial activity. In addition, we must understand the biogeographical diversity of microbes and the time-scales over which microbes adapt physiologically and adapt by changing population frequencies. As described below, we found several distinct patterns of microbial species and processing rates in terrestrial and aquatic ecosystems across the landscape.

Arctic terrestrial ecosystems are commonly defined by the dominant vegetation, such as hilltop heath or lowland wet sedge, and these ecosystems differ from each other consistently in landscape position, plant species composition, litter biochemistry, and biogeochemical cycling rates (see chapter 5). To test the idea that these

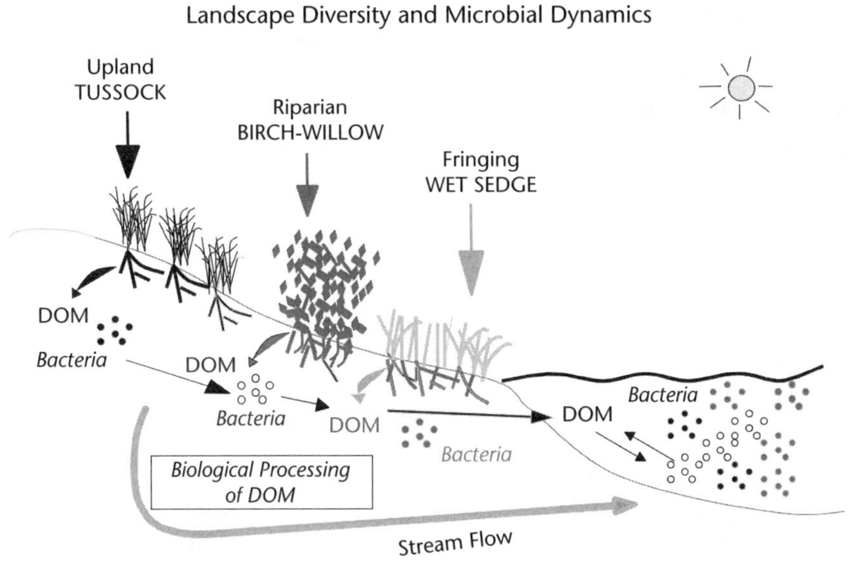

Figure 6.4 Conceptual model of landscape and microbial diversity showing fidelity of DOM character and soil bacterial communities to overlying vegetation, and mixing of DOM and bacteria moving down the landscape and into receiving waterbodies.

ecosystems contain distinct microbial communities that differentially transform DOM as it moves downslope from dry upland to wet lowland tundra, we studied soil microbial communities in upland tussock, stream-side birch-willow, and lake-side wet sedge tundra ecosystems (Figure 6.4).

One set of experimental results indicated that differences in the quality of the organic carbon were related to its place of origin on the landscape and to the time of season, and this controlled the rate of conversion of DOC to CO_2 (Michaelson et al. 1998). In a second set of experiments, phospholipid fatty acids and 16S rRNA analyses were used to examine differences in the microbial community composition among tundra ecosystems; we found that tussock tundra had a higher abundance and activity of fungi and a higher ratio of fungal:bacterial phospholipid fatty acids than did birch-willow or wet sedge ecosystems (Judd et al. 2006; Zak and Kling 2006). A third study added compound-specific ^{13}C isotope tracers and made measurements of extracellular enzymes to determine the potential rates of microbial activity during the degradation of organic molecules representing the common biochemical categories of cellulose, chitin, and lignin in soils. For example, the rates of chitin degradation were assessed by tracing the fate of ^{13}C-labeled N-acetylglucosamine added to soil microcosms. Although the majority of ^{13}C-labeled substrates rapidly moved into soil organic matter in all tundra soils (i.e., 50%–90% of applied ^{13}C), microbial respiration of labeled substrates in wet sedge tundra was lower than in tussock and birch-willow tundra (Figure 6.5). Despite these differences, wet sedge tundra exhibited the greatest potentials for extracellular enzyme activity.

Another critical control on microbial activities in tundra is soil moisture, and the effects of soil moisture also depend on the microbial community composition and DOC chemical characteristics. In soil incubation experiments we found, as expected, that respiration was lower in saturated soils than in soils at field capacity (Figure 6.6). However, differences in microbial assemblages and DOC characteristics among vegetation types controlled the respiration response to moisture.

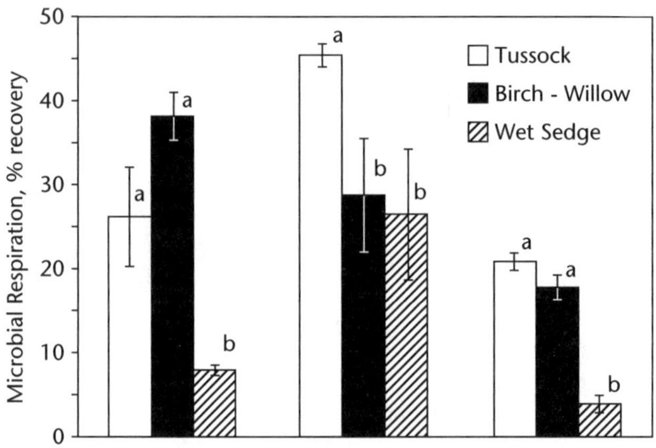

Figure 6.5 Microbial respiration of applied ^{13}C-labeled compounds representing, from left to right, cellulose, chitin, and lignin in soils. Processing rates varied by compound (not shown) and by the location on the landscape (vegetation communities). Used with permission from Zak and Kling (2006).

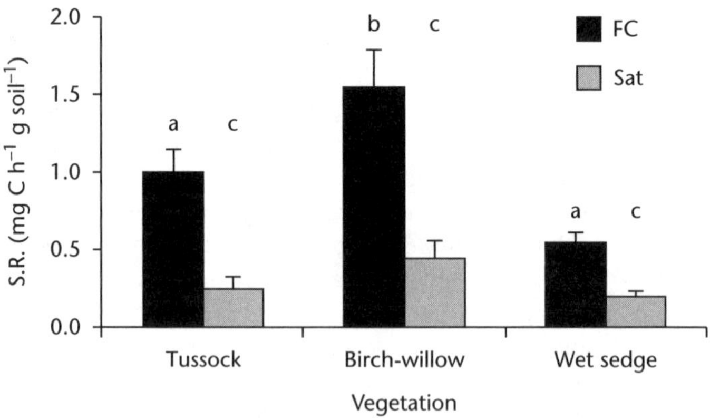

Figure 6.6 Effects of moisture on soil respiration (S.R.) in ~1L airtight glass jars. Black indicates soils at field capacity (FC) conditions and grey indicates saturated (Sat) conditions. Respiration (± standard error, N = 4) was measured by gas chromatography of headspace CO_2 from 20–30 g of soil with large roots and rhizomes removed added to the jars. Tussock and birch-willow soils received a "water addition" treatment (saturated; grey bars) and wet sedge soils received a "drying" treatment (to field capacity; black bar). (Data from K. Judd, unpublished.)

Path analysis revealed significant ($p < 0.05$) direct effects of microbial community composition on soil respiration, but the strength of the relationship and the aspect of community composition influencing respiration differed among vegetation types. For example, respiration in birch-willow soils was strongly influenced by factors related to microbial community composition, based on phospholipid fatty acid (PLFA) analysis, while soil moisture was the stronger control in tussock and wet sedge soils. Thus, topographic variation in plant litter biochemistry and soil drainage shape the metabolic capability of soil microbial communities, which, in turn, influence the chemical composition of DOM across the arctic tundra landscape.

In addition to discovering these patterns of community composition and potential activity, we tested an ongoing debate in ecology that revolves around how species composition and ecosystem function are related. To address the mechanistic controls of this relationship, manipulations in mesocosms of the composition of DOM fed to aquatic bacteria were used to determine effects on both bacterial activity and community composition. Sites along terrestrial to aquatic flowpaths were chosen to simulate movement of DOM through catchments (Figure 6.4), and DOM was fed to downslope and control bacterial communities. Bacterial production was measured and DOM chemistry and bacterial community composition, using denaturing gradient gel electrophoresis (DGGE) of 16S rRNA genes, were characterized following incubations. Bacterial production, DOC-specific bacterial production, and DOC consumption were greatest in mesocosms fed soil-water DOM; soil-water DOM enhanced lake and stream bacterial production by 320%–670% relative to lake and stream controls (Judd et al. 2006). But the novel finding was that adding upslope DOM to stream and lake bacterial communities resulted in significant changes in bacterial community composition relative to controls. In these experiments the bacterial community composition converged based on DOM source, regardless of the initial inoculum (Figure 6.7). In other words, when lake bacteria were fed soil or stream DOM, the lake community assemblage shifted to resemble the species present in the soil or the stream (dashed arrows in Figure 6.7). Clearly the soil and stream bacteria were already present in the lake in undetectable numbers, but when exposed to soil or stream DOM these populations had a metabolic advantage and grew to replace the originally dominant lake bacteria. These results demonstrate that shifts in the supply of natural DOM were followed by changes in both bacterial production and community composition, suggesting that changes in function are likely dependent on at least an initial change in the community composition in order to take advantage of the new DOM source.

In similar experiments we also examined how photo-oxidation of DOM affected microbial activity and DOM processing along these dominant hydrological flowpaths. The impacts of DOM photo-oxidation depended in part on the DOM source, but also on the relatively rapid shifts in bacterial community composition to groups better able to consume photo-products or tolerate harmful radicals (see Vignette 6.1; Judd et al. 2007). We have also shown, using mesocosms, bioassays, and monitoring of inflow events, that the DOM exported from land is accessible to lake bacteria and that the C use by bacteria depends on the time of year and the

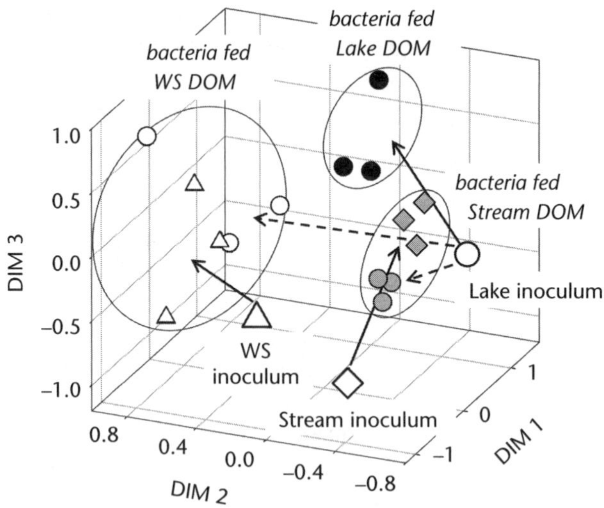

Figure 6.7 Changes in bacterial-community composition in experimental and control mesocosms. Multidimensional scaling shows the direction of change in bacterial-community composition from the beginning (initial inoculum, large symbols at arrow tail) to the end (small symbols at arrow head). Symbol shape represents the source of the bacterial inoculum (circle = lake; diamond = stream; triangle = wet sedge soil water), and symbol shading indicates the DOM source (black = lake; gray = stream; open = soil water). Note that the "control" community for comparisons was the ending, not the beginning, community. (Modified with permission from Judd et al. 2006.)

extent of preprocessing in soils. In early spring, when soils are frozen, the runoff water from snowmelt leaches and flushes DOM from the organic mat, but the flowpath short-circuits deeper soil microbes and the exported C is of relatively high quality for lake bacteria (Kling 1995; Crump et al. 2003). As soils thaw in summer, the water flowpath allows soil microbes the first chance to process high-quality C, and the reactivity of the remaining C reaching lakes is reduced. Our newest findings relate to the lability to microbes of soil carbon frozen for thousands of years and recently exposed by thermokarst activity (land-surface failures; see chapter 7). In general this deep, "thermokarst C" is as susceptible to bacterial attack as is soil C in the active layer, and on exposure to ultraviolet (UV) light the thermokarst C supports ~40% more bacterial respiration than does thermokarst C held in the dark (Cory et al. 2013). It is apparent that as thermokarst disturbances increase on the landscapes in the future (see chapter 10), more and more soil C will be exposed to light and processed at the surface rather than being processed *in situ* in the soils. Overall, these results on microbial processing of materials moving from land to surface waters indicate that (1) variation in DOM composition of soil and surface waters influences bacterial community dynamics, and (2) in turn different communities and exposure to light control the rates of carbon processing in set patterns across the landscape.

Vignette 6.1. DOM Photo-oxidation and Community-Level Bacterial Responses, by Kristin E. Judd

Terrestrial DOM is more chemically heterogeneous than aquatic-derived DOM and originates from the decomposition of soil organic matter and plant litter, plus leakage from and decomposition of living plant and microbial cells. When DOM is exposed to UV light, photo-oxidation reactions alter its chemical composition and bioavailability to aquatic bacteria. Because the amounts and chemistry of DOM vary spatially, seasonally, and with thaw depth in tundra soils, the response of aquatic bacteria to photo-exposed DOM affects the movement and export of DOM from land to surface waters and eventually the ocean.

The extent to which DOM fuels microbial growth depends on its chemical composition and on the metabolic capabilities of the microbial community. At the catchment scale, spatial patterns of DOM chemistry and microbial community composition are often related to the overlying vegetation. For example, water flowing downslope encounters distinct microbial communities that differ in their ability to consume particular mixtures of DOM, and bacterial communities respond to new (i.e., chemically distinct) mixtures of DOM from upslope by shifting their community composition to groups that can consume the new sources of DOM (Figure 6.4).

Ultraviolet light can break down large, plant-derived DOM compounds characteristic of terrestrial DOM into smaller molecules more readily available to bacteria. However, photo-oxidation reactions can also inhibit bacterial growth by producing harmful by-products (e.g., free radicals) and by producing partially oxidized molecules that are less labile to bacteria. The current paradigm is that the initial chemical composition of the DOM explains the direction and magnitude of photo-oxidation on DOM bioavailability because recalcitrant compounds become more labile, while labile compounds become more recalcitrant after UV exposure. However, these models fail to incorporate changes in bacterial communities in response to photo-products.

We addressed these questions by comparing short- and long-term responses of bacteria to photo-oxidized DOM. We found that contact time between photo-oxidized DOM and bacterial communities influenced the direction of bacterial responses, and that the direction and magnitude of response depended on the DOM source (Figure 6.V1). For example, photo-oxidation of soil DOM, not previously exposed to sunlight, greatly reduced bacterial growth at short contact times (hours) compared to dark controls. The effect was reversed at longer contact times (weeks), with light-exposed DOM significantly increasing bacterial growth. This shift from a negative to positive effect of light-exposed DOM on bacterial growth with increasing contact time between DOM and bacteria suggests that (1) harmful radical production initially inhibits bacterial growth, (2) bacterial communities require time to respond to the new DOM through either shifts in enzyme production or community composition, or (3) both processes are occurring. The same pattern in bacterial growth was seen with DOM from a first-order stream, although the magnitude of response at both short and long time-scales was lower. This response may occur because previous sunlight exposure of stream DOM reduces the degree to which photo-oxidation reactions occur. Lake water, on the other hand, which has likely had extensive sunlight exposure, showed the opposite pattern; short-term effects were slightly positive, but long-term effects were negative. These results from lake water suggest that low production of harmful radicals at short contact times account for the lack of a negative

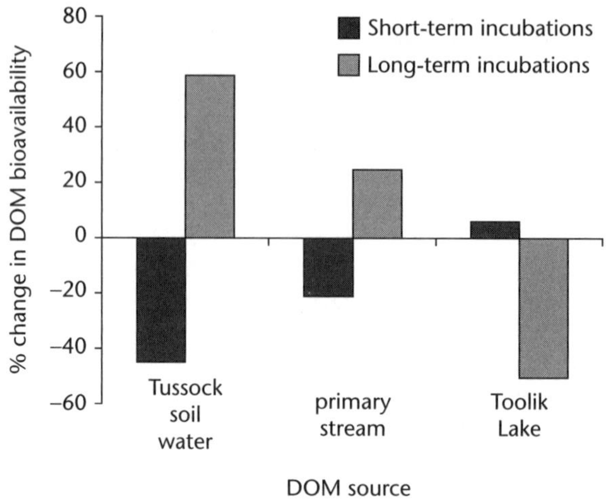

Figure 6.V1 Effect of photo-oxidation on DOM bioavailability. Bars indicate percent change in bacterial production ($\mu g\ C\ L^{-1}\ d^{-1}$) standardized by DOC concentration ($mg\ C\ L^{-1}$) in short-term (hours) and long-term (weeks). Values from data in Judd et al. (2007).

response at short time-scales, while photo-oxidation of labile aquatic DOM explains the negative effects over longer contact times.

DNA fingerprint analysis revealed that exposure to photo-oxidized DOM altered the bacterial community composition (Judd et al. 2007), indicating that specific bacterial groups are either resistant to harmful radicals or more efficient at using photo-oxidized forms of DOM. Either way, the community changed only in the longer incubations, which suggests that pulsed inputs of terrestrial DOM (e.g., during a rain event) may require several bacterial generations before these groups are present in high enough numbers to use the photo-products. These experimental studies indicate that the ultimate effect of photo-oxidation on DOM and its movement through catchments is a result of the interaction between DOM source and chemical composition, microbial community composition, and the contact time between DOM and microbial communities.

Ecosystem Interactions at Hillslope and Landscape Scales

One of the principles of ecology is that most descriptors of a system vary with the spatial or temporal scale of measurement. In biology, the spatial and temporal heterogeneity of organisms is a central component of several theories (e.g., evolution, succession, disturbance), and new advances in understanding the consequences of environmental heterogeneity on biological systems have been made in the last ~25 years (Turner and Gardner 1991; Kratz et al. 1991). In addition to the importance of spatial distribution or patchiness of resources and organisms, the movement of organisms and materials and the "connectivity" among patches is critical to system functioning (e.g., Reiners and Driese 2001), and this movement has been studied in detail at the Arctic LTER (e.g., Giblin et al. 1991). In addition to movement itself, in studies of temporal dynamics, organisms are considered to adapt and evolve, and characteristics of the environment (e.g., soil properties) are considered to change from one time period to another. However, in studies of system function across spatial scales, the environmental characteristics and organisms of interest are often considered to be static. In other words, the rates of movement of organisms or materials may be affected by the boundaries or ecotones (e.g., riparian zones) that separate different environmental patches, but there is little or no change or "processing" that occurs as materials and organisms move within a patch or even across the landscape.

The exceptions to this generalization of an assumed lack of processing of materials moving across a landscape include one derived from concepts in stream ecology (the "river continuum," "resource spiraling," and "serial discontinuity" concepts) and one derived from studies of landscape position of lakes (Magnuson et al. 1990; Kling et al. 2000). While these ideas about spatially dependent processes in rivers and in lakes have had utility in aquatic ecology, there has been very little empirical or conceptual integration of lakes and rivers at a landscape level. The result is that a large amount of "residual variation" remains among lakes and among streams that may only be explained through an understanding of ecosystem interactions at larger scales, including linkages at "hillslope" scales (similar to a soil catena or toposequence) and at broader, landscape scales. Our LTER research has contributed to this topic by studying the spatially dependent processing of materials in ecosystems that are linked in space.

Aquatic Ecosystem Interactions across the Landscape (1) Chemistry

Studies of the series of eight connected lakes that flow into Toolik Lake (the "inlet series," or "I-series"; see also chapter 8) have shown that there are distinct spatial patterns in chemical variables (Kling et al. 2000). These observed spatial patterns include mean concentrations of variables that increase, decrease, or remain constant along the lake chain from high to low altitude in the catchment (data for stream sites as yet show no strong spatial patterns with any variables). These patterns are determined in part by the effect of increasing catchment area feeding into lakes further downslope, and in part by the systematic processing of materials in lakes and in the stream segments between lakes. This processing in lakes and streams can be substantial for many limnological variables (Table 6.3). For example, lakes tend to "consume" alkalinity and calcium, while the stream segments produce these materials. These results, plus the more detailed analysis presented in Kling et al. (2000), illustrate that over small geographic areas, and somewhat independent of lake or stream morphometry, the consistent and directional (downslope) processing

Table 6.3 Effects of lake and stream processing for all sites of the Toolik I-series (~10 years of data for eight lakes connected by streams)

Lakes(Out—In)					Streams (Down—Up)				
	Mean	SE	N	p		Mean	SE	N	p
PCO_2 (μatm)	−374	60	86	**	PCO_2	338	60	139	**
PCH_4 (μatm)	−92	40	86	*	PCH_4	50	28	139	*
Alk (μEq L^{-1})	−35	9.9	102	**	Alk	35	7.6	136	**
DIC (μM)	−24	6.9	48	**	DIC	28	3.2	96	**
Ca^{2+} (μM)	−11	4.3	69	**	Ca^{2+}	9.9	2.7	75	**
PC (μM)	−5.7	4.4	54		PC	5.8	3.6	80	
DOC (μM)	17	12	103	**	DOC	−8.1	8.0	145	
NH_4^+ (μM)	0.11	0.04	88	**	NH_4^+	−0.12	0.05	128	**
Chl*a* (μg L^{-1})	0.17	0.08	93		Chl*a*	−0.06	0.08	111	

Mean values (± standard error, SE) are calculated for lakes by subtracting the inlet values from the outlet values, and for streams by subtracting the upstream values (at the lake outlet) from the downstream values (at the next lake's inlet). PCO_2 and PCH_4 are partial pressures of dissolved gases, Alk is total alkalinity; DIC is dissolved inorganic carbon; PC is particulate carbon; DOC is dissolved organic carbon; and Chl*a* is chlorophyll a. Positive values indicate production of materials, and negative values indicate consumption either in the lake or in the stream reach between adjacent lakes in the chain. N is the number of individual difference comparisons, and p indicates the significance level of a t-test (unequal variances) that the mean value is significantly different from zero; ** = $p < 0.01$ and * = $p < 0.05$. (Modified with permission from Kling et al. 2000.)

of materials helps produce spatial patterns that are coherent over time for many limnological variables.

Aquatic Ecosystem Interactions across the Landscape (2) Biology

The next stage in our research investigated whether biological variables responded in the same way that chemistry did to the landscape-level interactions between lakes and streams. In this research we showed that the processing of DOM by microbes, and the species of microbes present, vary consistently as water moves through a network of streams and lakes in the Toolik catchment. One study found that temperature is a critical factor in setting the level of bacterial activity (see Vignette 6.2; Adams et al. 2010). A second study found that in Toolik Lake itself the rate of bacterial activity was related to shifts in the source (terrestrial plants versus phytoplankton) and lability of DOM, and that bacterioplankton communities were composed of persistent populations present throughout the year and transient populations that appeared and disappeared (Figure 6.8). Shifts in community composition, measured by DGGE of 16S rRNA genes, were associated with an annual peak in bacterial productivity driven by the large influx of terrestrial DOM, some of which is labile to

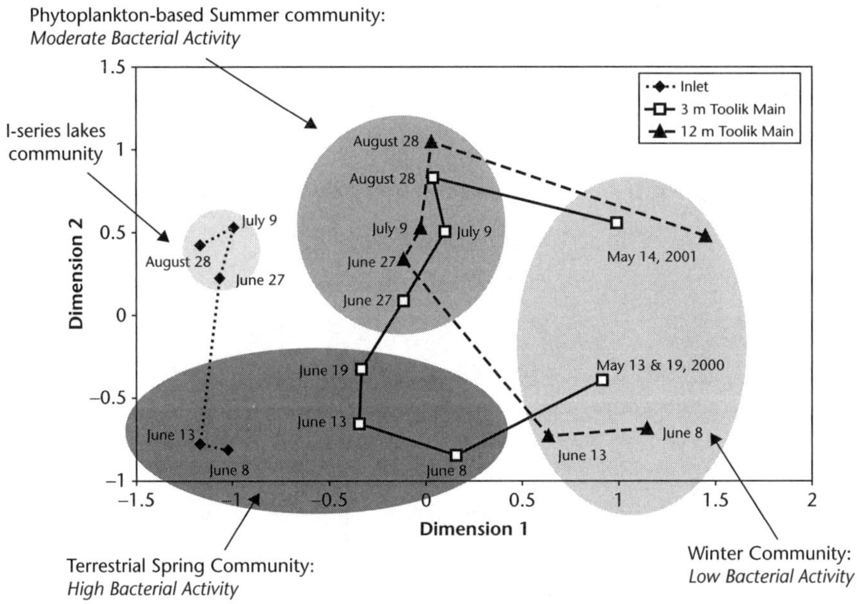

Figure 6.8 Multidimensional scaling calculated from DGGE banding patterns, showing consistent shifts in bacterial productivity and community composition in Toolik Lake throughout the annual cycle. (Modified with permission from Crump et al. 2003.)

bacteria, associated with spring snowmelt runoff. A second shift occurred after the terrestrial DOM flux declined and as the summer phytoplankton community developed.

Bacterioplankton community composition was also compared across 10 lakes and 14 streams within the Toolik catchment. Both lake and stream systems shared bacteria species (operational taxonomic units from DGGE analysis), and stream communities changed with distance from the upstream lake, suggesting a role of both dispersal of species between lakes and streams as well as inoculation and dilution with bacteria from soil waters or hyporheic zones (Crump et al. 2007). At the same time, similarity in lake and stream communities shifted gradually down the catchment (Figure 6.9). We found evidence that dispersal influences bacterioplankton communities via advection and dilution (mass effects) in streams, and via inoculation and subsequent growth in lakes, and that the spatial pattern of bacterioplankton community composition was strongly influenced by interactions among soil, stream, and lake environments. Overall these results reveal large differences in lake-specific and stream-specific bacterial community composition over restricted spatial scales (<10 km), and suggest that geographic distance and connectivity influence the distribution of bacterioplankton communities across a landscape.

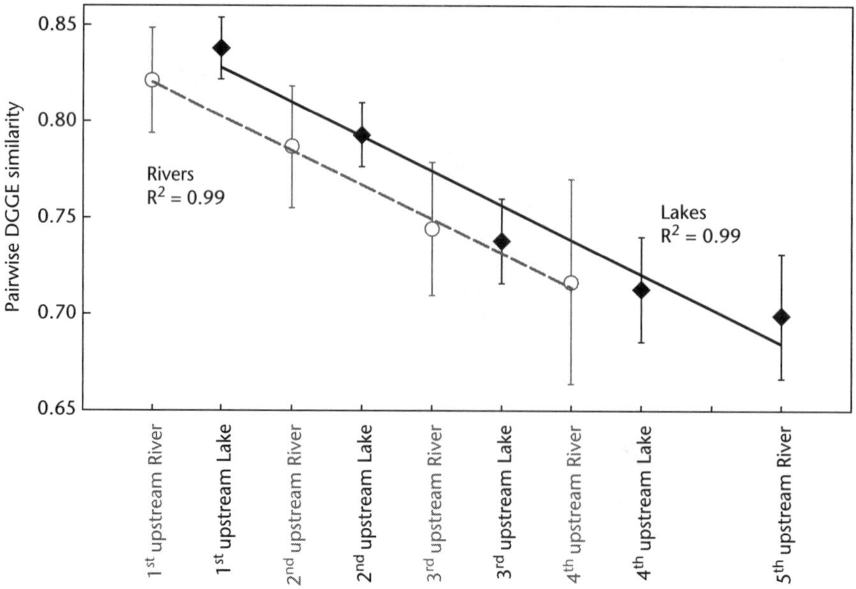

Figure 6.9 Average pairwise similarity values for DGGE banding patterns plotted against the degree of landscape separation. This separation is a categorical variable representing the number of upstream lakes or inlets separating two sites. X-axis values of "1st upstream lake" mean that comparisons are made between any lake and its closest upstream lake, and "1st upstream river" compares a lake inlet stream with its closest, nearby inlet stream. X-axis values of "2nd upstream lake or stream" represent a separation of two upstream lakes or two lake inlets. Error bars indicate standard error of all possible combinations within a category. Used with permission from Crump et al. (2007).

Vignette 6.2. Temperature Controls on Bacterial Activity and Communities, by Heather E. Adams

Bacterial activity is controlled primarily by temperature, carbon inputs, and community composition. Temperature impacts both bacterial metabolism and cell structure, and warmer temperatures generally increase reaction rates and also require more stable cellular membrane structures. Cold-adapted bacteria have more fluid cellular membranes and have enzymatic structures that allow reactions to proceed at colder temperatures. These structures result in bacterial populations adapting for particular temperature ranges as designated by the cardinal temperatures of minimum, optimal, and maximum temperatures for growth. Psychrophilic bacteria are defined as bacteria with optimal temperatures of <15°C and maximum temperatures <20°C, while psychrotolerant bacteria can tolerate cold waters down to 4°C but have a higher optimal growth temperature around 20°C (Morita 1975). Both psychrophilic and psychrotolerant bacteria are found in arctic lakes and streams.

Studies on the controls of bacterial activity at Toolik Lake include incubation experiments where bacteria have been held under different conditions or "fed" DOM from different habitats (e.g., lake-inlet water provides more terrestrial carbon while

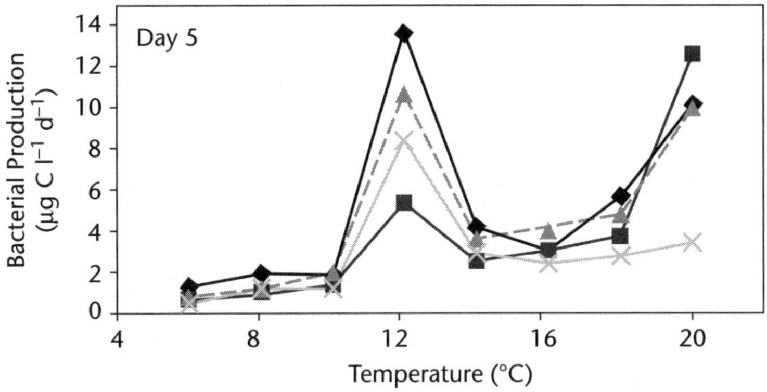

Figure 6.V2 Bacterial activity at different temperatures after five days of incubation with bacteria and DOM from Toolik Lake inlet and outlet, July 2004. Symbols indicate different locations and carbon sources: Diamonds = inlet community plus inlet DOM (control); Triangles = outlet community plus inlet DOM; Crosses (X) = outlet community plus outlet DOM (control); Squares = inlet community plus outlet DOM. Regardless of C source, two distinct temperature optima developed. (Modified with permission from Adams et al. 2010.)

lake-outlet water provides more aquatic carbon). Experiments run at temperatures ranging from 6°C to 20°C showed two peaks in activity at 12°C and 20°C, most strikingly after five days of incubation (Figure 6.V2). These two peaks correspond with the optimal temperatures for psychrophilic and psychrotolerant bacteria, and the peak temperatures were independent of the carbon source (terrestrial versus aquatic) for the bacteria. DNA fingerprinting of the bacterial communities using 16S rDNA verified that there was a shift in bacterial populations during the five-day incubation. The development of distinct peaks in activity indicates that the effect of temperature is community specific and can shift bacterial community structure. It is likely that differential growth of better-adapted populations within the community drives the community shift and also the substantial differences in activity rates at different temperatures.

Temperature can also interact with and constrain bacterial response to carbon inputs. In addition to having distinct optimal temperatures, bacterial populations also vary in their ability to break down different types of organic matter. Bacteria that are flushed into lakes from terrestrial and stream habitats are typically able to break down more heterogeneous forms of carbon (from land plants) than can lake bacteria, which are better able to process labile algal carbon. Heterogeneous, large-molecule terrestrial carbon requires greater enzyme action to be broken down, and low temperatures limit enzymatic activity. If temperatures are below the optimal range for an enzyme or bacterial species, temperature will constrain the bacterial response to available carbon. For example, in lake I-8, near Toolik, bacteria increased their activity in response to storm events that brought in terrestrial carbon subsidies. However, later in the summer season, bacteria did not respond to similar storm events that had colder water temperatures. This switch of the bacteria from carbon limitation early in the summer to temperature limitation later in the summer was also seen in the other lakes of the I-series catchment in 2003. Therefore, both carbon source and water temperature interact to govern the bacterial response in these lakes and streams.

Following on the results of Crump et al. (2007), a study of the distribution and biodiversity of microbes across the both terrestrial and aquatic habitats specifically examined the importance of dispersal. We know that microbes are transported in hydrological networks, but little is known about the composition and dynamics of underlying microbial metacommunities and the impact of dispersal on patterns of microbial diversity across landscapes. Using 454-pyrosequencing we found a general pattern of decreasing microbial diversity (Bacteria and Eukarya) along a hydrological continuum from soils to small streams to lakes, with highest species richness in soil waters and headwater streams and lowest species (OTU) richness in lake water (Crump et al. 2012). This is the first report of decreasing downslope diversity along a set of hydrologically connected ecosystems, but what was most surprising was the pattern of overlap in species distributions. For example, in Toolik Lake a substantial fraction of the taxa of Bacteria (58% of taxa) and Archaea (43% of taxa) were first observed in soil water or a headwater stream (Figure 6.10). This included

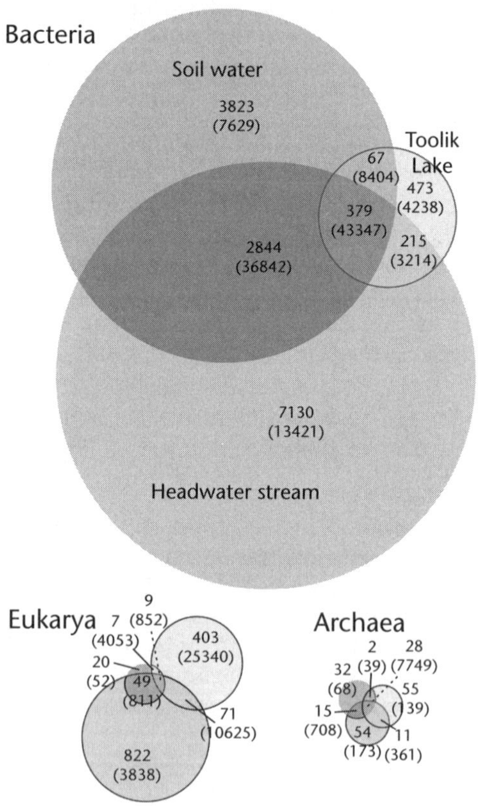

Figure 6.10 Venn diagram showing the number of shared bacterial, archaeal and eukaryotic taxa (in bold, defined by 97% sequence similarity) among soil water, headwater stream, and lake samples. The three circles indicating soil, stream, and lake are in the same relative position for each group shown, Bacteria, Archaea, and Eukarya. The number of sequences associated with taxa is shown in parentheses. Used with permission from Crump et al. (2012).

the 39 most common bacterial taxa in the lake, and accounted for the vast majority (89%) of the bacterial sequences that we collected from the lake. In addition, most of these abundant lake taxa were classified as "rare" in the upslope environments (<0.1% of sequences), which indicates that the rare taxa transferred into the lake must undergo species sorting processes (e.g., competition and predation) in order to form the resultant lake community. These results suggest that terrestrial environments serve as critical reservoirs of microbial diversity, and that the patterns of diversity in surface waters are structured by initial inoculation from upslope habitats. One implication of this conclusion is that environmental changes on land (e.g., permafrost thaw from climate warming) which affect microbes will propagate to surface waters; understanding freshwater microbial diversity and dynamics can only be accomplished by also studying the diversity and dispersal of terrestrial communities.

Hillslope Dynamics and Controls on Biogeochemistry and Catchment Export

One of the most critical issues in ecosystem research today is understanding how, exactly, do we apply our mechanistic or process-based knowledge of ecosystem function generated at small scales, such as a square-meter plot, to larger scales, such as an entire catchment, region, or biome. There are myriad concerns and approaches related to issues of "ecological scaling," but in the Arctic LTER we have focused on hillslopes as the "missing scale" required to transfer detailed process information to larger and larger areas (Figures 6.2, 6.4, and 6.11). The toposequence of a hillslope represents the major ecosystem types and landscape morphology of an entire catchment, yet can be studied in depth and cohesively (e.g., Giblin et al. 1991). For example, we can monitor soil water chemistry from the hilltop to valley bottom through time and relate the observed patterns to soil, plant, and microbial processes. One study by Yano et al. (2010) added a ^{15}N tracer to plots on the mid-slope and found that N retention was high and thus the movement of N downslope was slow. Another study in 2005 (Figure 6.12) showed that the pattern of DOC concentration in soil waters across the entire Imnavait Creek hillslope had early and late summer peaks at mid-slope, and slightly elevated concentrations at the footslope near the valley floor throughout the summer. Similar to the ^{15}N tracer results, there was no evidence of major transport of DOC from upslope to downslope during the summer. Our interpretation of this pattern is that most DOC production and consumption occurs *in situ*, which is consistent with the idea presented earlier that large amounts of DOC processing occur before DOC leaves the catchment. Because 2005 was a relatively wet year (compared to 2003 and 2004, for example), we assume that downslope transport of C, N, and P would be lower in drier years (see Vignette 6.3). Preliminary data also suggest that the same patterns (and interpretation) occur for other dissolved materials such as inorganic P and both inorganic and organic N; the next step is to examine the specific processes and rates at the landscape points where concentrations are high or where they change rapidly.

Although the mass of C or nutrients processed on the hillslope may be much greater than that transported downslope and into streams and lakes, the materials

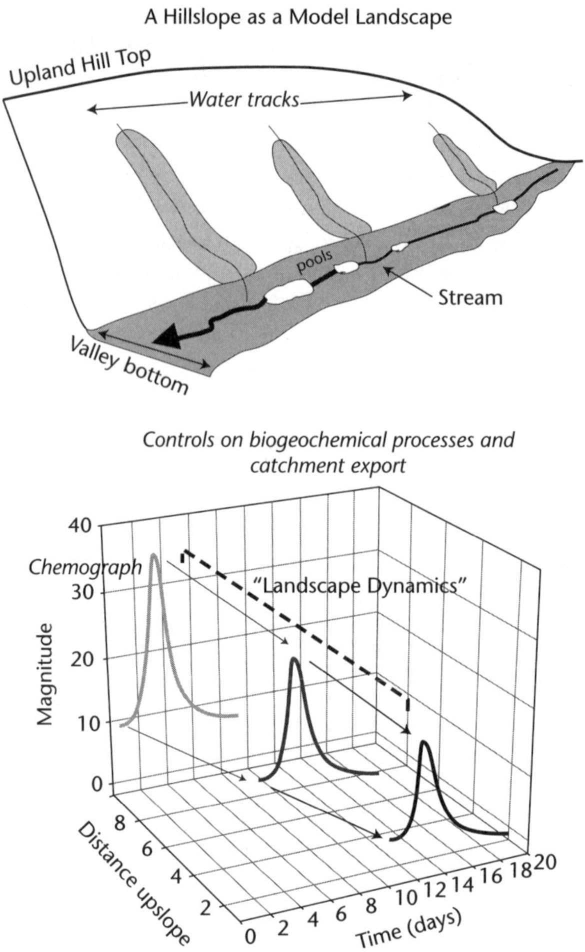

Figure 6.11 The hillslope (top) as a representative landscape model, with landscape components represented by the toposequence from upland heath to midslope tussock tundra to valley wet sedge vegetation, connected hydrologically by water tracks and riparian birch-willow vegetation. "Landscape dynamics" (bottom) in 3-D can be represented by changes in patterns or processes (e.g., a chemograph of soil water chemistry) moving from upslope to downslope and through time.

transported (1) have great impacts on the functioning of receiving surface waters and (2) can be substantial relative to the net C storage on land. Modeling of these landscape interactions based on a spatially linked, transect model indicates that hillslope interactions such as the downslope movement of nutrients and water may account for a ~40% increase in C sequestration in tundra ecosystems over the next century (Figure 6.13; Rastetter et al. 2004).

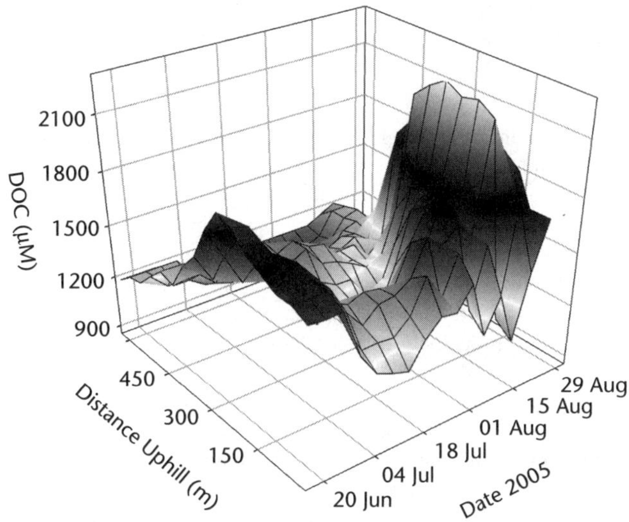

Figure 6.12 Plot showing the concentration of DOC (µM) in soil waters along the toposequence at Imnavait Creek through the summer of 2005 (x-axis). The stream is located at the bottom of the graph, and the hilltop is at the top of the graph (y-axis, distance uphill) following the layout in Figure 6.11.

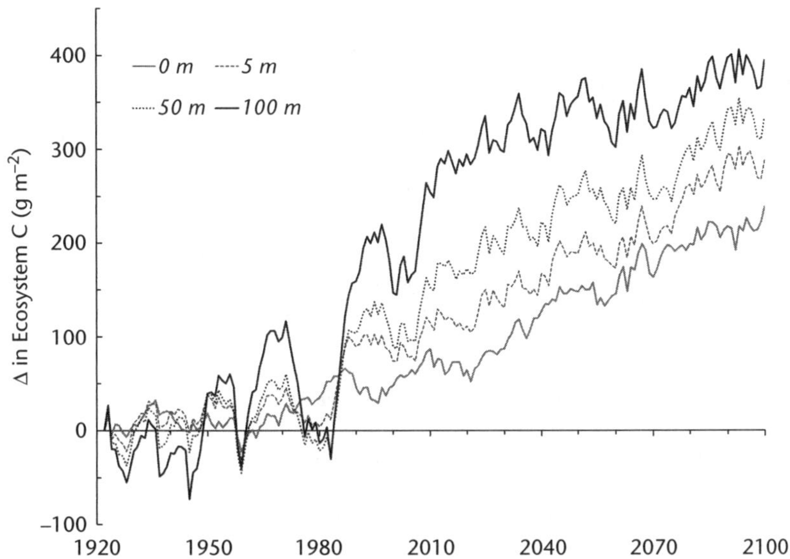

Figure 6.13 Model-simulated changes in ecosystem C stores (g m^{-2}) along a hillslope transect of tussock tundra with points ranging from 0 to 100 m downslope. Responses of net carbon storage to increasing atmospheric temperature and CO_2 concentrations over the next century are enhanced by the downslope movement of nutrients and water. Used with permission from Rastetter et al. (2004).

Vignette 6.3. Modeling Hillslope Dynamics, by Edward B. Rastetter

The downslope movement of water and materials on hillslopes produces strong linkages among hydrological, biogeochemical, and plant-community processes. The importance of these hillslope processes in the Arctic is clearly evident when moist tundra hillslopes are viewed from the distance of a kilometer or two, especially in early autumn. From that distance it is easy to see the zonation of vegetation from dry heath at the crest through moist tussock tundra and shrub-filled water tracks on the slopes to wet sedge meadows and tall riparian shrubs at the base. This striking pattern of vegetation composition and productivity arises because of the movement and accumulation of water and nutrients downslope, the effects of water and nutrients on plant growth and decomposition, and the interaction of these processes with soil freeze-thaw dynamics.

Rastetter et al. (2004) developed a hillslope model to assess the importance of these hillslope interactions on the responses of tundra ecosystems to changes in CO_2 and climate. The model was based on a model of linked C and N cycling through vegetation and soil organic matter on square-meter plots. These plot-scale models were linked to one another in series using a simple hydrology model that moved water and nutrients downslope. Thus, each plot received nutrients and water both from the atmosphere (nutrient deposition and rainfall) and from upslope so that at steady state the inputs of water and nutrients to each plot included the cumulative runoff from all plots upslope. Simulations then examined responses to the projected changes in climate and atmospheric CO_2 over the next century.

An increase in CO_2 increased N demand along the entire hillslope, causing downslope N losses to initially decline (Rastetter et al. 2004). Thus, vegetation at all slope locations, except the very top, must contend with both the increase in CO_2 and a decline in N supplied from upslope. Downslope locations cannot fully respond to the change in CO_2 until nutrient fluxes from upslope have come back to a steady state, causing the response to propagate downslope. This propagation is slow because nutrients moving downslope are repeatedly taken up, cycled through vegetation and soils, and released back into the soil solution. Thus nutrients spiral downslope in a way directly analogous to long-recognized nutrient spiraling in streams. Because of this spiraling, responses might take over 1000 years to fully propagate down a 100-m slope. Nevertheless, the simulations indicate that the base of the hillslope will store ~40% more C than the top over the next 100 years in response to elevated CO_2 and projected changes in climate (Figure 6.13). However, because of the long delays associated with nutrient spiraling, the magnitude of the difference between C storage at the top and bottom of the slope is likely to be much larger in the very long term.

Many of the anticipated responses of the Arctic to a warming climate are likely tied to these hillslope processes. For example, the nutrients necessary to fuel increases in shrub production and biomass must come either from mineralization of the organic matter on site or from mineralization of upslope organic matter and subsequent movement downslope (or from fixation, in the case of alder). In either case, hillslope processes are an important component of warming-induced changes in nutrient availability; if the nutrients are derived on site but are accompanied by downslope losses, production should eventually decline once the organic matter is mineralized. On the other hand, if the nutrients are derived from upslope, it is unclear how long an enhanced nutrient supply will persist once the upslope locations have equilibrated to climate warming.

Responses of Aquatic Ecosystems to Terrestrial Processes

Permafrost Thawing and Biogeochemical Impacts on Terrestrial and Aquatic Ecosystems

Despite clear evidence of arctic warming, from the loss of sea ice to shifts in vegetation and species ranges, many measurements throughout the Arctic, including at Toolik Lake, show a surprising lack of permafrost thawing in the soil (Figure 6.14). At the same time, we have measured a substantial increase (a doubling) in the alkalinity of Toolik Lake since 1975; these changes are unrelated to processes in the lake and appear instead to be caused by deeper hydrological flowpaths and increased weathering of mineral soils in the catchment. This is puzzling given the fact that the depth of summer thaw has not increased significantly over time to expose more mineral soil. However, these measurements of the maximum depth

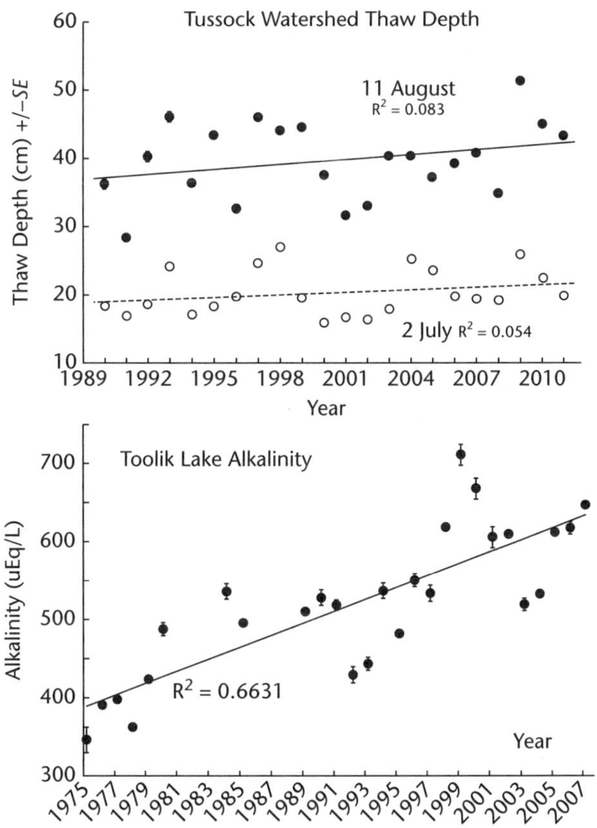

Figure 6.14 Summer thaw depth in July and August in the Tussock watershed near Toolik Lake (top), showing no statistically significant trends over time. The mean summer alkalinity in Toolik Lake (bottom) has increased over time. Standard error bars of mean values are often hidden by the points in both plots.

of summer thaw in soils are traditionally made at many hundreds of points along a set transect using steel probes to measure the depth to permafrost. This method is limited to upland terrestrial tundra, and has a relatively large coefficient of variation due to variable surface topography. To overcome these limitations we used a new approach with geochemical tracers instead of steel probes to show that changes in the geochemistry of surface waters must be related to changes in thaw depth, although the thawing may be confined to rocky hilltops with less vegetative insulation, and especially to the unfrozen zones underneath streams and lakes rather than to the entire upland catchment (Figure 6.2, top).

The first piece of evidence we have to build this conclusion is that the carbonate concentrations in the soils increase with depth because the deeper soils are more mineral rich, and they have been frozen so that less weathering has occurred (Keller et al. 2007). The second piece of evidence is that the ratio of strontium isotopes ($^{87}Sr/^{86}Sr$) in soils decreases with depth near the Arctic LTER (Figure 6.15). This means that as rainwater flows through deeper and deeper soils it will pick up the soil signature and the strontium isotopic ratio in the water will decrease. Finally, we observed just such a decrease in strontium isotope ratio in the stream entering Toolik Lake over a ten-year period (Figure 6.15). The implication is that water

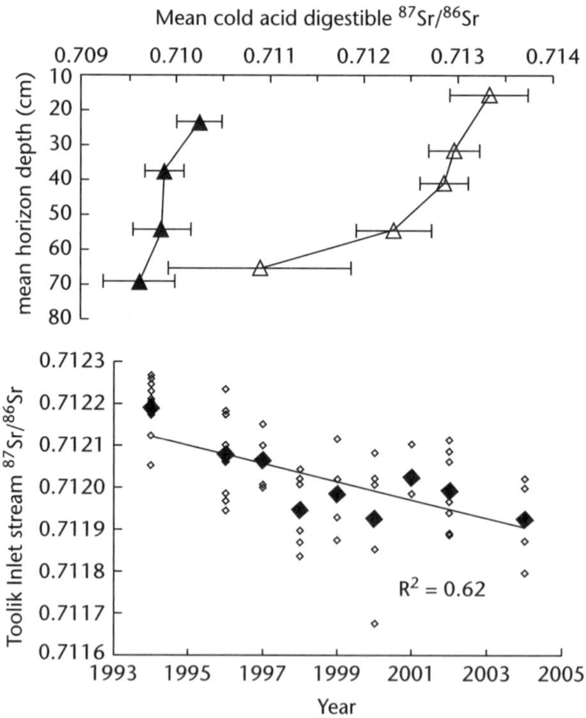

Figure 6.15 Strontium isotope ratios decrease with depth in older (filled) and younger (open) soils (top). Strontium isotope ratios in the inlet stream to Toolik Lake over time (bottom). (Modified with permission from Keller et al. 2007, 2010.)

flowpaths in the basin have progressively deepened and are now in contact with previously frozen soils of different chemistry. Because thaw depths of terrestrial sites have not changed during that time period (Figure 6.14, top), it is likely that the unfrozen thaw bulb (talik) found underneath streams and lakes has actually expanded, and the deeper thaw here contributes most to the altered chemistry. Our results also suggest that increasing thaw depth will lead to increasing Ca supply to soils and streams as well as to spatially variable increases in P and K supply (Keller et al. 2007). It may be that such changes in stream chemistry caused by permafrost thawing are more widespread in the Arctic than currently believed; this can be tested using this new geochemical method at other sites in arctic and boreal regions with permafrost.

Given the soil profiles of increasing concentrations of carbonates and other elements with depth (Keller et al. 2007), it is likely that as temperature warms or precipitation increases in the future (chapter 2) there will be increases in active layer depth and in weathering of these currently frozen materials. We expect that such weathering will continue to impact the amounts of materials transferred from land to water, and may have important effects on the chemistry of surface waters.

Summary

In summary, the main conclusions of land–water research at Toolik Lake thus far are related to the four major control steps presented in figure 6.1:

(1) The amount of C exported from land to surface waters and subsequently released to the atmosphere is important for tundra carbon budgets from local scales to the circumarctic. Despite this important C export, the amounts exported are dwarfed by the amount of C produced in the soil, especially by roots, which means that DOC processing by microbes in terrestrial ecosystems must be substantial;

(2) Microbial community composition and activity are strongly linked, and there are distinct and consistent patterns of microbial processing at key points as hydrological export moves materials along the continuum of terrestrial to aquatic ecosystems across the landscape;

(3) Interactions between different ecosystems along the toposequence of hillslopes are critical to the processing of carbon and nutrients and must be understood before plot-scale, process-level knowledge can be scaled-up to answer questions about catchment and regional biogeochemistry; and

(4) Despite no evidence of increasing thaw depth in terrestrial uplands near Toolik Lake, observed changes in the geochemistry of lakes and streams over time can only be explained by a thawing of permafrost and thus systematic changes in thaw depth in the basin due to global warming.

Acknowledgments

We thank NSF-DEB and NSF-OPP for funding support over the years, the staff of the Toolik Field Station, and many other assistants and students who have contributed greatly to the field and lab research, especially the senior LTER research associates for the land–water group, including Bryan Harper, Karen Riseng, Chris Wallace, Amanda Field, Jen Kostrzewski, Sara Fortin, and Jason Dobkowski. Lou Kaplan and Tim Kratz provided helpful reviews of the draft manuscript.

References

Adams, H. E., B. C. Crump, and G. W. Kling. 2010. Temperature controls on aquatic bacterial activity and community dynamics. *Environmental Microbiology* 12:1319–1333.

Aufdenkampe, A. K., E. Mayorga, P. A. Raymond, J. M. Melack, S. C. Doney, S. R. Alin, R. E. Aalto, and K. Yoo. 2011. Riverine coupling of biogeochemical cycles between land, oceans, and atmosphere. *Frontiers Ecology Environment* 9: 53–60.

Battin T. J., S. Luyssaert, L. A. Kaplan, A. K. Aufdenkampe, A. Richter, and L. J. Tranvik. 2009. The boundless carbon cycle. *Nature Geoscience* 2:598–600.

Binkley, D., P. Sollins, R. Bell, D. Sachs, and D. Myrold. 1992. Biogeochemistry of adjacent conifer and alder-conifer stands. *Ecology* 73:2022–2033.

Bormann, F. H., G. E. Likens, T. G. Siccama, R. S. Pierce, and J. S. Eaton. 1974. The export of nutrients and recovery of stable conditions following deforestation at Hubbard Brook. *Ecological Monographs* 44:255–277.

Canadell, J. G., C. Le Que´re´, M. R. Raupach, C. B. Field, E. T. Buitenhuis, P. Ciais, T. J. Conway, N. P. Gillett, R. A. Houghton, and G. Marland. 2007. Contributions to accelerating atmospheric CO_2 growth from economic activity, carbon intensity, and efficiency of natural sinks. *Proceedings National Academy of Sciences* 104:18866–18870.

Cole, J. J., N. Caraco, G. W. Kling, and T. Kratz. 1994. Carbon dioxide supersaturation in the surface waters of lakes. *Science* 265:1568–1570.

Cory, R. M., and L. A. Kaplan. 2012. Biological lability of streamwater fluorescent dissolved organic matter. *Limnology and Oceanography* 57:1347–1360.

Cory, R. M., B. C. Crump, J. A. Dobkowski, and G. W. Kling. 2013. Surface exposure to sunlight stimulates CO_2 release from permafrost soil carbon in the Arctic. *Proceedings of the National Academy of Sciences* 110:3429–3434.

Crump, B. C., G. W. Kling, M. Bahr, and J. E. Hobbie. 2003. Bacterioplankton community shifts in an arctic lake correlate with seasonal changes in organic matter source. *Applied Environmental Microbiology* 69:2253–2268.

Crump, B. C., H. E. Adams, J. E. Hobbie, and G. W. Kling. 2007. Biogeography of bacterioplankton in lakes and streams of an Arctic tundra catchment. *Ecology* 88:1365–1378.

Crump, B. C., L. A. Amaral-Zettler, and G. W. Kling. 2012. Microbial diversity in arctic freshwaters is structured by inoculation of microbes from soils. ISME Journal 6:1629–1639.

Giblin, A. E., K. J. Nadelhoffer, G. R. Shaver, J. A. Laundre, and A. J. McKerrow. 1991. Biogeochemical diversity along a riverside toposequence in arctic Alaska. *Ecological Monographs* 61:415–435.

Hope, D., M. F. Billett, and M. S. Cresser. 1994. A review of the export of carbon in river water: Fluxes and processes. *Environmental Pollution* 84:301–324.

Judd, K. E., and G. W. Kling. 2002. Production and export of dissolved C in arctic tundra mesocosms: The roles of vegetation and water flow. *Biogeochemistry* 60:213–234.

Judd, K. E., B. C. Crump, and G. W. Kling. 2006. Environmental drivers control ecosystem function in bacteria through changes in community composition. *Ecology* 87:2068–2079.

Judd, K. E., B. C. Crump, and G. W. Kling. 2007. Bacterial responses in activity and community composition to photo-oxidation of dissolved organic matter from soil and surface waters. *Aquatic Sciences* 69:96–107.

Keller, K., J. Blum, and G. W. Kling. 2007. Geochemistry of soils and streams on surfaces of varying ages in arctic Alaska. *Arctic, Antarctic, and Alpine Research* 39:84–98.

Keller, K., J. D. Blum, and G. W. Kling. 2010. Stream geochemistry as an indicator of increasing thaw depth in an arctic watershed. *Chemical Geology* 273:76–81.

King, J. Y., W. S. Reeburgh, K. K. Thieler, G. W. Kling, W. M. Loya, L. C. Johnson, and K. J. Nadelhoffer. 2002. Pulse-labeling studies of carbon cycling in arctic tundra ecosystems: The contribution of photosynthates to methane emission. *Global Biogeochemical Cycles* 16.4:1062. doi:10.1029/2001GB001456.

Kling, G. W., G. W. Kipphut, and M. C. Miller. 1991. Arctic lakes and rivers as gas conduits to the atmosphere: Implications for tundra carbon budgets. *Science* 251:298–301.

Kling, G. W., G. W. Kipphut, and M. C. Miller. 1992. The flux of carbon dioxide and methane from lakes and rivers in arctic Alaska. *Hydrobiologia* 240:23–36.

Kling, G. W. 1995. Land-water linkages: The influence of terrestrial diversity on aquatic systems. Pages 297–310 in F. S. Chapin and C. Korner, editors, *The role of biodiversity in arctic and alpine tundra ecosystems.* Berlin: Springer-Verlag.

Kling, G. W., G. W. Kipphut, M. C. Miller, and W. J. O'Brien. 2000. Integration of lakes and streams in a landscape perspective: the importance of material processing on spatial patterns and temporal coherence. *Freshwater Biology* 43:477–497.

Kratz, T. K., B. J. Benson, E. R. Blood, G. L. Cunningham, and R. A. Dahlgren.1991. The influence of landscape position on temporal variability in four North American ecosystems. *The American Naturalist* 138: 355–378.

Lewis, W. M., Jr., and J. F. Saunders, III. 1989. Concentration and transport of dissolved and suspended substances in the Orinoco River. *Biogeochemistry* 7:203–240.

Likens, G. E., F. H. Bormann, N. M. Johnson, D. W. Fisher, and R. S. Pierce. 1967. The calcium, magnesium, potassium and sodium budgets for a small forested ecosystem. *Ecology* 48:772–785.

Magnuson, J. J., B. J. Benson, and T. K. Kratz. 1990. Temporal coherence in the limnology of a suite of lakes in Wisconsin, U.S.A. *Freshwater Biology* 23:145–159.

McDowell, W. H., and G. E. Likens. 1988. Origin, composition, and flux of dissolved organic carbon in the Hubbard Brook valley. *Ecological Monographs* 58:177–195.

McGuire, A. D., L. G. Anderson, T. R. Christensen, S. Dallimore, L. Guo, D. J. Hayes, M. Heimann, T. D. Lorenson, R. W. Macdonald, and N. Roulet. 2009. Sensitivity of the carbon cycle in the arctic to climate change. *Ecological Monographs* 79:523–555.

McKane, R. B., E. B. Rastetter, G. R. Shaver, K. J. Nadelhoffer, A. E. Giblin, J. A. Laundre, and F. S. Chapin. 1997. Reconstruction and analysis of historical changes in carbon storage in arctic tundra. *Ecology* 78:1188–1198.

Michaelson, G. J., C. L. Ping, G. W. Kling, and J. E. Hobbie. 1998. The character and bioactivity of dissolved organic matter at thaw and in the spring runoff waters of the arctic tundra North Slope, Alaska. *Journal of Geophysical Research* 103:28939–28946.

Michalzik, B., K. Kalbitz, J. H. Park, S. Solinger, and E. Matzner. 2001. Fluxes and concentrations of dissolved organic carbon and nitrogen—A synthesis for temperate forests. *Biogeochemistry* 52:173–205.

Morita, R. Y. 1975. Psychrophilic bacteria. *Bacteriological Reviews* 39:144–167.

Rastetter, E. B., B. L. Kwiatkowski, S. Le Dizès, and J. E. Hobbie. 2004. The role of down-slope water and nutrient fluxes in the response of arctic hill slopes to climate change. *Biogeochemistry* 69:37–62.

Reiners, W. A., and K. L. Driese. 2001. The propagation of ecological influences through heterogeneous environmental space. *BioScience* 51:939–950.

Shaver, G. R., K. J. Nadelhoffer, and A. E. Giblin. 1991. Biogeochemical diversity and element transport in a heterogeneous landscape, the North Slope of Alaska. Pages 105–126 in M. G. Turner and R. H. Gardner, editors. *Quantitative methods in landscape ecology.* New York: Springer-Verlag.

Stieglitz, M., A. Giblin, J. Hobbie, M. Williams, and G. Kling. 2000. Simulating the effects of climate change and climate variability on carbon dynamics in arctic tundra. *Global Biogeochemical Cycles* 14:1123–1136.

Stieglitz, M., J. Hobbie, A. Giblin, and G. Kling. 1999. Hydrologic modeling of an arctic watershed: Towards Pan-Arctic predictions. *Journal of Geophysical Research* 104:27507–27518.

Turner M. G., and R. H. Gardner. 1991. *Quantitative methods in landscape ecology.* New York: Springer-Verlag,.

Yano, Y., G.R. Shaver, A. E. Giblin, E. B. Rastetter, and K. J. Nadelhoffer. 2010. Nitrogen dynamics in a small arctic watershed—Retention and downhill movement of ^{15}N. *Ecological Monographs* 80:330–351.

Zak, D. R., and G. W. Kling. 2006. Microbial community composition and function across an arctic tundra landscape. *Ecology* 87:1659–1670.

7

Ecology of Streams of the Toolik Region

William B. Bowden
Bruce J. Peterson
Linda A. Deegan
Alex D. Huryn
Jonathan P. Benstead
Heidi Golden
Michael Kendrick
Stephanie M. Parker
Elissa Schuett
Joseph J. Vallino
John E. Hobbie

Introduction

The original goal of the stream research was to describe the basic ecology of streams near Toolik, including the organisms of the streams, stream flow, water chemistry, and selected ecological processes such as primary productivity, respiration, and fish growth rates. Monitoring continues today to quantify year-to-year variability and to identify long-term changes in the ecology of streams. Long-term fertilization experiments have been a critically important tool used to understand the role of nutrients in limiting and structuring stream ecosystems. Today the overall theme of stream research is to integrate past work to inform future expected changes driven by climate.

This chapter first describes the ecological characteristics of the five types of arctic streams in the Toolik region, including biogeochemistry, primary producers, microheterotrophs, and secondary producers. Next, we describe the details of three stream-fertilization experiments that provide the basis for much of our past and current research. Finally, we synthesize our research results under six themes: resource limitation (bottom-up control); recovery from nutrient enrichment; trophic cascades (top-down control); hyporheic processes; disturbance as a driver; and how warming leads to changes in permafrost and thermokarst.

How do arctic streams differ from their boreal, temperate, and tropical counterparts? First, and perhaps most significantly, the vast majority of arctic streams less than a depth of 2 m freeze solid during winter. This results in communities with low taxonomic richness and a strong pan-Arctic element. Second, small arctic streams are underlain by permafrost (described in chapter 1). This restricts hyporheic processes to seasons during which water can flow and affects catchment-level patterns of runoff. Third, biotic communities receive nearly continuous light in the summer months, affecting patterns of primary production and the behavior of consumers. Fourth, arctic streams lack surrounding forest canopies; they have little shade and receive lesser amounts of coarse particulate organic matter than do streams in forested regions or in riparian forests of grasslands. Finally, arctic streams and their watersheds are undergoing extensive warming that will continue for a century or more (chapter 2), which will affect organisms, especially fish, and will increase their nutrient supply.

Streams of the Toolik Lake Region

Types of Streams

The landscapes of the Toolik Lake region have five general types of streams—tundra, mountain, glacier, tundra spring, and mountain spring. These stream types are characterized on the basis of their primary water sources and have been described in detail (Craig and McCart 1975; Oswood et al. 1996; Huryn et al. 2005). Briefly, the stream types have the following characteristics.

Tundra, Mountain, and Glacier Streams

Tundra streams flow from the foothills and across the coastal plain (Figure 7.1A). The source of water to tundra streams is snowmelt and precipitation. Their catchments are composed of a relatively thin layer of peat (often only 25–30 cm deep) underlain by a largely impermeable layer formed by permafrost. Thus, tundra streams are particularly responsive to rainfall events. Tundra streams that drain areas with ice-rich permafrost often have a "beaded" channel form—a series of pools (the "beads") connected by relatively short connector reaches (Figure 7.1B). The pools are roughly circular and deep (1–2 m) relative to their width (typically <10 m) and form by thermal erosion of ice wedges in the permafrost that form the borders of polygons (Hopkins et al. 1955). Consequently the beads and their connector reaches often have a zigzag pattern rather than the looping meanders commonly expected for stream channels. In terms of overall stream length, tundra streams are by far the most abundant stream type on the North Slope, where they contribute 82% of the total stream length.

Mountain and glacier streams (Figure 7.1C, D) drain high-elevation catchments of the Brooks Range. Like tundra streams, the source of water to mountain streams is runoff from snowmelt and precipitation. However, unlike tundra streams, these streams drain deep deposits of talus and rubble. In addition, glacial streams receive melt water from cirque glaciers. The snow and ice that persists in these bowl-like depressions on mountains are the most common form of glacier in the Toolik region. The channels of mountain and glacier streams tend to have high gradients, with habitats consisting of a series of cascades and pools structured by cobble and

Figure 7.1 Stream types in the Toolik Lake Region. (A) Meandering tundra stream, Kuparuk River. (B) Beaded tundra stream, Imnavait Creek. (C) Mountain stream, Ivishak Stream. (D) Glacial stream, Gates Glacier. (E) Mountain spring stream, Ivishak Spring. (F) Aufeis from a mountain spring, Flood Creek area. Photo credits: W. B. Bowden, except (D) by B. Peterson.

boulders. The channels of glacier streams differ from mountain streams in having angular stones rather than stones rounded by alluvial erosion, large amounts of fine sediment derived from glacial flour, and frequent shifts in channel location. Together, mountain and glacier streams contribute 16% to total stream length on the North Slope.

The peak flows in tundra, mountain, and glacier streams typically occur during spring and early summer, with low flows during late summer. Mountain and tundra streams also have relatively "flashy" hydrographs characterized by rapid increases in discharge in response to storms (Figure 7.2; Parker 2008). Unlike mountain and tundra streams, the hydrographs of glacier streams respond relatively slowly to storms. They are characterized by a regular fluctuation in flow driven by the daily

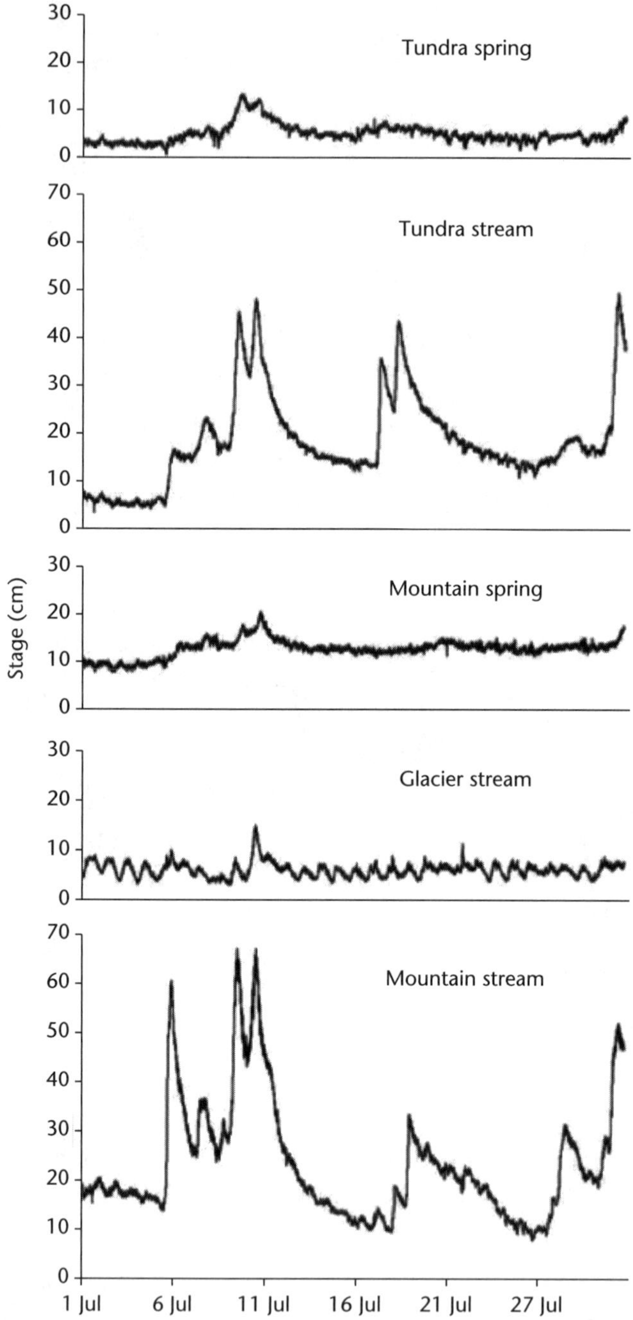

Figure 7.2 Seasonal variability of stage (water depth) in five types of arctic streams. Discharge is directly related to stage, but the relationship is different for each of these stream types (from Parker 2008).

cycle of melt water release from headwater glaciers; the highest discharge occurs late in the afternoon and the lowest discharge early in the morning (Figure 7.2). With the onset of winter, runoff to mountain, glacier, and tundra streams generally ceases, channel volume reduces, and the water column freezes, although unfrozen water may persist at the bottom of deep pools (>2 m).

Spring Streams

Flowing water may persist during the arctic winter in rare locations where thawed ground water feeds perennial springs. Spring streams typically arise in tundra and mountain environments. These springs receive discharge from deep limestone beds (the Lisburne Limestone Group) that underlie parts of the eastern North Slope and Brooks Range (Craig and McCart 1975; Yoshikawa et al. 2007) and in some places also from deep lakes and suprapermafrost sources associated with Tertiary sediments.

The streams that arise from springs (Figure 7.1E) differ from tundra, mountain, and glacier streams because they have perennial flow and water temperatures that exceed 0°C year-round, despite air temperatures below –20° to –30°C during winter. In addition, they have relatively uniform depths and well-sorted particle distributions usually consisting of gravel to cobble-sized particles. This lack of spatial complexity is due to the absence of scouring storm flows. Because a large proportion of the discharge of tundra spring streams and mountain spring streams is derived from groundwater infiltration, their hydrographs are relatively flat and typically they do not respond dramatically to storm runoff (Figure 7.2). The water discharged from springs eventually cools and will freeze at some point downstream. This ice often accumulates downstream of the spring source and may cover many hundreds of hectares to a thickness of many meters, depending on the size of the spring. This ice formed from spring water is one source of aufeis, the overflow ice that accumulates in streams and rivers overwinter (Figure 7.1F). Some of the aufeis in certain locations may persist throughout the year; the larger formations are visible in summertime satellite imagery. Tundra and mountain spring streams together contribute <1% of total stream length on the North Slope.

Characteristics of the Streams near Toolik Lake

Chemistry of Nitrogen, Phosphorus, and Organic Carbon

Comparisons among the stream types on the basis of soluble reactive phosphorus and ammonium are generally difficult because the concentrations of these nutrients are at or below the detection limits of standard analytical methods. However, concentrations of nitrate, dissolved organic nitrogen (DON), total dissolved nitrogen (TDN), and total dissolved phosphorus (TDP) are all at concentrations that can be measured readily. Although these forms of nitrogen and phosphorus are less readily available to algae, they may serve as an important supplemental source of these essential nutrients.

The summer-season concentrations of nitrate are highest in glacier streams, intermediate in mountain and spring streams, and lowest in tundra streams (Figure 7.3). Conversely the DON concentrations are lowest in glacier streams, intermediate in mountain and spring streams, and highest in tundra streams. This pattern appears to reflect the amounts of terrestrial primary production and soil organic matter in the catchments of the contrasting stream types.

The absolute concentrations of TDP are low in all stream types with slightly higher average concentrations in springs and glacier streams and lower concentrations in mountain and tundra streams (Figure 7.3). In addition, the relative concentrations of TDP compared to concentrations of TDN are also quite low in the same streams. The TDN:TDP molar ratios average over 100 in mountain and tundra streams and over 40 in spring and glacier streams. Compared to the molar Redfield ratio of about 15:1, which represents the expected composition of algae, it appears that all of these stream types are phosphorus limited—that is, they have on average a nitrogen surplus relative to phosphorus.

A survey of 20 headwater streams in the Toolik region showed a wide range of DOC concentrations (Table 7.1). Springs and streams in the mountains had the lowest concentrations (the means ranged from 109 to 193 μM), while tundra springs and tundra streams had the highest concentrations (538 and 584 μM, respectively).

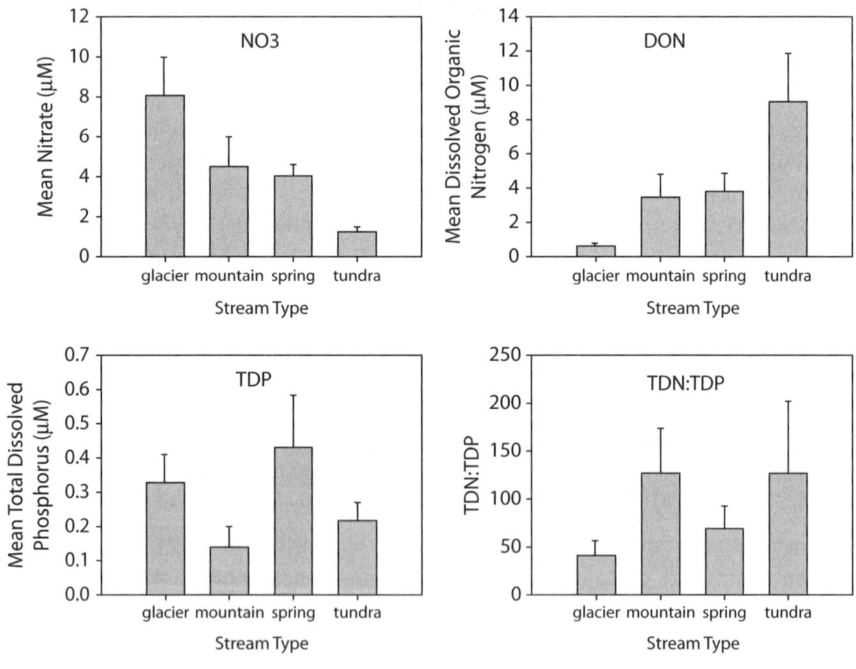

Figure 7.3 Concentrations of nitrate (NO_3), dissolved organic nitrogen (DON), total dissolved phosphorus (TDP), and the ratio TDN:TDP for streams in the region of Toolik Lake. TDN is effectively the sum of NO_3 plus DON.

Table 7.1 Water quality characteristics of streams in the Toolik region.

Stream type	Slope (%)	pH	Conductivity (µS cm^{-1})	SRP (µmol L^{-1})	NH_4 (µmol L^{-1})	NO_3+ NO_2 (µmol L^{-1})	DOC (µmol L^{-1})
Tundra	2.0 ± 0.0 (3)	7.2 ± 0.4 (3)	57.7 ± 55.3 (7)	0.13 ± 0.15 (3)	0.18 ± 0.1 (3)	2.8 ± 2.9 (3)	584 ± 167 (4)
Tundra spring	1.0 ± 0.0 (4)	7.1 ± 0.8 (4)	129.2 ± 112.5 (4)	0.48 ± 0.16 (4)	0.13 ± 0.1 (4)	3.2 ± 2.0 (4)	538 ± 441 (2)
Mountain	3.4 ± 3.6 (7)	8.0 ± 0.4 (12)	235.8 ± 78.8 (8)	0.21 ± 0.23 (11)	0.28 ± 0.21 (11)	4.0 ± 4.1 (12)	193 ± 214 (5)
Mountain spring	2.7 ± 1.5 (3)	8.2 ± 0.2 (6)	289.0 ± 57.1 (5)	0.53 ± 0.57 (6)	0.32 ± 0.26 (6)	4.0 ± 1.0 (6)	109 ± 46 (5)
Glacier	5.7 ± 4.6 (3)	7.4 ± 0.6 (3)	104.2 ± 38.3 (6)	0.12 ± 0.11 (3)	0.22 ± 0.24 (3)	7.8 ± 3.2 (3)	112 ± 53 (2)

Mean values ± standard deviations with number of replicate streams noted in parentheses. All data except conductivity and DOC are from Huryn et al. (2005) in a survey of 32 headwater streams. The conductivity and DOC data are from Peterson from 20 streams (LTER, unpublished data). Note that the nitrate+nitrite values reported here are from a different data set than those reported in Figure 7.3. The two data sets are similar, but not identical.

Other chemical measures were highly variable within and among catchments, making it difficult to identify consistent differences among stream types. For example, in a related survey of 32 headwater streams in the Toolik Lake region (Table 7.1), there was a ten-fold difference in conductivity within springs, a pH range from 6.4 to 8.6 in springs, and a many-fold range in nutrient concentrations among stream types. In general all chemical constituents were higher in concentrations in the tundra springs and streams than in mountain waters.

Particulate organic matter in the form of detritus is an important source of carbon and energy to stream food webs. Major sources of detritus include senescent aquatic macrophytes, epilithic biofilms, and riparian vegetation, particularly deciduous tree leaves (willows, poplar, and birch) that enter stream channels during autumn. Erosion of peat banks in tundra streams is another source. Compared with forested streams in temperate regions, arctic streams, including those in the Toolik Lake region, tend to receive and store relatively low amounts of detritus because the shrubby vegetation that dominates in the riparian zones of these streams does not overhang the channel as would be the case for canopy-forming trees in the riparian zones of other biomes (Benstead et al. 2005). The larger amounts of coarse particulate organic matter (CPOM) stored in small tundra streams than in other stream types is likely due to greater inputs of riparian litter per unit streambed area, as well as higher retention of CPOM in these small streams (Table 7.2).

Table 7.2 Masses in important autotrophic and detrital pools in streams in the Toolik region.

Stream type	Chlorophyll *a* (mg DW m^{-2})	Bryophytes (g DW m^{-2})	FBOM (g DW m^{-2})	CBOM (g DW m^{-2})
Tundra	4.8 ± 3.1 (3)	0.0 ± 0.0 (4)	~10[a]	45–151
Tundra spring	6.3 ± 3.9 (4)	84.2 ± 26.2 (5)	More	More
Mountain	3.4 ± 5.1 (7)	0.0 ± 0.0 (11)	Less	Less
Mountain spring	11.6 ± 7.7 (6)	64.0 ± 27.4 (7)	More	More
Glacier	5.6 ± 6.9 (3)	3.0 ± 5.2 (3)	Least	Least

[a] From Peterson et al. (1986).

FBOM is fine benthic organic matter, CBOM is coarse benthic organic matter, and DW is dry weight. Approximate organic content or ash-free dry weight (AFDW) is 50% of DW. Note that chlorophyll *a* is in different units. The dry weight of epilithic algal mass can be estimated as ~100x the reported chlorophyll *a* values (Peterson et al. 1986 after Strickland 1960). Note that the resulting values would be in mg DW m^{-2}. Data from Huryn et al. (2005), unless noted otherwise.

Breakdown rates of leaf detritus in the Kuparuk River and nearby Hershey Creek are relatively slow (0.004–0.006 d^{-1}) compared with rates measured in temperate streams (Benstead et al. 2005). Breakdown rates are driven by the initial colonization of leaf tissues by aquatic hyphomycete fungi ("conditioning," Suberkropp 2001) followed by consumption of such conditioned detritus by macroinvertebrates. The spores of aquatic hyphomycete fungi were common in samples of surface foam taken from streams in the Toolik Lake region, and changes in concentrations of ergosterol (a component specific to fungi and found in all fungal cell walls) in leaf litter incubated in a tundra stream indicated rapid rates of fungal colonization and conditioning. Nevertheless, ergosterol concentrations were consistently lower in willow and sedge leaf detritus incubated in an arctic tundra stream compared to concentrations generally reported for tree leaves decomposing in temperate streams. Once conditioned, willow and sedge leaf detritus in arctic tundra streams is readily consumed by "shredders," a group of macroinvertebrates that are defined functionally by their preference for feeding on CPOM.

Primary Producers

Most is known about autotrophic communities in alluvial streams that have cobble substrate. In comparison, peat- or silt-bottomed streams are common but little studied. In alluvial streams the cobbles are covered with a biofilm that contains a diverse community of autotrophic organisms dominated by diatoms (Table 7.3). In the Kuparuk River, the four most common genera, out of 187 found, accounted for about 90% of the total diatom abundance. In 24 headwater streams, 120 taxa of periphyton were found, including two widespread cyanobacteria. Differences in periphyton community structure between streams and stream types are likely

Table 7.3 Representative primary producers in streams near Toolik Lake

Primary	Study area	Representative organisms	Comments	Reference
Epilithic algae	Kuparuk River	*Achnanthes spp.* *Cymbella spp.* *Hannaea arcus* *Diatoma spp.*	90% of biofilm mass is diatoms. Of 187 reported species, 137 were at <1% abundance.	Miller et al. (1992)
	24 headwater streams in Toolik region	*Achnanthidium biasolettianum* *A. minutissima* *Encyonema minutum* *Fragilaria vaucheriae* *Hannaea arcus* Cyanobacteria *Oscillatoria* spp. *Schizothrix calcicola*	120 taxa found across all stream types. Only 5 taxa in glacier streams versus 42 in tundra streams.	Huryn et al. (2005)
Bryophytes	31 streams in Toolik region	*Hygrohypnum ochraceum* *H. alpestre* *Schistidium agassizii* *S. alpicola* *Fontinalis neomexicana* *Cratoneuron filicinum* *Drepanocladus* spp. *Caliergon* spp. *Philnotis* spp. *Solensotma* spp.	The species present in arctic streams can be found in other high-latitude and high-elevation environments.	Bowden et al. (1994) and Bowden (unpublished data)

attributed primarily to frequency of scouring during storm flows as described below in the section "Disturbance as a Driver in Arctic Streams."

Aquatic bryophytes can be an important component of the autotrophic community and may be responsible for the majority of primary production and nutrient uptake where they are abundant (Arscott et al. 1998; Stream Bryophyte Group 1999). Bryophyte biomass in tundra spring streams and mountain spring streams was significantly greater than in other stream types (Table 7.2). The ecophysiology of aquatic bryophytes is generally not well known. Arscott et al. (2000) found significant differences in the responses of two groups of bryophytes commonly found in the Kuparuk River to changes in light, temperature, nutrients, and desiccation. Two species of *Hygrohypnum* (*H. alpestre and H. ochraceum*) increased net primary production as light, temperature, and nutrient levels increased. In contrast, *Schistidium agassizii* was relatively unresponsive to changes in these variables. These differences explain why *S. agassizii* is able to persist in the extremely oligotrophic waters characteristic of tundra streams and why the *Hygrohypnum* spp. invade and overtake *S. agassizii* when nutrient levels increase.

Other aquatic macrophytes are found occasionally in arctic streams. Canadian waterweed (*Elodea canadensis*), Eurasian watermilfoil (*Myriophyllum spicatum*), and *Sparganium hypoboreous* are sometimes abundant in peat-bottomed streams, especially in the deep and fairly protected pools of beaded streams.

Microhetertrophs

Relatively little is known about microheterotrophs in streams in the Toolik Lake region. Lock et al. (1990) studied the effect of phosphorus fertilization on selected autotrophic and heterotrophic processes in the Kuparuk River and found that phosphorus stimulated both. Peterson et al. (1985) concluded that the stimulation of heterotrophic activity observed in the Kuparuk River was an indirect effect of greater algal production which led to greater production of algal exudates which then directly stimulated heterotrophic activity. Heterotrophs presented with phosphorus in the dark where algal exudates were absent did not exhibit greater uptake or production.

Crump et al. (2007) studied the planktonic bacteria in a chain of connected lakes and streams in the major catchment that discharges into Toolik Lake (the "Inlet Series," see details in chapter 6). This study was based on analysis of fragments of DNA to indicate when different taxonomic units were present. In the I-series the bacterial communities were most similar in connected lakes and streams (see Figure 6.9). The similarity between stream segments decreased as the water moved downstream and the number of stream or lake elements between the comparison points increased.

Larouche et al. (2012) used similar methods to show that bacterial community composition in the epilithic biofilm of tundra and mountain streams in the Feniak Lake region (about 200 km southwest of Toolik Lake) differed somewhat on the basis of local geology but differed strongly between epilithic diatom communities and interstitial sediment communities only a few centimeters away. Based on phylogenetic analyses they determined the presence of 32 families of bacteria belonging to 14 classes. The most common bacterial families in interstitial sediment samples were Enterobacteriaceae (25.1%), Paenibacillaceae (28.3%), Pseudomonadaceae (9.4%), and Xanthomonadaceae (11.6%). Epilithic biofilm samples were dominated by species of Cyanobacteria (43.5%), Flexibacteraceae (17.7%), Comamonadaceae (12.9%), and Deinococcaceae (5.5%).

Secondary Producers

Seventy-two macroinvertebrate taxa were identified in a survey of 32 streams in the Toolik Lake region (Huryn et al. 2005, Table 7.4). The regional pool of taxa (β-richness) is represented by 70+ taxa (approximates generic richness), which is relatively low when compared with streams in interior Alaska (Oswood 1989). The taxonomic richness of individual stream communities (α-richness) ranged from a high of 42 taxa (tundra stream) to a low of 12 taxa (two glacier streams). The

Table 7.4 Representative benthic macroinvertebrates in streams of the Toolik region (from Huryn et al. 2005)

Functional	Representative taxa	Comments
Filter feeders	Simuliidae Brachycentridae some Tanytarsini (Chironomidae)	Biomass ranges over five orders of magnitude among stream types (0.09 to 374 mg DM m^{-2}). Contribute 36% to total BMI biomass in tundra streams but <2% in other stream types.
Predators	Tanypodinae (Chironomidae) Empididae (Diptera) Chloroperlidae (Plecoptera) Perlodidae (Plecoptera) *Rhyacophila* (Trichoptera) Acarina (water mites)	The Empididae and Chloroperlidae are widespread, occurring in most stream types. Predator biomass is greatest in mountain springs (1881 mg DM m^{-2}) compared to other stream types (17–247 mg DM m^{-2}). Predators are <6% of total BMI biomass in other stream types
Scrapers	*Diamesa* spp. (Chironomidae) *Baetis* (Ephemeroptera: Baetidae), *Orthocladius* spp. (Chironomidae)	The contribution of scraper biomass ranges from 41% of total biomass in tundra streams to 71% in tundra spring streams.
Shredders	*Nemoura* spp. (Plecoptera) *Podmosta* spp. (Plecoptera) *Zapada* spp. (Plecoptera)	The contribution of shredders to community biomass ranges from 1% in tundra springs in to 27% in mountain streams.
Collector-Gathers	Oligochaeta *Orthocladius* spp. (Chironomidae) some Tanytarsini (Chironomidae) *Eukiefferiella* spp. (Chironomidae)	The contribution of collector-gatherers ranges from 4% in tundra and mountain spring streams to 21% in tundra spring streams.

Oligochaeta and the chironomid genera *Diamesa* and *Eukiefferiella* were the most widespread taxa, with both occurring in almost all streams sampled. Acarina (water mites), *Baetis* (Ephemeroptera: Baetidae), and the chironomid genera *Orthocladius, Corynoneura,* and *Tvetinia* were also widespread. Ordination analysis indicated that regional macroinvertebrate community structure of the Toolik region is comprised of five assemblages structured by the position of a stream within the landscape (tundra, tundra spring, mountain, mountain spring, and glacial). Differences in community structure among stream types is attributed to differences in physical disturbance regime due to landscape position (see "The Landscape Disturbance-Habitat Template," below).

Macroinvertebrate abundance ranges over three orders of magnitude among stream types near Toolik. Tundra spring streams (1.2×10^5 individuals m^{-2}) and mountain spring streams (9.2×10^4 individuals m^{-2}) have the greatest mean abundance; mountain streams have the lowest (4.2×10^3 individuals m^{-2}). Mean macroinvertebrate biomass, which ranges over two orders of magnitude among streams, is greatest in mountain spring streams (5800 mg dry matter [DM] m^{-2})

and lowest in glacier (442 mg DM m^{-2}) and mountain streams (399 mg DM m^{-2}). The range of community biomass among streams is similar to that documented worldwide (Benke 1993), which is somewhat surprising for this harsh arctic environment.

Five functional feeding groups are generally recognized among freshwater stream macroinvertebrates: filter-feeders, predators, shredders, scrapers, and collector-gatherers. Near Toolik only the filter-feeders and predators show significant differences in contribution to total macroinvertebrate biomass among stream types (Table 7.4). Compared with filter-feeders and predators, the distribution of biomass among scrapers, shredders, and collector-gatherers is relatively uniform among stream types, with no significant differences in their proportional contributions to total macroinvertebrate biomass (Huryn et al. 2005).

Three species of fish can be abundant in headwater streams near Toolik: Dolly Varden char (*Salvelinus malma*), arctic grayling (*Thymallus arcticus*), and slimy sculpin (*Cottus cognatus*). All are opportunistic predators that feed primarily on benthic macroinvertebrates. Sculpin are commonly found in spring streams and inlet and outlet streams of deep lakes. Dolly Varden char are year-round residents of spring streams. Following hatching, most populations of arctic Dolly Varden remain within their natal spring streams for about three years, after which they migrate to the Beaufort Sea during summer and then return during early autumn to spawn and overwinter (Richardson et al. 2001). There is also a "residual" form of Dolly Varden that is nonmigratory. These consist of males that remain within the spring stream for their entire lives. Grayling are widely distributed in mountain and tundra streams, as well as in the lower reaches of spring streams. Grayling require overwintering habitat of deep lakes and rivers to which they migrate from headwater streams to overwinter. For more information on arctic grayling, see the section "Life History and Feeding Ecology of Arctic Grayling" and Vignette 7.2, "Forecasting the Future for Fish," later in this chapter.

A number of bird species use stream habitats in the Toolik Lakes region. Most are piscivorous, such as the Arctic tern (*Sterna paradisaea*) and glaucous gull (*Larus hyperboreus*). Two other species, the dipper (*Cinclus mexicanus*) and harlequin ducks (*Histrionicus histrionicus*), feed at several trophic levels and thus are more intimately integrated with stream food webs. The dipper nests along several mountain spring streams in the Toolik Lake region, but is absent from other stream types (Parker and Huryn 2006). Dippers feed on larvae of stream macroinvertebrates (stoneflies, black flies, mayflies) and young-of-the-year Dolly Varden char. Presumably their trophic focus on spring streams, where they are apparently year-round residents, is related to the presence of perennially flowing water combined with high levels of invertebrate productivity that remain accessible through the winter. Harlequin ducks are summer residents of North Slope tundra streams and near Toolik they nest along headwater streams of the Kuparuk River (Huryn, personal observation). Like dippers, harlequin ducks enter streams to feed on benthic macroinvertebrates, including caddisflies, black flies, stoneflies, and mayflies.

Food-Web Characteristics

Detailed connectance food webs have been produced for three stream communities representing three stream types (Parker and Huryn 2006): specifically, a tundra stream (Figure 7.4, Kuparuk River), a mountain stream, and a mountain spring stream in the Ivishak River drainage. Differences in food-web size (i.e., number of taxa participating in a food web) were relatively minor: 38 in the mountain stream, 42 in the mountain spring stream, and 52 in the tundra stream. Food-web connectance (i.e., the proportion of realized links relative to all possible links) was also similar: 0.12 for tundra stream and 0.19 for the mountain stream and the mountain spring stream.

There were, however, major differences in mean food-chain length, total macroinvertebrate biomass, and the apportionment of this biomass among trophic positions. Mean food-chain length jumped from 1.7 in the mountain stream and 1.9 in the tundra stream to 3.1 in the mountain spring stream, indicating the presence of an additional trophic level in the spring stream (~three trophic levels) compared with the tundra and mountain streams (~two trophic levels). The third trophic level in the mountain spring stream is occupied by the dipper *C. mexicanus* (Parker and Huryn 2006). Given the high energy requirements of birds, dippers can be significant predators of stream invertebrates. As noted above and in Table 7.4, the mountain spring stream also had the greatest total macroinvertebrate biomass and the highest proportion of biomass attributed to predators. Such differences in food-web structure indicate major functional differences (e.g., productivity, nutrient cycling) among the communities in these different stream types. These differences are ultimately

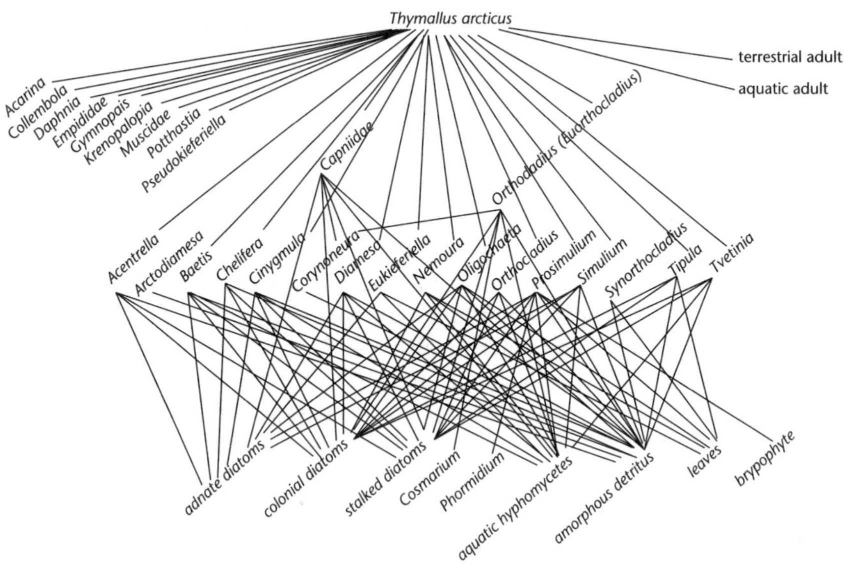

Figure 7.4 Food-web structure in the Kuparuk River, a typical tundra stream (from Parker 2008).

driven by factors related to landscape position, particularly those affecting patterns of channel morphology and disturbance regime.

Stream Research Themes

Over the past three decades several major themes have guided research on streams in the Toolik Lake region. Studies of three tundra streams, in particular, have demonstrated that the community structure and function is responsive to changes in nutrient inputs and to changes in physical factors such as discharge. The focus of the longest and most thorough investigation is the Kuparuk River near the Dalton Highway where a 30-year stream fertilization experiment is continuing. Oksrukuyik Creek also near the Dalton Highway was the site of a shorter (six-year) fertilization experiment, and Hershey Creek, a tributary of the Kuparuk, was the site for a one-year fertilization experiment. While all three streams are in the tundra-stream class, they differ substantially in size and morphology. The climate in the area of these study streams is described in chapter 2, and the types of vegetation found in the watersheds are described in chapter 3. In the following sections we first describe the general ecology of these three main study streams and then summarize the key findings from six research themes that have defined our research on streams in the Toolik area.

General Ecology of Detailed Study Streams

The Kuparuk River

The 5-km study reach is centered on the Dalton Highway (68°38' N, 149°24' W), where the Kuparuk is a meandering, fourth-order stream (Figure 7.5) with a catchment area of 143 km^2 of alpine and moist tundra communities. Dwarf birch and willows line sections of the stream bank but do not shade the channel. For a more complete description of the Kuparuk watershed see Kriet et al. (1992), Harvey et al. (1997), and Hershey et al. (1997).

This headwater reach ceases to flow in late September or early October, and pools are frozen solid from October to May; the open-water growing season is approximately four months. Discharge during the flowing season is highly variable and ranges from less than 1 $m^3 s^{-1}$ to around 5 $m^3 s^{-1}$ (Figure 7.6). Peak flows during the spring freshet regularly range from 10 to 30 $m^3 s^{-1}$ and can deliver 25% of the total seasonal runoff (McNamara et al. 1998). However, the two largest runoff events ever recorded on the upper Kuparuk River occurred during summer rain events in July 1999 and August 2002 and produced estimated peak flows in excess of 100 $m^3 s^{-1}$ (Kane et al. 2008). In contrast, drought conditions prevailed in 2005, 2007, 2009, and 2011, during which some reaches of the upper Kuparuk River became "hyporheic," with no surface water flowing.

Light inputs are high during most of the growing season due to the open canopy and 24-hour photoperiod. Summer water temperature in the Kuparuk River

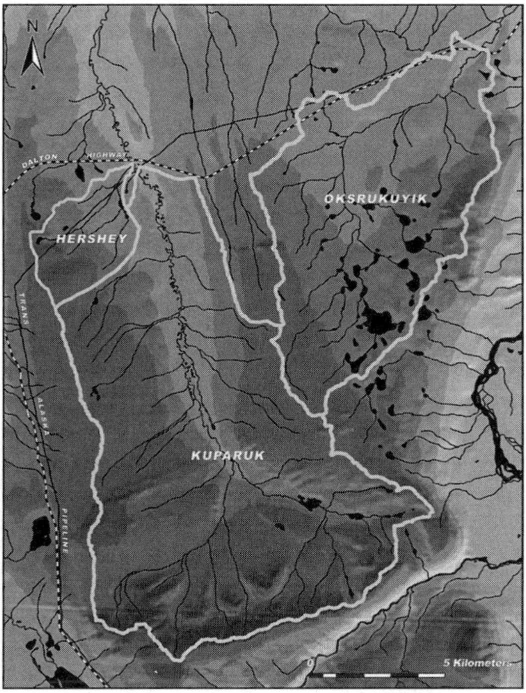

Figure 7.5 Map of the upper Kuparuk River and Oksrukuyik watersheds (maps from J. Stuckey and R. Fulweber, Toolik Field Station GIS and Remote Sensing Facility).

averages 7°C–13°C (Figure 7.7). Maximum precipitation occurs in July and August (Kane et al. 1989); however, the peak discharge usually occurs in late May or early June because of snowmelt. The Kuparuk River is oligotrophic and primary production here is most frequently phosphorus limited (Peterson et al. 1993).

Autochthonous production is dominated by benthic net primary production but is relatively low compared to temperate streams, ranging from 15 to 30 g C m^{-2} y^{-1} (Bowden et al. 1992). Early estimates of benthic primary production in the Kuparuk River were made using closed, recirculating chambers (Figure 7.8). Cappelletti (2006) adapted the widely used whole-stream metabolism methodology (e.g., Bott 2007), which is dependent on a nighttime "dark" period to identify daily respiration values; at the bottom of streams light intensity drops to near-zero levels even at the summer solstice (Figure 7.9). Cappelletti (2006) concluded that estimates of net and gross primary production from the whole-stream method were similar to the historical database of estimates based on closed-chamber methods (Figure 7.10). Estimates for ecosystem respiration by the whole-stream method were much larger than estimates of community (epilithic) respiration from the closed chambers, primarily because the whole-stream method includes hyporheic respiration and the closed-chamber method does not (Fellows et al. 2001).

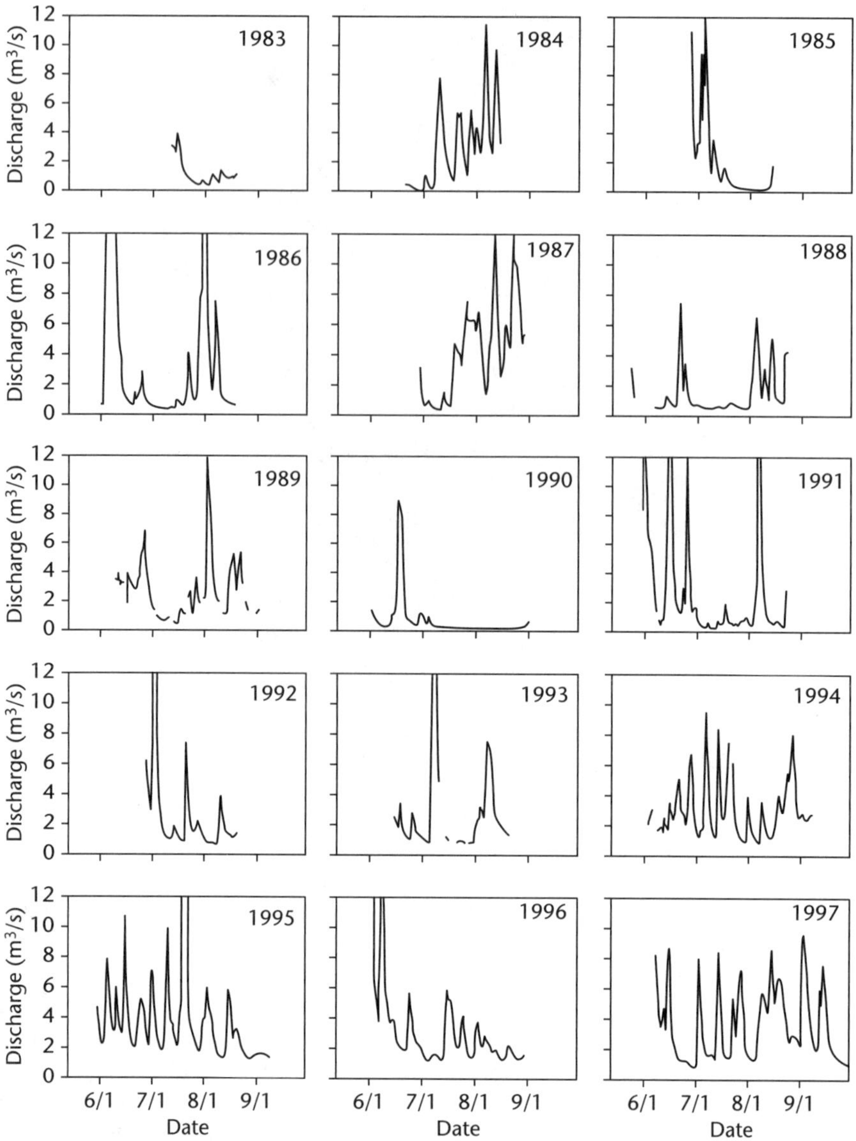

Figure 7.6 Annual discharge regime of the Kuparuk River, 1983–2010. Updated with permission from the Ecological Society of America.

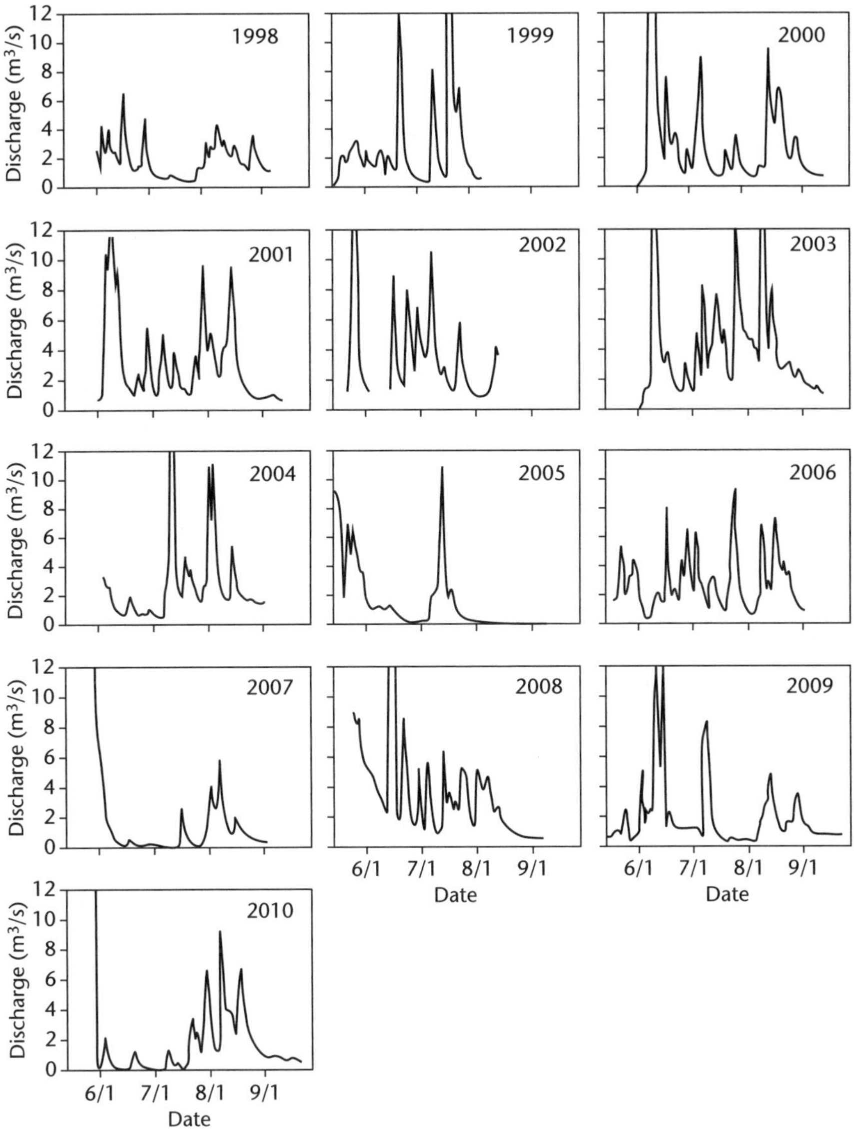

Figure 7.6 (*Continued*)

Peterson et al. (1986) and Harvey et al. (1997) assessed the relative importance of allochthonous versus autochthonous contributions to the carbon in the Kuparuk River (Figure 7.11). Allochthonous organic matter input is comparable to temperate, closed-canopy forest streams, but of different composition. Peat from eroding banks—rather than leaves from riparian vegetation—is the major source of particulate organic matter loading to the river (~200–300 g C m^{-2} y^{-1}). Dissolved organic

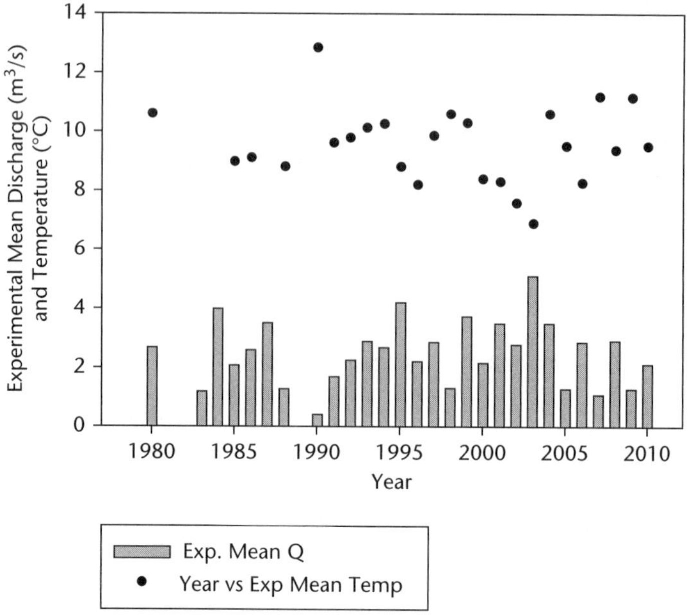

Figure 7.7 Mean summer discharge and temperature during the fertilization period (June 25–August 16) in the Kuparuk River. Note: Dates included in these means may vary slightly year to year, due to technical issues with equipment.

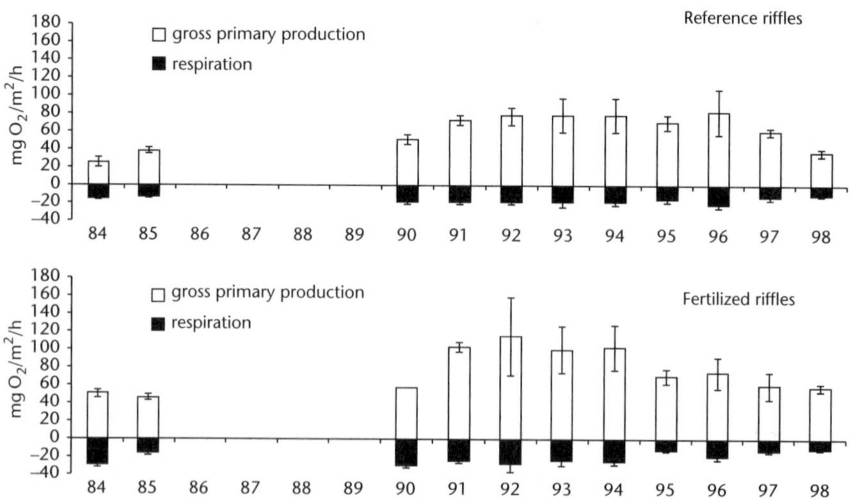

Figure 7.8 Production and respiration from chamber experiments in the Kuparuk River. From Slavik (2004), used with permission from the Ecological Society of America.

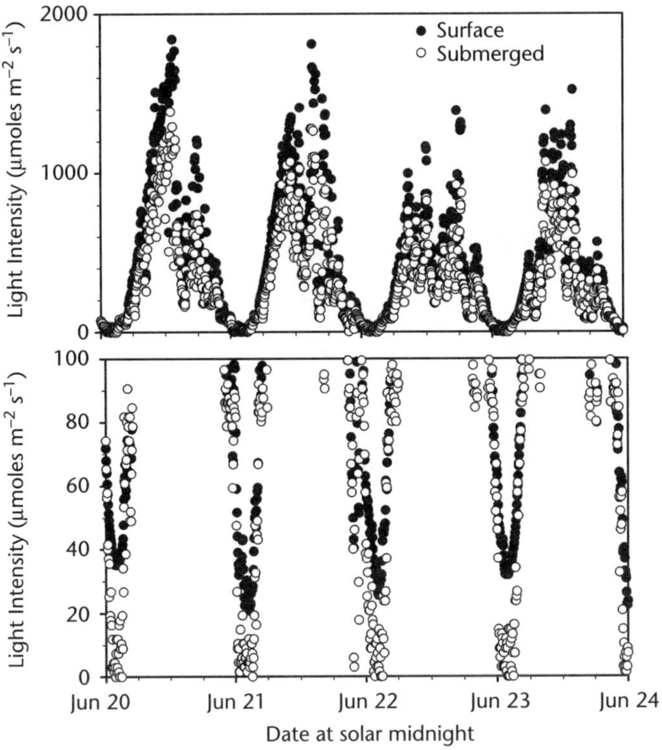

Figure 7.9 Typical surface and subsurface light levels in the Kuparuk River. (A) Five-minute light recordings over three days near the summer solstice. (B) Magnification of midnight light values (from Cappelletti 2006).

carbon comprises over 90% of the total organic carbon (mean TOC = 6.8 mg C L^{-1}) export from the watershed.

The Kuparuk River is a tundra stream and so reflects that community type, as described above. Filter-feeding black flies *Stegopterna mutata* and *Prosimulium martini* are abundant and emerge in mid-July and in early August (Hiltner 1985). *Baetis* and *Acentrella* spp. are the dominant grazing mayflies. The mayfly *Ephemerella* sp. is also present, but is far less abundant. There are several species of chironomids present, including the univoltine, tube-dwelling grazer *Orthocladius rivulorum,* which is one of the largest and most abundant (Hershey et al. 1988). The caddisfly *Brachycentrus americanus* is both a filter-feeder and a grazer and is represented by at least two cohorts during the summer season because this organism has a three-year life cycle (Peterson et al. 1993).

Arctic grayling (*Thymallus arcticus*) is the only fish species found in the study reach of the Kuparuk River. The grayling reside in the river from late June through mid-August, and then migrate upstream to overwinter in headwater lakes. There are approximately 50–100 adult grayling (20–40 cm total length, Deegan et al. 1997) per river kilometer during the summer. Juveniles (5–20 cm) are present, but

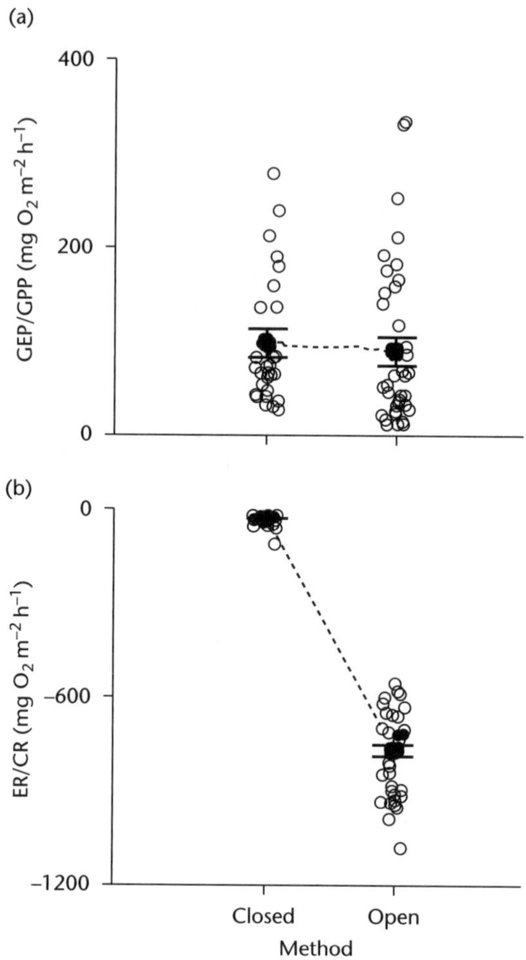

Figure 7.10 Closed-chamber versus open whole-stream estimates of metabolism in the Kuparuk River: (A) Comparison of closed-system gross primary production versus open-system gross ecosystem production (mean $\pm$ 1 SE). (B) Comparison of closed-system community respiration versus open-system respiration (mean $\pm$1 SE) (from Cappelletti 2006.)

in very low abundance (<10 per km). Mean estimates of age-0 (young-of-the-year, YOY) grayling density range from 50 to 100 per pool, and they are seldom found in riffles (Deegan and Peterson 1992).

Oksrukuyik Creek

Oksrukuyik Creek is a tundra stream that drains a cluster of lakes north of the Brooks Range and flows northeast into the Sagavanirktok River (Figure 7.5). Average discharge is 1.0 $m^3\ s^{-1}$ which may increase to more than 5 $m^3\ s^{-1}$ during

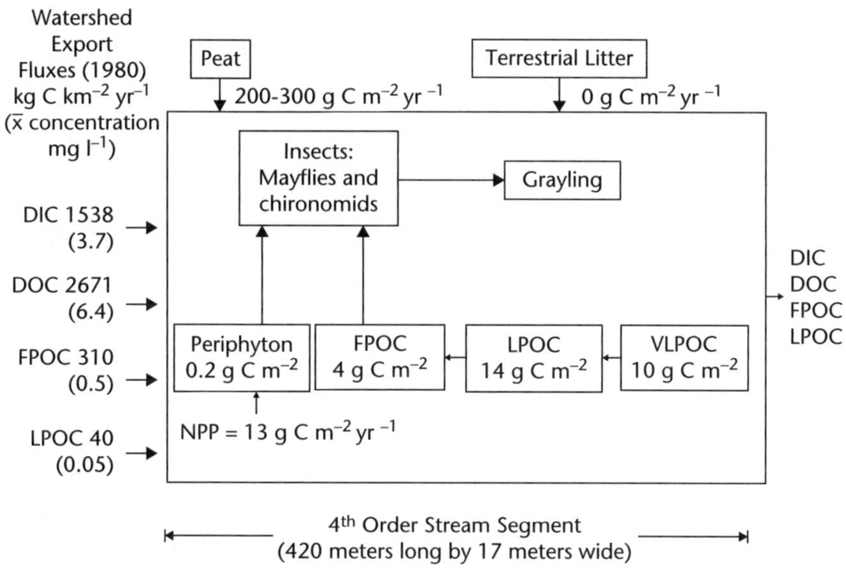

Figure 7.11 Carbon stocks and fluxes in the Kuparuk River ecosystem. From Peterson et al. (1986) with permission from the Canadian Journal of Fisheries and Aquatic Sciences.

summer rains, although the headwater lakes provide water storage and dampen discharge fluctuations. The benthic invertebrate biomass is dominated by insect larvae, especially *Orthocladius rivulorum.* Several species of the mayfly *Baetis* occur, all of which are univoltine, generalist collectors. The main filter-feeders are a species complex of univoltine black flies, notably *Prosimulium perspicuum.* Unlike the Kuparuk River, Oksrukuyik Creek lacks a large population of the caddis fly *Brachycentrus americanus.*

One fish species, the arctic grayling *Thymallus arcticus*, occurs in the study area. The creek provides spring spawning habitat and summer feeding areas for adult grayling (>30 cm) that probably overwinter in the Sagavanirktok River. Juvenile grayling occur infrequently. YOY grayling hatch in late June or early July and inhabit the margins of the stream throughout the summer.

Three experimental reaches were established for a short-term fertilization study on Oksrukuyik Creek (Harvey et al. 1998): reference, pre-fertilized, and fertilized. The pre-fertilized and fertilized zones are the same physical reach; pre-fertilized refers to the reach in 1989 and 1990, before fertilization began, and fertilized refers to the reach from 1991 to 1994, when nutrients were added.

Hershey Creek

Hershey Creek is a short stream that flows into the Kuparuk River at a point where the Trans-Alaska Pipeline crosses the river (Figure 7.5). CPOM storage is higher in small tundra streams like Hershey Creek than in the larger streams and rivers. The upper 2-km reach of Hershey Creek was used for as short-term phosphorus-enrichment experiment (Benstead et al. 2005). The study reach was bounded at its upstream and downstream limits by the pipeline and forded at both points by the pipeline's gravel access road (Figure 7.5). The most common riparian plants in this relatively low-gradient reach are *Salix* spp. and *Carex* spp.; shading of the channel by the dwarfed vegetation is limited to narrowest (<0.5 m width) sections of the stream. With the possible exception of Hershey Creek's deepest pools, >1.5 m in depth, its channel is frozen solid from October to May.

Synthesis Theme (1): Resource Limitation (Bottom-up Control)

The Kuparuk River Long-Term Fertilization Experiment

The Kuparuk River ecosystem has been studied intensively from 1978 until the present. The first years of this study were focused on the basic hydrology, nutrient biogeochemistry, primary productivity, and organic matter budgets. In 1980, nutrient enrichment bioassays showed that the low levels of dissolved phosphorus were limiting to primary production of the epilithic algal community. Starting in 1983 a long-term nutrient fertilization study began that continues to the present (results in Bowden et al. 1992 and Slavik et al. 2004).

Starting in the summer of 1983, the Kuparuk River was fertilized with a low-level, constant-drip addition of phosphoric acid, which raised the average P concentration by ~0.3 μM at a nominal flow rate of 2 m^3 s^{-1}. This experimental manipulation of the system stimulated algal production by 1.5 to 2 fold. The fertilization also increased bacterial activity, rates of respiration of the epilithic biofilm, and rates of decomposition of recalcitrant substrates such as lignin and lignocellulose. The bottom-up impacts of phosphorus addition increased the abundance of insects such as *Brachycentrus* and *Baetis*. In the first few summers of fertilization, there was evidence that at times insect grazers limited chlorophyll accumulation on the bottom rocks (Figure 7.12). While this can occur when mayfly and especially chironomid populations are at their maxima, subsequent studies with tracers (Wollheim et al. 1999) indicated that energy flow through grazers was on average far less than required to control algal biomass. However, the increased insect abundance stimulated the growth of both adult and YOY grayling in the fertilized reach relative to the reference reach (Deegan and Peterson 1992).

Although there was a consistent response to fertilization, the populations of algae and insects and the rates of fish growth were highly variable (~10 fold) from year to year. This was true in the reference and fertilized reaches (Hershey et al. 1997; Deegan et al. 1999; Slavik et al. 2004). Much of this variability is correlated with year-to-year variation in discharge. Under high discharge conditions (and presumably high scouring) algal biomass was only half as great as under low discharge conditions. Adult fish thrived when discharge was high, but the YOY grew poorly under high-flow conditions that are associated with low temperatures (Deegan et al. 1999). These impacts of discharge appear to be equally strong in the reference and

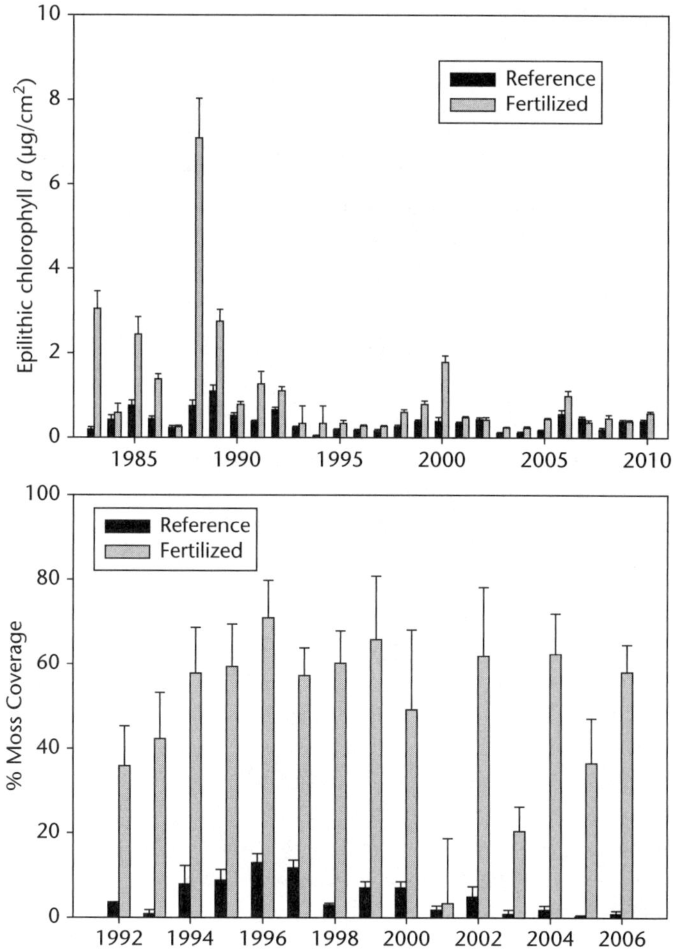

Figure 7.12 Epilithic total chlorophyll during fertilization (top) and percentage of total bryophyte cover (bottom) in the reference and fertilized reach riffles of the Kuparuk River. Chlorophyll means for 1987 and 1994 only include August values. In 1988, fertilized reach chlorophyll values are inflated due to contamination by green algal filaments. Bryophyte cover was negligible prior to 1992. Data are means ± 1 SE. Graphs are updated versions from Slavik et al. (2004).

fertilized reaches, although the higher biomass and productivity in the phosphorus fertilized reach made the effects easier to measure.

After nearly a decade of continuous summer fertilization, an important transition occurred. A carpet of two closely related aquatic moss species, *Hygrohypnum alpestre* and *H. ochraceum* rapidly covered the riffle habitat for several kilometers downstream of the phosphorus addition point (Figure 7.12). This was a surprise because neither of these species was observed in this reach of the Kuparuk prior to fertilization. A less obvious and slower growing species, *Schistidium agassizii*, was always present in relatively low abundance but had not responded to the added phosphorus. Between 1990, when the presence of *Hygrohypnum* spp. was first noted, and 1997, the bryophyte biomass in the treatment reach (riffles only) reached ~200 g ash free dry matter (AFDM) m^{-2}, while bryophyte biomass in the reference reach was only ~25 g AFDM m^{-2} (Arscott et al. 1998). Subsequent studies showed that the *Hygrohypnum* species are present in nutrient-rich seeps and springs in the headwaters of the Kuparuk River and that these species are capable of rapid growth under the conditions of nutrient enrichment.

The impact of this moss invasion on the physics, biogeochemistry, and biota of the Kuparuk River has been profound (Stream Bryophyte Group 1999). *Hygrohypnum* spp. provide a large amount of surface area for algal epiphytes, and the moss community (moss plus epiphytic algae) is several-fold more productive than the epilithic biofilm in the reference reach (Arscott et al. 1998). The biomass of moss dwarfs by orders of magnitude the biomass of epilithic algae. Consequently, biomass in the experimentally enriched reach is significantly higher than biomass in the reference reach. The moss fronds create a matrix containing slow-moving water, and the matrix traps and stores a large amount of fine particulate matter and may serve as an important locus of nutrient recycling. The moss matrix habitat is host to an insect community quite different in abundance and species composition than found in the rocky-bottom reference reach (Lee and Hershey 2000). Chironomids, *Brachycentrus*, and a large mayfly (*Ephemerella*) are more abundant by an order of magnitude in the mossy fertilized reach (Figure 7.13). In contrast, other common insects including baetid mayflies, black flies, and *Orthocladius* are less abundant in the mossy reach than in the reference reach. Despite higher insect biomass and secondary production in the fertilized reach, the growth of adult, drift-feeding grayling was sometimes greater and sometimes less in the fertilized reach compared to the reference reach (Figure 7.14). Some aspect of the link between insect production and fish growth was changed by the phosphorus-induced moss invasion, but further work is required to understand these results.

Oksrukuyik Creek Fertilization Experiment

The Kuparuk River fertilization experiment was repeated on a smaller tundra stream, Oksrukuyik Creek, to determine if the responses to nutrient enrichments were repeatable (Harvey et al. 1998). Additions of phosphorus or phosphorus plus nitrogen were continued for six years, and once again nutrients stimulated production at all trophic levels. Nitrogen was added in addition to phosphorus to avoid depletion of available nitrogen downstream of the P-addition point. Epilithic chlorophyll, primary production, and respiration increased in the fertilized reach. The biomass of filamentous algae and epilithic diatoms also increased in the fertilizer reach relative to stocks in the reference reach. The densities of the mayfly *Baetis* spp. and the chironomid *Orthocladius rivulorum* were increased by fertilization, whereas the abundance of black flies was not significantly affected. Fertilization stimulated the growth of both YOY and adult grayling.

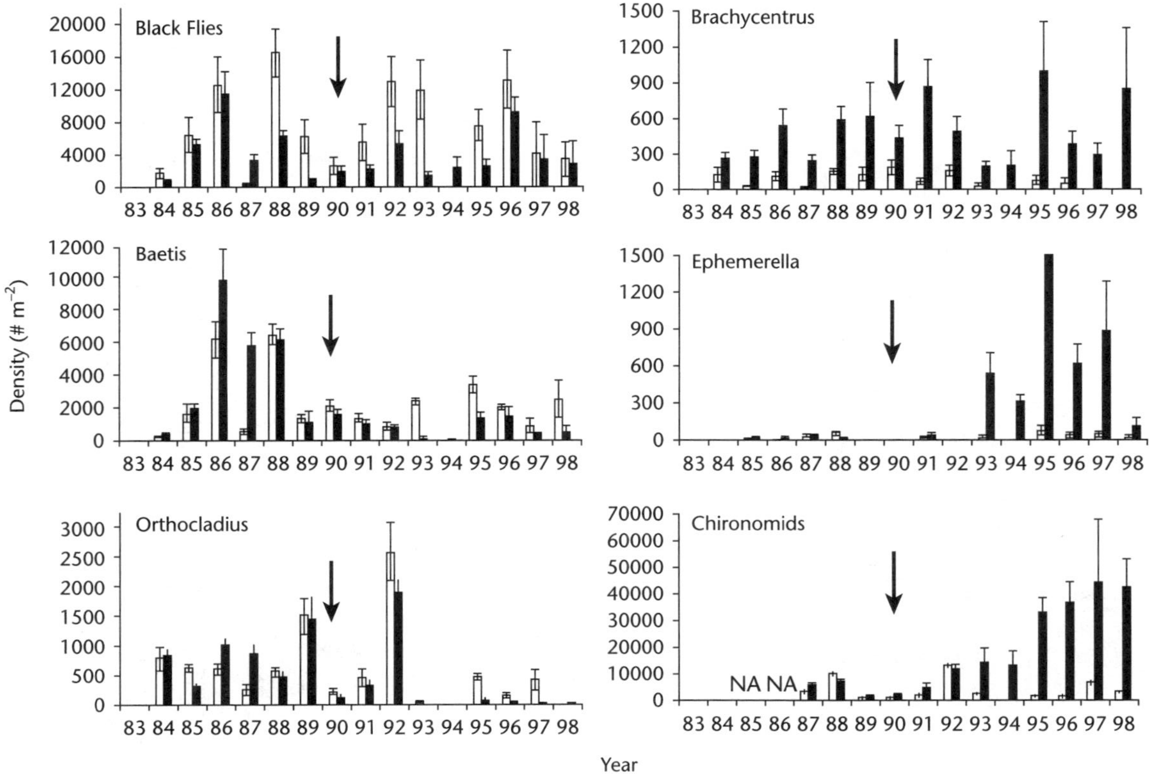

Figure 7.13 Mean July densities of six major benthic insect taxa in the Kuparuk River from 1983 to 1998. The shift to a moss-covered fertilized reach caused different changes in insect densities. The species on the left were disadvantaged by the moss; the species on the right benefited from the moss. The mosses appeared in 1990 (indicated by the arrows).

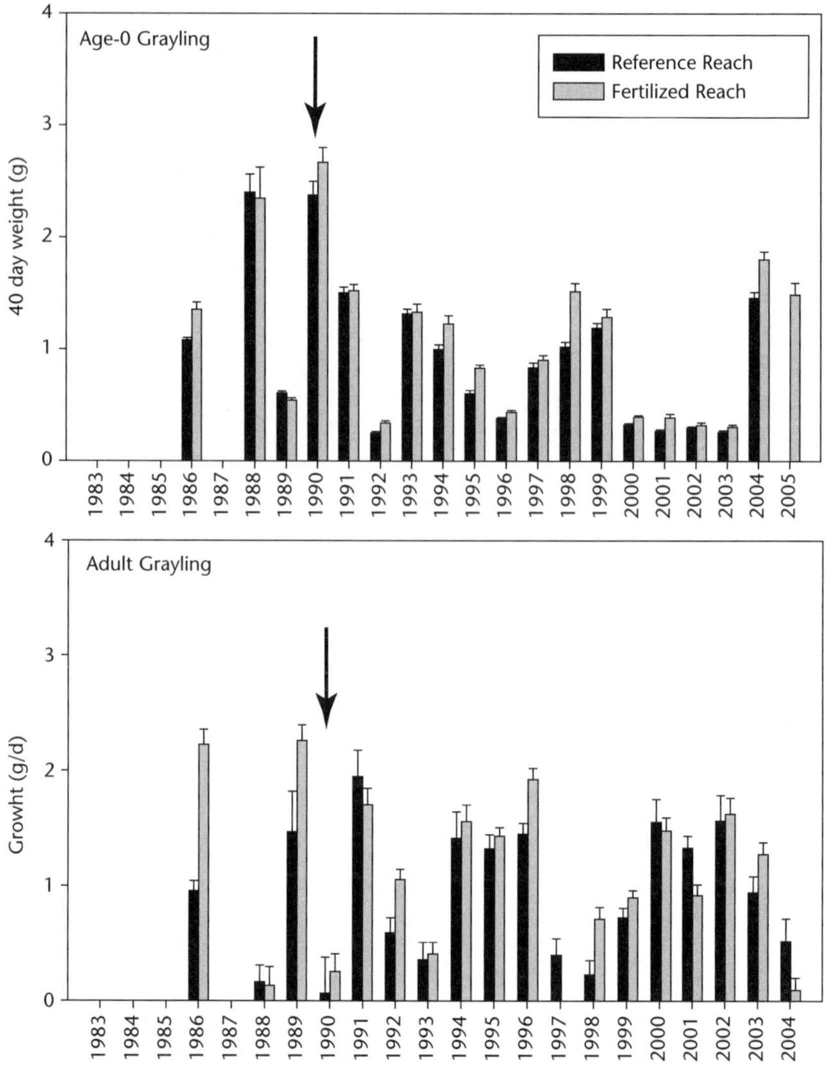

Figure 7.14 Long-term fish response. (A) Mean age-0 grayling 40-day weight (g ± 1 SE) in the reference and fertilized reaches of the Kuparuk River. (B) Adult grayling growth (mean g d^{-1}, ± 1 SE) in the reference and fertilized reaches of the Kuparuk River. N> 5 for all years; the mosses appeared in 1990 (arrows). From Deegan et al. (1999), used with permission from the American Fisheries Society.

Comparison with the Kuparuk River results showed that of nine measures of biotic responses to fertilization six were the same in the two rivers. Fertilization of the Kuparuk caused *Brachycentrus* abundance to increase significantly, but in comparison *Brachycentrus* was too scarce in Oksrukuyik Creek for abundance measurements. Black fly production did not change in Oksrukuyik but decreased in the Kuparuk when nutrients were added. The decrease in black fly abundance in the Kuparuk, where *Brachycentrus* increased, and the lack of a response by black flies in Oksrukuyik, where *Brachycentrus* was essentially absent, suggests a negative interaction between these two organisms. *Orthocladius* production increased in Oksrukuyik Creek but did not change in the Kuparuk. As in the Kuparuk, summer discharge interacted significantly with nutrient fertilization to modulate biotic activity. For example, in both rivers nutrients stimulated YOY and adult grayling growth, but discharge had an even larger impact. The YOY grew fastest in low flow years in both reference and fertilized reaches, whereas adults grew best in moderate to high flow years.

Hershey Creek Fertilization Experiment

Ammonium-N and phosphoric acid were added to a second-order beaded, tundra stream to investigate the consequences of nutrient enrichment for low-order arctic streams (Benstead et al. 2005). Responses in nutrient chemistry, chlorophyll *a* standing crop, and in the breakdown and macroinvertebrate colonization of leaf litter were measured over a 38-day summer period. During the time of P addition, nutrient concentrations immediately downstream of the dripper averaged 6.4 μM NH_4-N and 0.45 μM soluble reactive P (SRP). Concentrations upstream of the dripper averaged 0.54 μM NH_4-N and 0.03 μM SRP. Uptake of both nutrients was rapid. Concentrations were reduced on average to 28% (NH_4-N) and 15% (SRP) of maximum values within 1500 m. Standing crops of chlorophyll *a* on standardized substrates were significantly higher by the end of the experiment. Breakdown rates of senescent willow (*Salix* sp.) and sedge (*Carex* sp.) litter and associated fungal biomass were also significantly increased by nutrient addition. Fertilization resulted in four- to seven-fold higher macroinvertebrate abundance and two- to four-fold higher macroinvertebrate biomass in litter bags, as well as an increase in late-summer body mass of larval *Nemoura* stoneflies.

Synthesis Theme (2): Recovery from Experimental Nutrient Enrichment

Community recovery from experimental fertilization (i.e., relative return to reference conditions) has been studied in three stream reaches at the Arctic LTER (Benstead et al. 2007). The three stream reaches (Figures 7.15 and 7.16) were created either by successive movement of nutrient addition points ("drippers") downstream from their original locations (Kuparuk River, two reaches) or by complete cessation of experimental nutrient addition (Oksrukuyik Creek). Specifically, in 1985 the original phosphorus dripper on the Kuparuk River was moved 0.59 km downstream from its original position at 0.0 km, forming the first Kuparuk recovery reach (enriched for the two previous summers, 1983–1984) between 0.0 km and 0.59 km. In 1996 the dripper was moved downstream a second time, to 1.4 km below the original dripper site. The reach between 0.59 km and 1.4 km became the second Kuparuk recovery reach (prior enrichment for 13 consecutive summers, 1983–1995). The third recovery reach was on Oksrukuyik Creek and was created

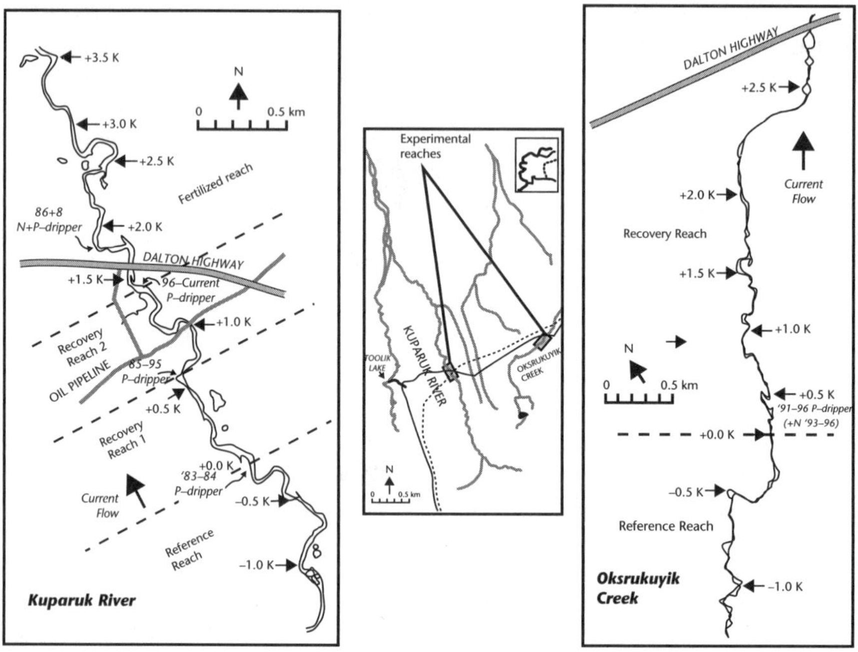

Figure 7.15 Maps showing the reference and recovery reaches of the Kuparuk River (left) and Oksrukuyik Creek (right), and the location of the two rivers on Alaska's North Slope (center). The phosphorus addition site on the Kuparuk River was moved downstream in 1985 and again in 1996, forming the two recovery reaches: Recovery 1 (fertilized for 2 years) and recovery 2 (fertilized for 13 years). Oksrukuyik Creek (right) was fertilized for 6 years. From Benstead et al. (2007), used with permission from *Freshwater Biology.*

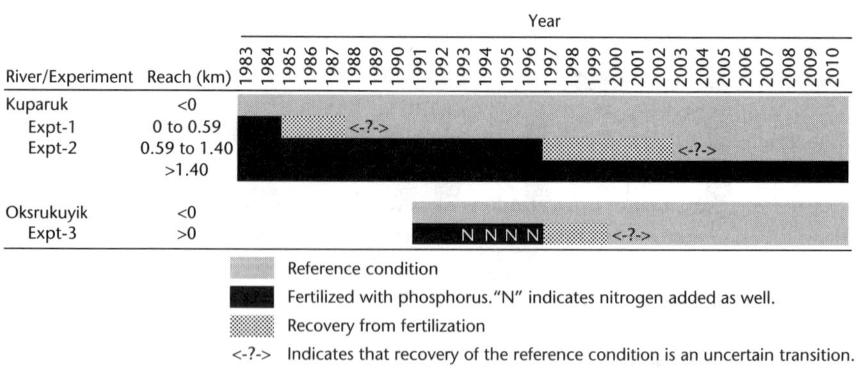

Figure 7.16 Time course of fertilization experiments on the Kuparuk River and Oksrukuyik Creek near Toolik Lake, Alaska, indicating the locations (reaches) and time intervals of the recovery process.

when the original experiment was terminated after six consecutive summers of fertilization (Harvey et al. 1998). During the first two summers of fertilization (1991–1992), Oksrukuyik Creek was fertilized with phosphorus only. For the last four summers (1993–1996), NH_4-N was also added as ammonium sulphate to prevent the possibility of severe N-limitation. The reference reach, upstream of the dripper, was not treated in any way throughout the entire experiment.

In all three of the recovery reaches, return of epilithic algal biomass to reference levels occurred within a year after cessation of fertilization (Figure 7.17).

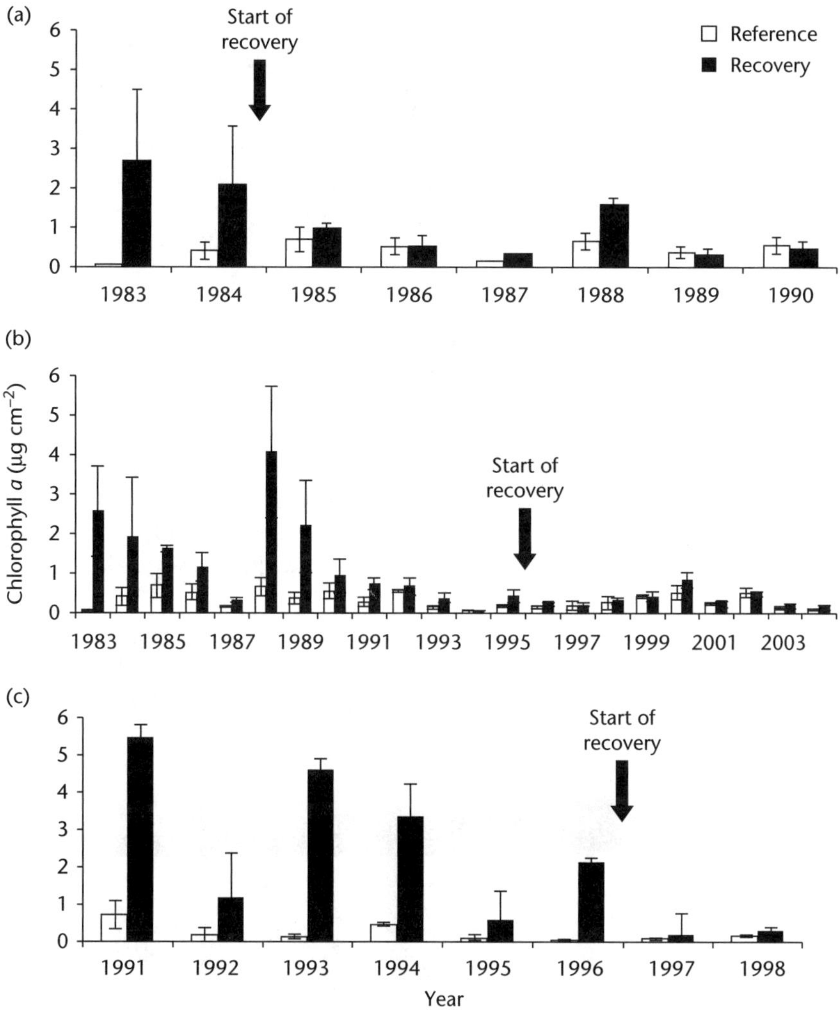

Figure 7.17 Mean epilithic chlorophyll *a* standing crop (±1 SE) in reference and recovery reaches of the (A) Kuparuk River, 1983–1990 (first recovery reach), (B) Kuparuk River, 1983–2003 (second recovery reach), and (C) Oksrukuyik Creek, 1991–1998. From Benstead et al. (2007), used with permission from *Freshwater Biology*.

Rapid shifts in algal standing crop were likely due to high turnover of algal cells combined with a return to P-limited growth after cessation of phosphorus addition. In contrast, bryophytes, which became the dominant primary producer in the Kuparuk River's fertilized reach only after seven years of long-term nutrient addition (Bowden et al. 1994), persisted for seven years after the phosphorus nutrient subsidy had been removed (Figure 7.18). Bryophyte cover eventually declined to 7% by 2003, but annual decreases prior to that year were highly variable with large relative decreases in bryophyte cover occurring only in 2001 and 2003. The relative persistence of bryophyte cover was presumably related to the longevity of individual bryophyte plants combined with secure attachment to cobbles and boulders by tough rhizoids (Stream Bryophyte Group 1999). The taxon largely responsible for increases in bryophyte cover, *Hygrohypnum* spp., has been shown to be strongly P-limited under ambient nutrient conditions in the Kuparuk River. It therefore seems likely that the gradual attrition of *Hygrohypnum* cover in the Kuparuk's second recovery reach was driven by senescence of individual plants stressed by increasing nutrient limitation, combined with storm-associated scour.

The benthic macroinvertebrates in the Kuparuk River differed in community structure among the unfertilized reference reach, the second recovery reach, and the fertilized reach during the years 1992–1998 (i.e., the first seven years of recovery of the second recovery reach; Figure 7.19). Density data for six numerically dominant taxa could be divided into three groups: taxa that were positively affected by bryophytes and that persisted at relatively high densities in the recovery reach (brachycentrid caddis flies, the mayfly *Ephemerella,* and most midge taxa); taxa that were negatively affected by bryophytes and that showed depressed densities (the tube-building midge *Orthocladius rivulorum*); and taxa that showed variable or neutral responses (simuliids and baetid mayflies). The presence of the bryophytes, which persisted longer than expected, tended to slow the recovery of the benthic

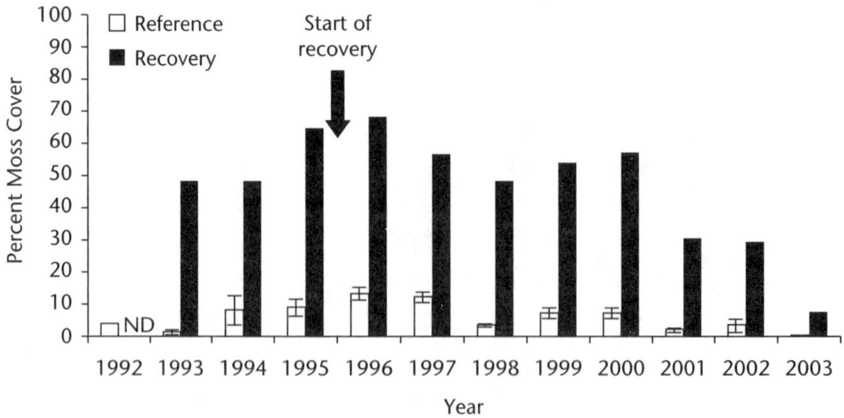

Figure 7.18 Time course of bryophyte recovery from fertilization in the Kuparuk River. Reproduced with permission from *Freshwater Biology.*

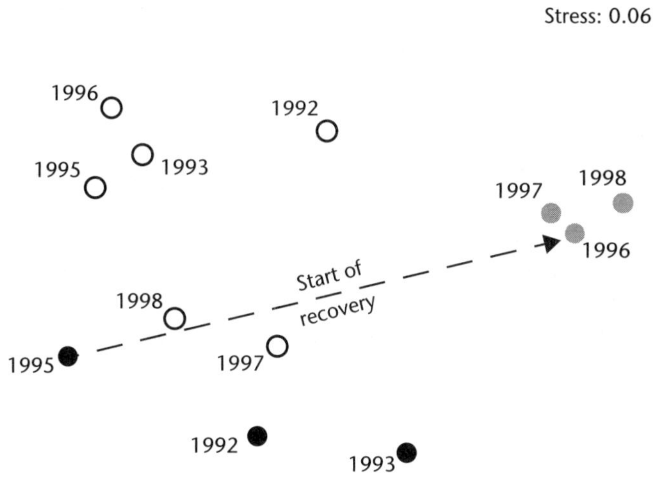

Figure 7.19 Nonmetric multidimensional scaling ordination plot of benthic macroinvertebrate samples based on Bray-Curtis dissimilarities among untransformed density data. Solid symbols are pre-recovery (fertilized) reach, grey symbols are recovery reach, and open symbols are reference reach. Arrow shows transition from fertilized to recovery conditions. No data were available for 1994. Ordination using log-transformed data showed similar patterns of recovery with an identical stress value. From Benstead et al. (2007), used with permission from *Freshwater Biology.*

macroinvertebrate community structure to a reference condition, compared to the recovery rates for the first recovery reach (Peterson et al. 1993, <2 years), where the *Hygrohypnum* spp. had not yet established.

Growth rates of adult and YOY grayling initially showed clear positive responses to N and P addition in Oksrukuyik Creek, presumably because of the observed increases in availability of important prey taxa (Harvey et al. 1998). Growth of both age-classes returned to reference values rapidly (within one to two years) after nutrient addition ceased. Arctic grayling feed predominantly on drifting benthic insects (Deegan and Peterson 1992), so it is likely that rapid return of grayling growth to reference rates was linked to correspondingly fast return of important prey taxa to lower levels of availability. However, no insect data are available for Oksrukuyik Creek's recovery reach.

Differences in recovery from nutrient enrichment among the three stream reaches appeared to be driven by the duration of nutrient enrichment and by the effects of bryophyte colonization. Short-term nutrient enrichment of the Kuparuk's first recovery reach (two years prior fertilization) and Oksrukuyik Creek (six years) did not result in bryophyte colonization. These two reaches showed a rapid, community-wide return to reference levels of biomass and biological activity. The main effect of bryophyte colonization on recovery patterns was the establishment of a high-biomass and physically complex benthic environment that was more stable, relatively resistant to scouring, and consequently persistent in the long term (Stream Bryophyte Group 1999). In turn, this allowed some dominant "moss-loving" primary consumer taxa to remain at high levels relative to other taxa that were negatively affected by bryophytes.

Nutrient enrichment of oligotrophic arctic stream ecosystems is likely to increase in the future due to permafrost thawing associated with climate change, as well as local anthropogenic impacts such as mining and drilling. The results of the recovery studies at the Arctic LTER site show that streams can recover rapidly from short-term, low-concentration nutrient enrichment, as has been seen in studies of many different perturbations to stream ecosystems (Niemi et al. 1990). Longer enrichment can lead to shifts in the primary producer community, however, with persistent effects on benthic habitat and community recovery (see Benstead et al. 2007 for more information on recovery in the Kuparuk River and Oksrukuyik Creek).

Synthesis Theme (3): Trophic Cascades (Top-down Controls)

Life History and Feeding Ecology of Arctic Grayling

Arctic grayling are an attractive sport fish, taken primarily by fly-fishing. They are distributed throughout the Arctic and are symbolic of clear, cold streams in the northern wilderness. Grayling are considered an obligate fluvial species, although some populations have been isolated from streams with suitable summer habitat and have adapted to lake environments. These populations are still obligate stream spawners and move into lake outlets to spawn and then return immediately to the lake. Away from lakes, grayling are often the only fish species in tundra rivers; this is the case for over 200 km of the Kuparuk River. Individual grayling can range widely, moving as much as hundreds of kilometers on a seasonal or annual basis between spawning, rearing, and sheltering habitats (Parkinson et al. 1999; West et al. 1992).

Grayling make an annual migration to deep-water lakes or flowing spring habitats, driven by the fall freeze-up of tundra streams (West et al. 1992). In late May they reenter the river, spawn, and then move to summer territories that they defend against other grayling. Grayling faithfully return to the same spawning and feeding areas (Buzby and Deegan 2004). Juvenile age classes (1- to 5-year-old fish; 5–20 cm total length; 2–150 g) use smaller streams, while adult grayling (6- to 18-year-old fish; 20–40 cm total length; 150–450 g) tend to use larger rivers. Thus, there is a distinct separation in size classes between small and large streams (Figure 7.20). An important consequence of this observation is that it clearly takes an interconnected landscape composed of lakes and large and small rivers to support a viable arctic grayling population.

Within the three months of June, July, and August, grayling must obtain enough energy through feeding on insects in the stream to grow, migrate, survive the winter, and spawn the following spring. Grayling are generalists in their food habits, but drifting aquatic insects, especially mayflies, stoneflies, and caddis flies are their primary food items (Figure 7.21). Despite differences in the size of fish occupying each river, diets of grayling in Hershey Creek and the Kuparuk River are similar. Early in the summer, fish in both streams rely heavily on Chironomidae, although

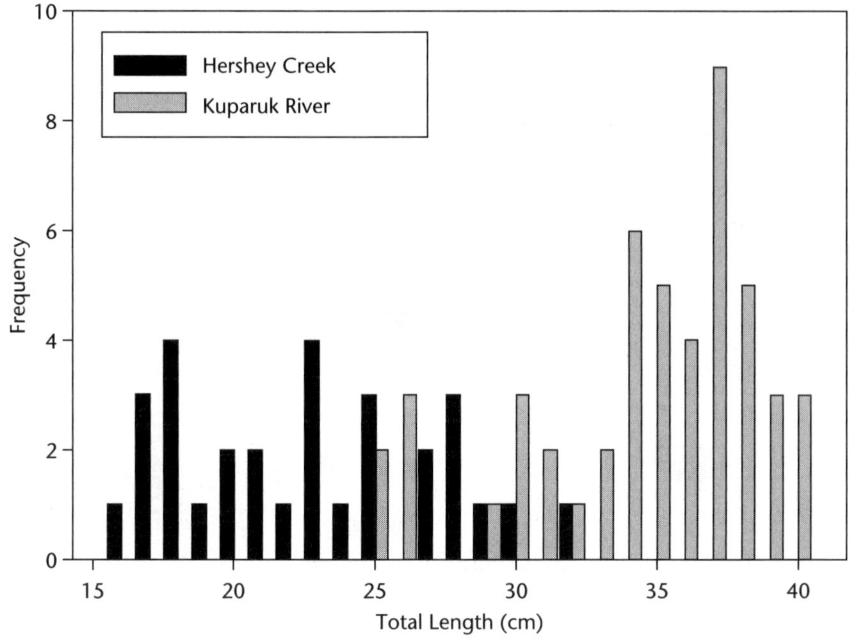

Figure 7.20 Size distribution of arctic grayling in small (Hershey Creek) and large (Kuparuk River) streams.

the Kuparuk diet was slightly more diverse. The numerical dominance of the Chironomidae was somewhat less when diets were viewed on a biomass basis due to the small size of Chironomidae and the much larger sizes of Trichoptera, Tipulidae, and terrestrial insects. Fish in the Kuparuk River consumed more prey; when results are normalized for fish size, however, there were no differences between streams. Diets were less similar late in the summer when fish in the Kuparuk River began consuming Ephemeroptera almost exclusively while diets of fish in Hershey Creek were dominated by Chironomidae. Fish in the Kuparuk River consumed four times more biomass than fish in Hershey Creek; and, unlike earlier in the summer, significant differences were maintained even after prey biomass was normalized for fish size. In addition to consuming more prey, fish in the Kuparuk River consumed larger prey late in the summer thereby reducing apparent foraging costs.

Terrestrial insects made up only a small fraction of the prey items in Hershey Creek and in the Kuparuk River (Figure 7.22). Less than 10% of the abundance and less than 20% of the biomass consumed by fish in both streams came from terrestrial sources. Fish in Hershey Creek tended to have the same fraction of terrestrial insects in their diet over the season, while fish in the Kuparuk River substantially reduced their reliance on terrestrial insects later in the season. The low reliance on terrestrial insects by fish in Hershey Creek contrasts with the predictions made by the river continuum concept (Vannote et al. 1980), which suggests that grayling diets in smaller creeks should reflect a greater terrestrial component because small

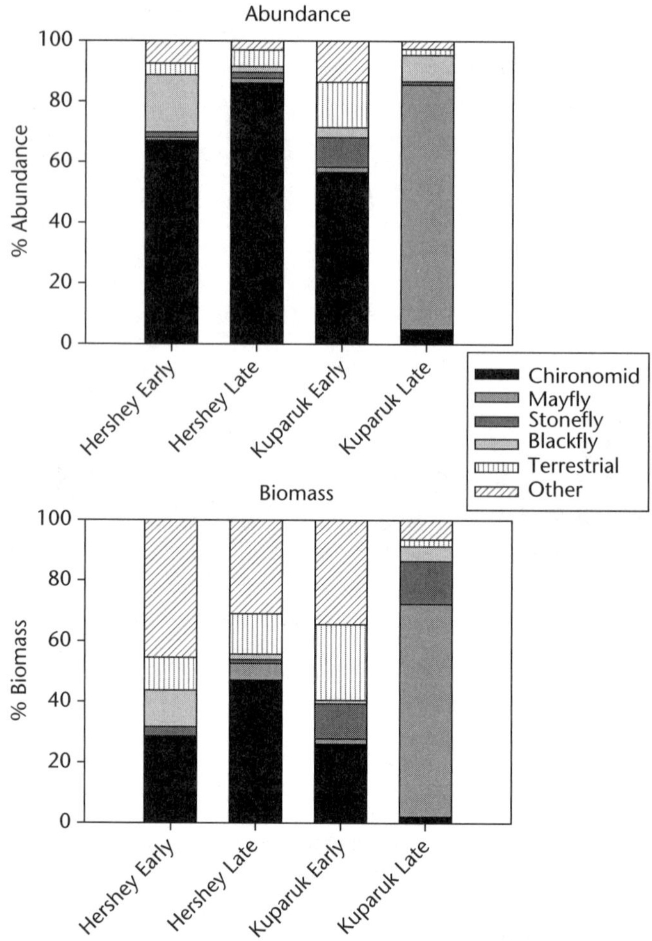

Figure 7.21 Diets of arctic grayling in Hershey Creek (n = 58 fish) and the Kuparuk River (n = 51 fish) from 1999 to 2001. "Early" refers to late June–early July, while "late" refers to early August. Abundance is based on number of prey; biomass is based on weight of prey. Stomach contents were obtained by gastric lavage and all insects from the contents were identified. Data from unpublished manuscript of C. MacKenzie and L. Deegan.

streams are expected to have a greater connectivity to the surrounding terrestrial environment. Drift data were unavailable so it was not possible to determine if the low consumption of aerial prey was due to low quantities in the drift or because grayling did not actively select them. These results were comparable to those found by Huryn (1998) in New Zealand streams. However, they were lower than the 15%–45% of terrestrial prey found in salmonid stomachs from forested streams in southeastern Alaska (Wipfli 1997).

The degree to which grayling feed in overwintering lakes during the arctic fall and winter is not known. Arctic grayling are visual feeders (O'Brien and Showalter

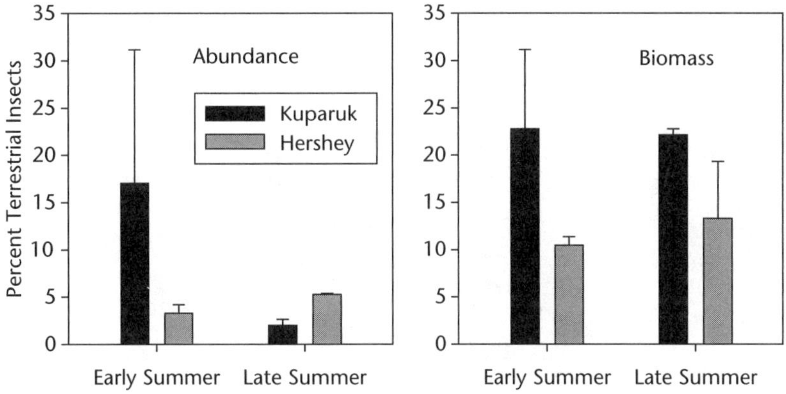

Figure 7.22 Contribution of terrestrial insects to diets of arctic grayling in small streams and large rivers. Means values are from the same fish described in Figure 7.21. The total number of fish represented in each category of this figure across all three seasons are Hershey early n = 21, Hershey late n = 37, Kuparuk early n = 24, and Kuparuk late n = 27. There are no data for Kuparuk early fish in 1999.

1993) and short day length, thick lake ice, and snow greatly reduce the light available for feeding during midwinter. During the fall and spring, arctic grayling may feed in overwintering lakes, however. Lake-adapted grayling are zooplanktivores, while fluvial grayling consume drifting mayflies and other benthic food. It is unknown if fluvial grayling that enter lakes in the fall continue to focus on insects and benthic invertebrates or switch to zooplankton. A few arctic grayling caught in March under the ice in central Alaska, had mayfly and stonefly larvae in their guts (Ziemann 1986).

Response to Food Changes and Other Environmental Variables

The rate of grayling growth is food limited. When fertilization with phosphorus stimulated the underlying food web (Figure 7.23), there was enhanced growth of adult and YOY grayling; in some years, this was more than two-fold over reference habitats, although in most years the difference was less enhanced (Figure 7.23). Overall, growth in the reference and fertilized reaches was linearly related with high and low years of growth in the same years in both reaches, and with about a 20% increase in the fertilized reach.

However, the year-to-year variation in the growth of grayling driven by environmental variability exceeded the differences caused by nutrient enhancement. Between 1986 and 2004 a 15-fold difference in growth among years (range 0.1 to 1.7 g d^{-1}) was measured, while nutrient enrichment provided only a 1.2 times increase in growth. Thus, it was not possible to identify a reliable growth benefit from the phosphorus enrichment. There is a correlation between discharge and growth for adult and YOY fish in the Kuparuk River and in Oksrukuyik Creek

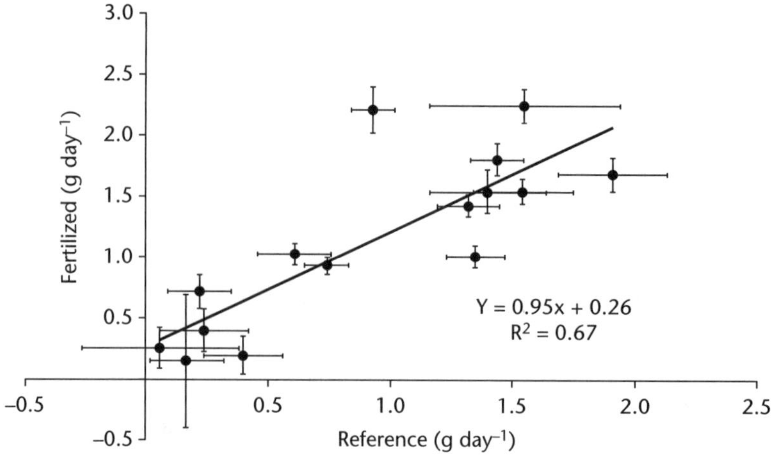

Figure 7.23 Growth rate of adult arctic grayling in reference and phosphorus enriched reaches of the Kuparuk River between 1986 and 2004. The solid line is the regression between reference and fertilized reaches.

(Deegan et al. 1999), but this was evident only after many years of study because of the high variability in this system (Figure 7.24). Fifty-six percent of the variance in adult grayling growth rate in both rivers was associated with nutrient level and mean summer discharge, while 60% of the variance in YOY grayling growth was associated with nutrient level, mean summer discharge, mean summer temperature, and differences between the two study streams. As noted previously, it appears that adult fish do better when discharge is high but that YOY grow poorly under high-flow conditions that are associated with low temperatures. Superimposed on the relationship between growth and discharge is the stimulatory effect of altering the nutrient regime of the river.

In these arctic tundra streams, because of the opposite response of adult and young grayling to the same environmental conditions, interannual sequences of environmental conditions can influence population structure. Optimal conditions for growth include a combination of beneficial temperature and discharge regimes superimposed on basic levels of stream productivity set by nutrient input. Recruitment may be maximized when environmental conditions in sequential years first favor adult (high flow, low temperature), then young (low flow, high temperature) fish. Adults would grow well in the high-flow year, resulting in high egg production the following year; a low-flow year would lead to high growth and survivorship of YOY grayling, resulting in a strong year-class the following year. Conversely, consecutive years of high discharge, while producing a healthy adult population, could result in little to no recruitment because of poor YOY growth and survivorship. Grayling appear to have evolved a life history in response to the variability of the arctic environment, including a relatively long life span, annual reproduction, and relatively few offspring per spawning event, which increases

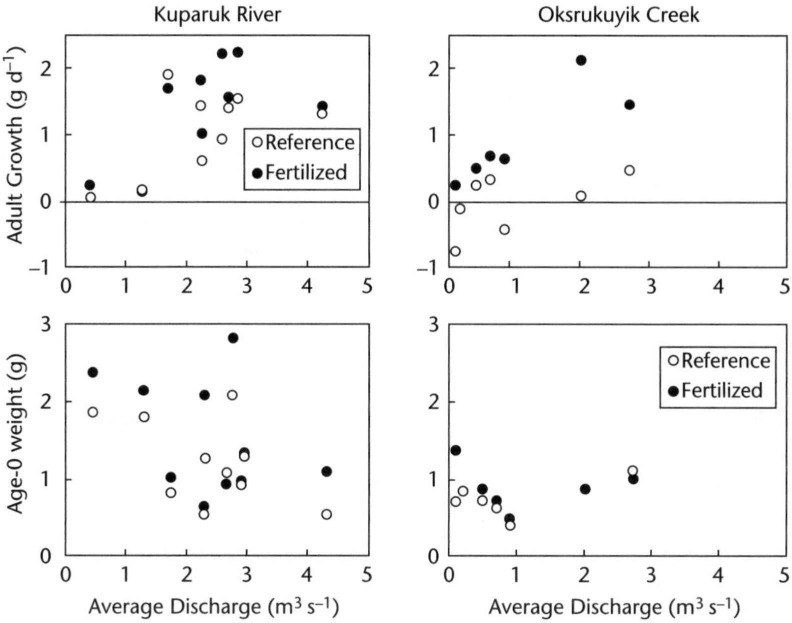

Figure 7.24 Relationship between growth of adult and young-of-the-year (age-0) grayling and river discharge in the Kuparuk River and Oksrukuyik Creek. From Deegan et al. (1999), used with permission from the American Fisheries Society.

the likelihood that some offspring will encounter exceptionally good conditions in some years.

Synthesis Theme (4): Hyporheic Processes

Vignette 7.1. The Unseen River Below, by William B. Bowden

The bottoms of most streams are composed of porous materials like sand, gravel, and cobble that allow water in the open stream channel to move into and out of the stream substrate. The subsurface zone in which this exchange occurs is called the "hyporheic zone." The thickness of this zone may range from centimeters in low-gradient streams dominated by small particle sizes (silt and sand) to many meters in high-gradient streams dominated by large particle sizes (gravel and cobble). Water in the hyporheic zone typically flows at speeds that are intermediate between the speed of groundwater in the soil (which may be less than a meter in a year) and stream water in the channel (which may be a meter per second). Thus, water that enters the hyporheic zone from the stream channel will move more slowly than it would in the open channel and is in intimate contact with surfaces of the particles that make up the stream substrate. These particles are covered by a diverse and active microbial biofilm. Essentially, the hyporheic zone behaves like a flow-through bioreactor that can process large quantities of carbon and nutrients.

All streams with porous substrates (i.e., excluding streams that flow over bedrock) have a hyporheic zone of some size that remains relatively fixed. However, in frigid environments like the Arctic, streams can freeze solid during the winter and so the hyporheic zone ceases to function. During the summer, the stream sediments thaw, and the hyporheic zone becomes thicker. Thus, these frigid streams are special in that they have a transition zone in space—from groundwater to open-channel flow (like all other streams)—but they also have a transition zone in time from frozen winter conditions to flowing summer conditions and back again.

Prior to the Arctic LTER research program, virtually nothing was known about hyporheic dynamics in arctic streams. Substantial evidence from temperate streams suggests that the hyporheic zone can be an important—perhaps critical—area of organic matter turnover and source of nutrient regeneration. Results from field studies on streams and rivers around the Toolik region reported by Edwardson (1997) and Edwardson et al. (2003) are consistent with results from these temperate studies. In particular, this early research on arctic hyporheic processes found that geomorphic profiles of stream channels (longitudinal more so than lateral) provide the necessary hydraulic gradients to drive hyporheic exchange through the streambed, as shown by Kasahara and Wondzell (2003). It also showed that biogeochemical processing could potentially supply from 14% to 162% of the benthic nitrogen uptake requirements in the Kuparuk River.

Hyporheic Dynamics in Seasonally Thawed Streams

A variety of mechanisms detain water as it moves down a stream channel, including eddies, backwaters, and hyporheic storage. Collectively these detention mechanisms are referred to as "transient storage," and they are thought to significantly influence biological and biogeochemical dynamics in stream ecosystems. Conservative tracer additions are used widely to calculate transient storage indicators such as mean storage residence time, storage zone area, hydraulic retention, and storage exchange rates. However, an important limitation of this approach is that it does not distinguish among different storage modes within streams. In particular, it does not distinguish between in-channel storage locations such as eddies and backwaters and below-channel storage locations such as the hyporheic zone. It is important to estimate the relative influences of in-channel and below-channel storage because the two environments are very different and will likely have very different influences on biogeochemical dynamics in streams.

It was found that ground-penetrating radar (GPR) could be used to follow changes in the depth of the thawed zone below arctic stream channels (Bradford et al. 2005). Because frozen water and liquid water have very different dielectric values, GPR can easily distinguish between the two. The depth to the liquid-frozen interface was interpreted from GPR data by hammering a metal probe into the stream substrate to the point of refusal and recording the temperature. Brosten et al. (2006) used this approach to acquire a series of thaw profiles at five sites from May to September 2004 in fast-flowing, cobble-bottomed (alluvial) streams as well as slow-flowing, peat-bottomed streams. Development of the thawed zone beneath the two stream environments was distinctly different (Figure 7.25). Within cobble-bottom streams thaw depths increased in thickness up to 1 m within the first four weeks after flow began but only up to 32 cm in the peat-bottomed streams. In August and September the cobble-bottomed sites began to refreeze, while the peat-bottom sites continued to thaw; thaw depths had not receded at the peat-bottom sites by the last site visit in September. These results indicate that rapid heat absorption and loss occurs in the cobble-bottom stream sites, while peat insulates the permafrost and introduces a lag in the seasonal thermal profile.

Conservative tracer additions were used to calculate transient storage indicators. Zarnetske et al. (2008) showed that these indicators were sensitive to discharge and strongly correlated with total stream power (Figure 7.26). However, the relationship between transient storage and extent of thaw was less clear, suggesting that hyporheic storage might not be a large proportion of overall transient storage. Transient storage indicators increased with increasing thaw depths under base- and low-flow conditions, but as expected these relationships diminished at high flow. There were good correlations between easily measured hydrologic variables of these channels (e.g., the Darcy-Weisbach friction factor) and several metrics of transient storage, in agreement with previous studies (e.g., Harvey and Wagner 2000). Notably, this research showed that stream power is a particularly good predictor of transient storage characteristics because it normalizes simple characteristics of hydraulics and morphology, thereby

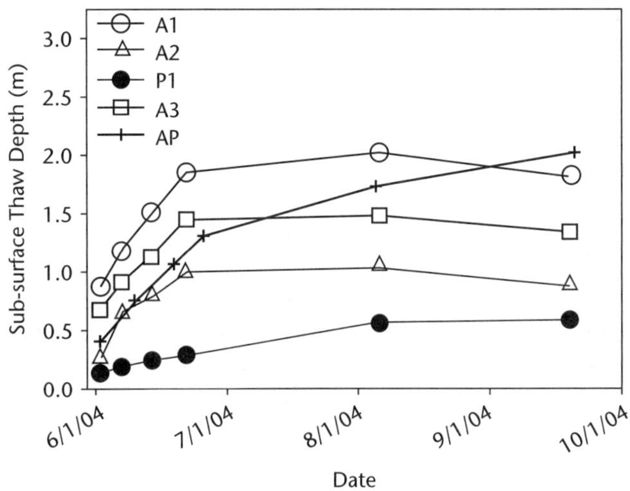

Figure 7.25 Subsurface thaw depth determined by ground penetrating radar data for fast-flowing cobble-bottom streams (A1, A2, A3, AP) and one slow-moving, peat-bottomed stream (P). From Zarnetske et al. (2007), used with permission from the American Geophysical Union.

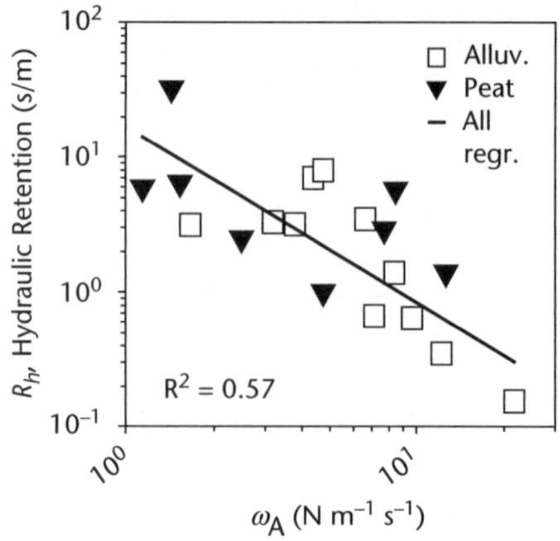

Figure 7.26 Relationship between hydraulic retention, representing the time spent in storage per meter length of stream (seconds per meter), and stream power (x-axis), as determined by numerous conservative solute tracer experiments in arctic streams. From Zarnetske et al. (2007), used with permission from the American Geophysical Union.

allowing better comparisons across streams that differ widely in these characteristics. Thus, it may be possible to estimate transient storage in these streams using metrics that are very simple to derive rather than methods that are laborious and expensive. It is likely that these results are transferable to non-arctic streams.

Water samples taken at fixed depths in the hyporheic zones of the cobble-bottomed and peat-bottomed streams during conservative solute tracer experiments described above (Figure 7.27) showed clearly that hyporheic exchange depths in the cobble-bottom and peat-bottomed streams were constrained by the depths of thaw beneath these two streams. In the cobble-bottom stream—which had a greater depth of thaw—stream water penetrated the hyporheic zone to a depth of up to 54 cm. In the peat-bottom stream, which had a much shallower depth of thaw, the actively functioning hyporheic zone was limited to a depth of 10 cm or less. The greatest depth of stream water penetration into the hyporheic zone in the cobble-bottom stream was seen at riffle head locations, while hyporheic exchange at riffle tail and pool features was less extensive. In the peat-bottom stream, a distinct boundary of active hyporheic exchange was observed at 10 cm, regardless of the stream feature sampled. In both streams the maximum extent of hyporheic penetration was much less than the maximum extent of thaw.

What would happen to the hyporheic zone in the future if the arctic environment warms? The finding described above, that the hyporheic zone occupies only a portion of the thaw basin, indicates that even if the depth of thaw increases, the hyporheic zone will not expand. However, in a future warmer environment the flowing season of the year will be longer and so the hyporheic zone will be active for a greater portion of the year. Thus, changing seasonality of the arctic system is more likely to have important influences on organic matter processing and nutrient dynamics in arctic streams by extending the period

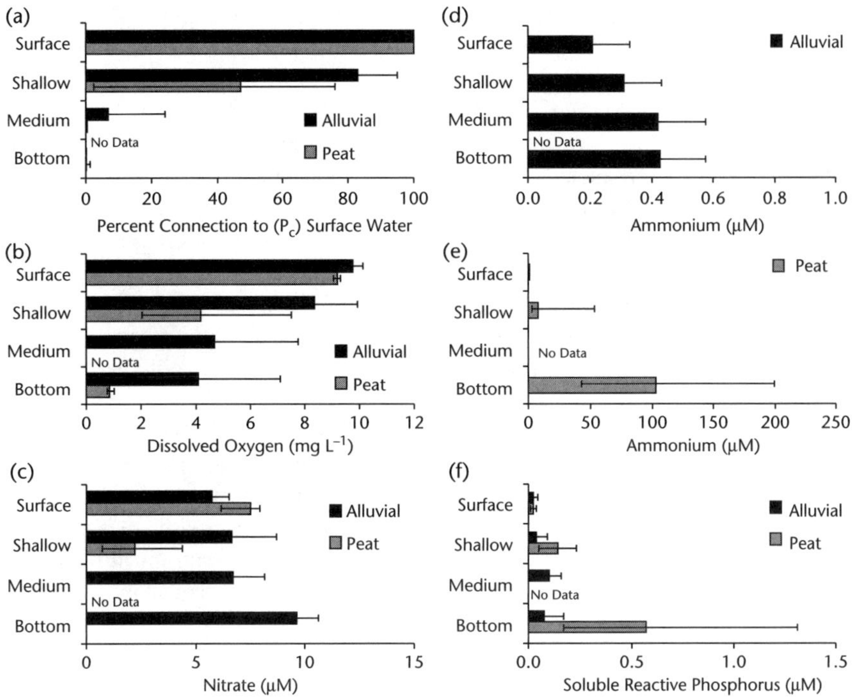

Figure 7.27 Characteristics of surface waters and hyporheic waters from mini-piezometers installed at three depths in an alluvial stream and two depths in a peat stream, as determined from conservative tracer addition experiments. Note that "bottom" in the alluvial stream is much deeper (110 cm) than "bottom" in the peat stream (33 cm). (A) Median percentage connectedness (P_C) of hyporheic waters to surface water at the same location. (B) Median dissolved oxygen concentrations. (C) Median nitrate concentrations. (D) and (E) Median ammonium concentrations. Note the change in scale between the alluvial (D) and peat (E) sites. (F) Median phosphate concentrations. From Greenwald et al. (2008), used with permission from the American Geophysical Union.

of time that the hyporheic zone is active rather than extending the volume of thawed soil or substrate it occupies. This is an important finding that will guide future models of how and why biogeochemical dynamics in arctic streams may change in future.

Synthesis Theme (5): Disturbance as a Driver in Arctic Streams

General Context

In the Toolik Lake region the spatial variability of stream ecosystem structure is, to a substantial degree, determined by the interaction between channel scouring, water-column freezing, and nutrient supply (Huryn et al. 2005; Parker and Huryn 2011). Channel scouring, which is a form of disturbance caused by sediment transport during storm flows, has been identified as a major type of stream disturbance worldwide (Scarsbrook and Townsend 1993). Streams with channels that are frequently scoured will be inhabited by taxa with life-history attributes allowing them to persist in such a physically unstable habitat.

Similarly, streams that regularly freeze solid will be inhabited only by taxa with life-history stages able to survive periods without liquid water. Water-column freezing is a form of disturbance that is assumed to contribute to the relative simplicity of arctic stream communities (Hershey et al. 1995) and that interacts with channel scouring to provide an additional level of physical control on stream community structure (*sensu* Poff 1997). Streams that are frequently scoured and that freeze during winter are expected to have extremely low levels of taxonomic richness, being inhabited by taxa with life-history attributes allowing them to persist in such a physically unstable habitat while also enabling them to survive cold periods without liquid water.

Nutrient supply will interact with a stream's physical-disturbance regime to affect the dynamics of community recovery following disturbance and will ultimately control levels of stream ecosystem productivity. An extremely oligotrophic stream, for example, may be expected to show long-term decreases in biofilm and bryophyte biomass following a scouring disturbance due to nutrient limitation of rates of net primary production, while a nutrient-rich stream may show more rapid recovery (Huryn et al. 2005).

The Landscape Disturbance-Habitat Template

The combination of physical disturbance, freezing, and nutrient scarcity creates an environmental template that explains community structure and complexity in the region around Toolik Lake. January benthic temperatures of different stream types in this region range from well below zero for mountain and glacier streams (Figure 7.28), to temperatures near zero for tundra streams, to well above freezing for mountain springs and some tundra springs (Parker and Huryn 2011). Given the overall stream length for different stream types described earlier, the winter freezing of streams in this region indicates that only the spring streams, or <1% of total stream length, remain flowing year-round. Although spring streams provide a negligible amount of habitat area, they may have large consequences for biodiversity and food-web dynamics of the North Slope landscape. During winter, for example, springs provide 100% of flowing headwater habitat and critical winter refugia for predators that are unable to tolerate freezing (e.g., perlodid stoneflies, Dolly Varden char, and American dippers).

Parker and Huryn (2006) measured channel scouring of streams in the mountain and mountain spring streams associated with the Ivishak spring. These measurements indicate that rates of movement of stones on the stream bottom range from a low of 4% measured over a ~6 week summer period in Ivishak mountain spring-stream, to a high of 66% measured over the same period in Ivishak mountain stream. Although this study was limited to only two streams and stream types, these data approximate the extremes of stream bottom displacement among the five different stream types of the Toolik region, as subsequently measured in 19 different

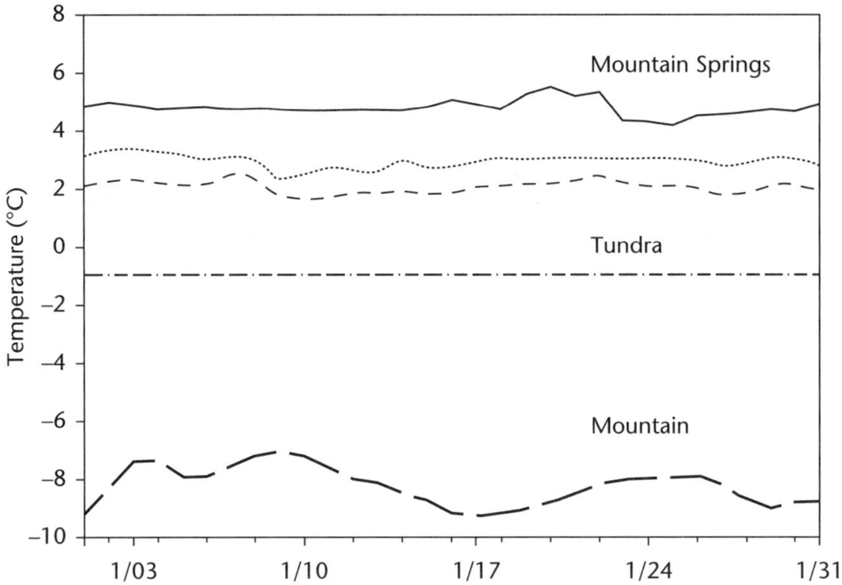

Figure 7.28 Midwinter (January) temperatures in different types of arctic streams (from Parker 2008).

streams by Parker and Huryn (2011). Surprisingly, given the large differences in the intensity of channel scouring between Ivishak mountain stream and Ivishak mountain spring-stream, the effect of disturbance on the macroinvertebrate communities appeared to be manifested primarily by differences in food-web function rather than food-web architecture (e.g., Figure 7.4). Although essentially the same macroinvertebrate taxa occurred in both streams, total macroinvertebrate biomass was 635 mg DM m^{-2} in Ivishak mountain stream and 4617 mg DM m^{-2} in Ivishak mountain spring-stream, and predators contributed 25% to biomass in the mountain spring-stream but only 7% in the mountain stream.

Although compared with other regions the concentrations of solutes tend to be uniformly low in North Slope streams (Kling et al. 1992), variability in the abundance of many solutes is high among individual streams within stream types. This variability may be explained, in large part, by differences in catchment geology, soil, and vegetation. Tundra streams, which have catchments consisting of layers of peat overlying permafrost and consequently little mineral material, tend to have extremely low concentrations of dissolved ions and are consequently very oligotrophic. These systems are similar to perched lakes that receive the majority of their solutes from the atmosphere. Compared to tundra streams, it was expected that other stream types would have greater concentrations of dissolved ions due to their proximity to less-weathered mineral sources. However, as noted earlier, differences in concentrations for many solutes are so variable that it is not possible to establish clear differences among stream types.

Nonetheless, each of the five stream types has a discrete periphyton and macroinvertebrate community type that is related to the characteristic disturbance regime of the stream type (Figure 7.29; Huryn et al. 2005; Parker and Huryn 2011). Although many factors likely contribute to the spatial variation of stream periphyton and macroinvertebrate community structure across the Toolik Lake landscape, the major factors can be conceptualized as a two-dimensional habitat template structured by a gradient in conductivity–substratum instability for periphyton, and a gradient in substratum instability–freezing probability for macroinvertebrates.

The domain of the landscape template is roughly bounded by the characteristics of mountain spring-streams, glacier streams, and tundra streams (figure 7.29).

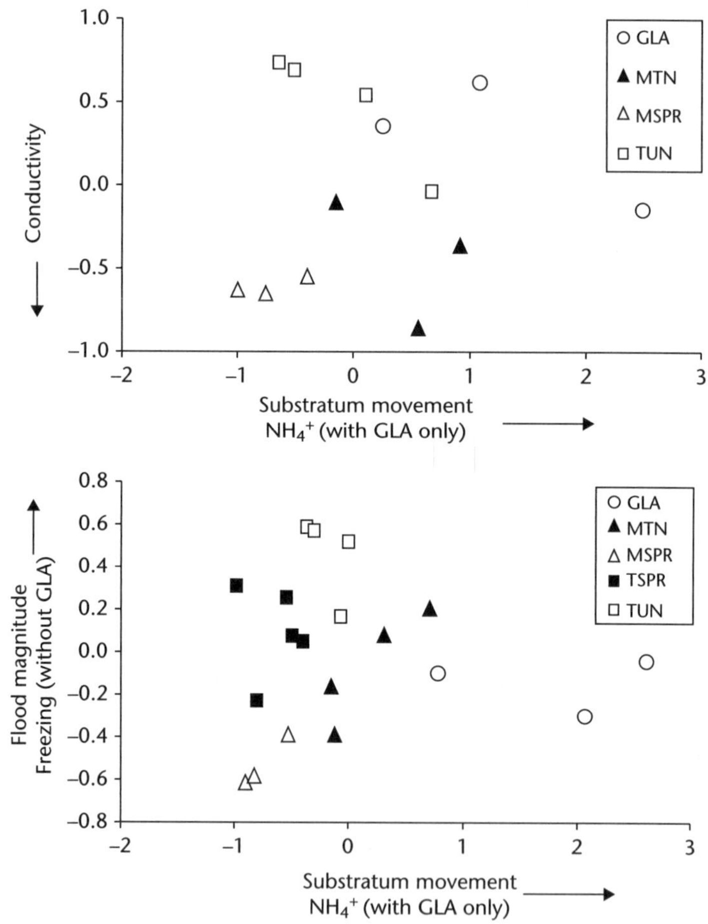

Figure 7.29 (A) Multidimensional scaling (MDS) analysis based on periphyton biovolume in 17 streams (2D Stress = 0.12). (B) MDS analysis based on invertebrate biomass (2D Stress = 0.08). GLA = glacier stream, MTN = mountain stream, MSPR = mountain spring, TSPR = tundra spring, TUN = tundra stream. From Parker and Huryn (2011), used with permission from *Freshwater Biology.*

Each of these stream communities has unique structural and functional attributes that are responses to its position within the Toolik Lake landscape. Mountain spring-streams have macroinvertebrate communities that are structured by a habitat with perennial flow and stable substrata. Mountain spring-streams also have, by far, the greatest total macroinvertebrate biomass. Communities of glacier streams offer habitats that freeze during winter, have unstable substrata, and the highest nutrient concentrations. Glacier streams have the lowest total macroinvertebrate biomass, indicating low productivity, and the lowest taxonomic richness. Communities of tundra streams are structured by a habitat that freezes during winter and has relatively stable substrata. Tundra streams also have the highest taxonomic richness.

Synthesis Theme (6): Changes in the Land—Permafrost and Thermokarst

Permafrost is an important feature of northern Alaska (chapter 1) that shapes the ecology through soil moisture and drainage, stream flow patterns, weathering of soil material, distribution of plants, the sequestration of large stores of organic matter, and even the location of animal burrows. Permafrost is defined by its temperature, always at or below 0°C, and it may or may not contain ice (van Everdingen 1988). Ice in permafrost occurs as massive ice (ice wedges, pingo ice), as ice within microscopic soil pores, and as lenses of segregated ice of various sizes.

When permafrost thaws and the contained ice melts, the land surface may subside, forming irregular features that are collectively referred to as "thermokarst" terrain. After the soil is destabilized, an event such as a large rainfall may cause mass movement of the soil downslope. This movement can take several forms, including glacial thermokarst, retrogressive thaw slumps, gully thermokarst, and active layer detachments (Figure 7.30). These thermokarst features have the potential to impact small lakes and tundra streams like the Kuparuk River by increasing sediment and nutrient loading. The sediments and nutrients that come from newly thawed and exposed minerals and soils may affect all aspects of the food web in arctic streams, from primary production to fish growth. Even when streams are not involved, the land surface will change because of downslope redistribution of soil organic matter; increased availability of organic matter to microbes and subsequent increases in CO_2, methane, and nitrous oxide; alterations of drainage and snow accumulations; and better growth sites for shrubs.

Thermokarst terrain is a natural feature of the arctic landscape, but it is relevant to question whether the rate of thermokarst formation has increased in recent decades along with climate warming in the Arctic. Unfortunately, most of these features are too small to be discerned reliably from satellite imagery. However, comparisons of recent aerial photos with aerial photos of the same areas taken in the late 1970s document a significant increase in the number of thermokarst hillslope failures in the region around Toolik Lake at a time when permafrost in the area has warmed (Figure 2.15). In 2006 a semiquantitative, low-level aerial survey identified at least 34 active thermokarst features in a ~25 km × ~25 km area between Toolik Lake and areas east of Happy Valley to the north (Bowden et al. 2008; Gooseff et al. 2009). The climate of the south slope of the Brooks Range is warmer than that

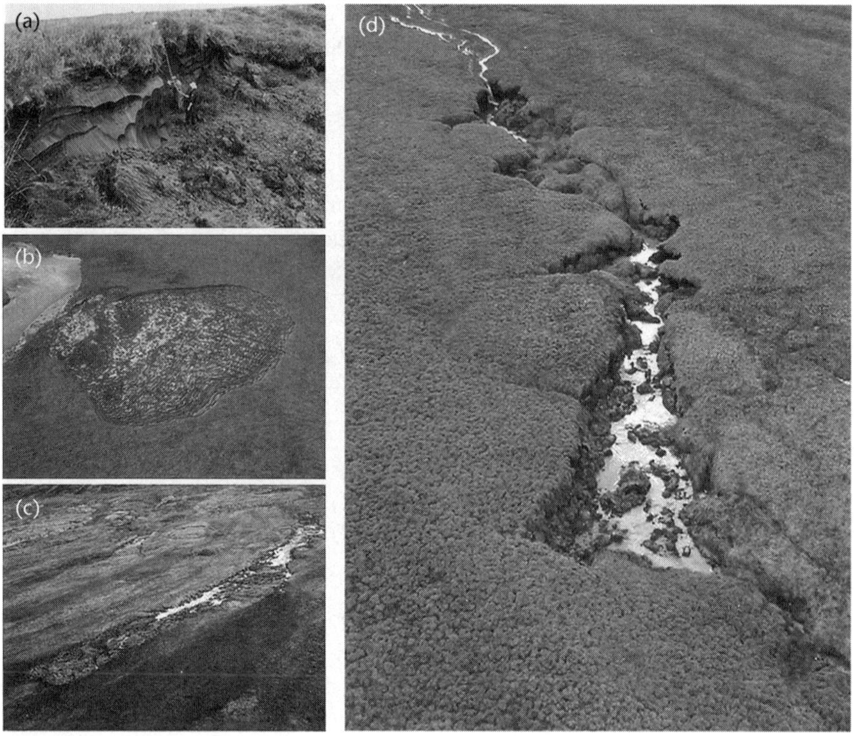

Figure 7.30 Examples of hillslope failures induced by thermokarst terrain. (A) Glacial thermokarst; note the exposed massive ice and people for scale. (B) Retrogressive thaw slump; widest dimension is about 100 m. (C) An active layer detachments slide; length is about 800 m. (D) Gully thermokarst; length is about 150 m. Photo credits W. B. Bowden, except (C) by M. Gooseff (reproduced with permission from the American Geophysical Union). [REFER COLOR IMAGE]

of Toolik; in subsequent, quantitative surveys in the regions around Feniak Lake (central) and Kelly River (west) within the Noatak National Preserve, we enumerated over 800 active features. We concluded on the basis of aerial photos taken ~25 years apart that the number of active features may have increased by 200% and that thaw slumps on lake shores have expanded over this period (Figure 7.31; Gooseff et al. 2009).

Several thermokarst features near the Toolik Field Station have been studied to determine the impacts of thermokarst formation on the arctic tundra landscape and environment. One of these is a gully thermokarst that formed rapidly in late July 2003 in the headwaters of the Toolik River, immediately after a period of heavy rainfall (Figure 7.32). This thermokarst feature occurred along a small drainage track on a gentle hillslope in an area characterized by Hamilton (1978) as ice-rich silt in the Sagavanirktok River glaciation. This type of feature tends to form when an ice wedge in the permafrost melts, leaving a large void through which water begins to move. At some point the overlying soil fails, creating a sharply defined

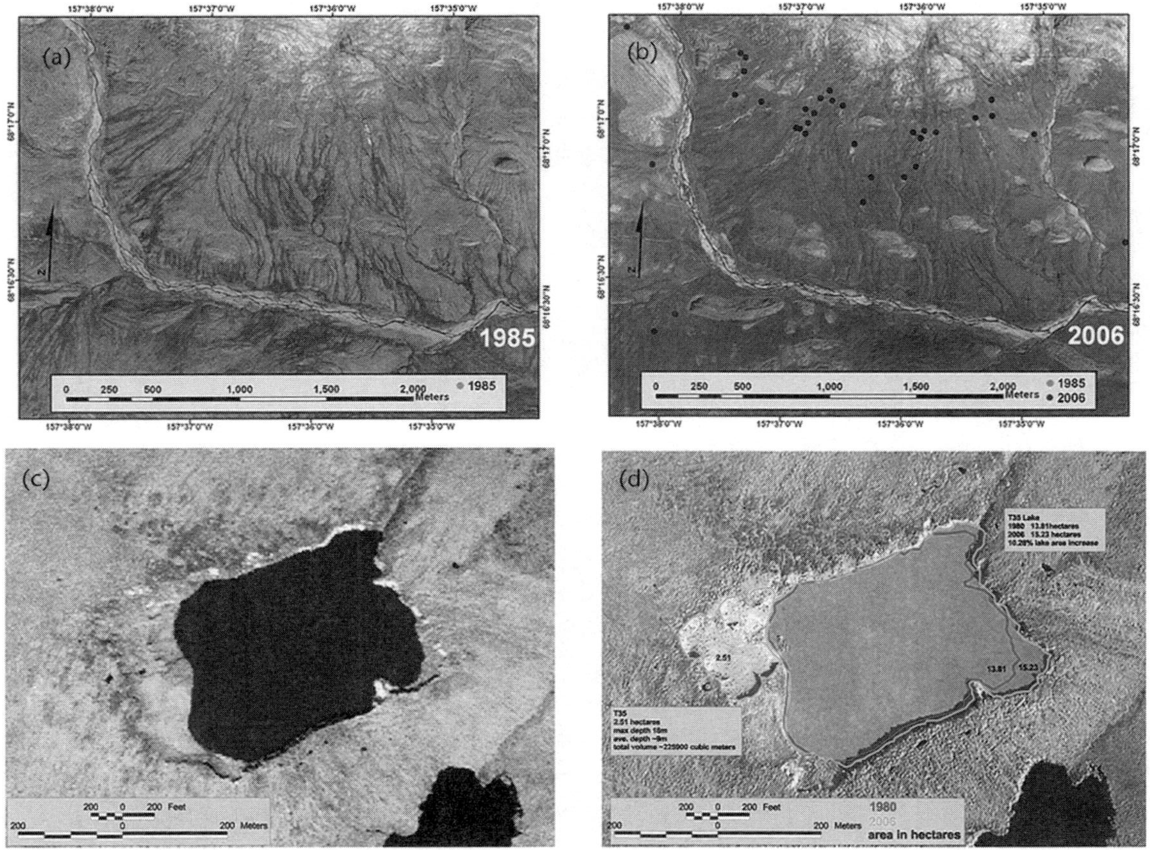

Figure 7.31 Thermokarst distribution near Feniak Lake: Changes in the past 25 years (A) 1985 Alaska High-Altitude Photography (AHAP) photo (color-infrared) showing 1 active thermokarst feature in 1985 (green dot) and 27 active thermokarst features in 2006 (yellow dots, see panel B). Note the dark red strips of riparian vegetation (mainly *Salix* spp.) along water tracks in this image. (B) 2006 aerial photo (real-color) showing the same location and features as in panel A. (C) 1980 AHAP photo of lake in process of expansion. (D) 2006 air photo with graphics showing 26 years of substantial change (1980–2006). From Gooseff et al. (2009), used with permission from the American Geophysical Union. [REFER COLOR IMAGE]

Figure 7.32 Toolik River looking north on July 28, 2003. Flow is toward top of the picture. The thermokarst is just off of the image to the left, on the small drainage that can be seen joining the Toolik River at the center right of the image. The change in brightness in the upper impacted reach of the Toolik River is due to sediment input from the small tributary. On this date the sediment impact could be seen from the air at least 20 km downstream. For scale, the main river at this point is about 5–10 m wide. Photo credit: W. B. Bowden.

gully. Initially the gully may expand due to localized permafrost thaw and further soil erosion. However, over time the side slopes of this form of thermokarst lessen and begin to revegetate, stabilizing the feature (Figure 7.33).

During the early rapid development and active expansion, sediment and nutrient export from thermokarst features can add significantly to the background load in receiving streams. In the case of the gully thermokarst on the Toolik River tributary, the total suspended solids (TSS) concentration in the stream below the thermokarst drainage confluence were two to three orders of magnitude higher than just above the thermokarst (Figure 7.34). It is instructive to consider the impact of this single gully thermokarst on the Toolik River in the context of the background levels of sediment generation and transport in the Kuparuk River, located 5 km to the west. Water and sediment yield from the Kuparuk River were measured from 1978 to 1980, including one of the wettest years on record (1980; Kriet et al. 1992). Discharge ranged from 0.3 to 28.3 $m^3\ s^{-1}$ and sediment concentrations ranged from 0.4 to 35 $mg\ L^{-1}$. Average water yields over the three years (late May to mid-August) were 15.7, 29.7, and 33.2 cm with specific sediment yields of 0.5, 1.1, and 3.5 metric tons

Figure 7.33 Images of the gully thermokarst on a tributary near the headwaters of the Toolik River shortly after it formed in 2003 and in 2004 and 2007. Note how the feature grows rapidly from 2003 to 2004 and how the edge features collapse and "soften" between 2004 and 2007. From Bowden et al. (2008), used with permission from the American Geophysical Union.

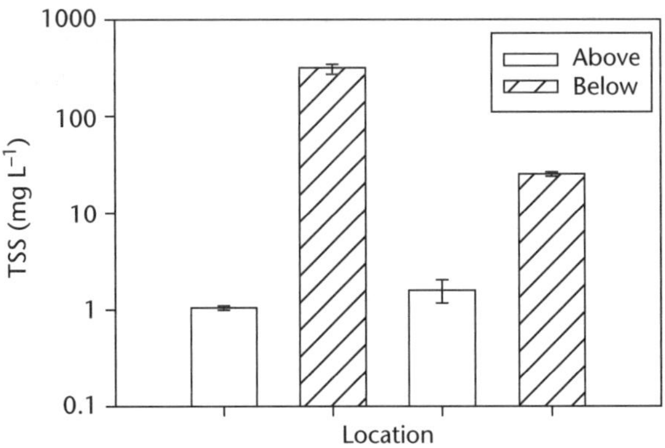

Figure 7.34 Total suspended sediment concentrations in the small drainage above and below the thermokarst near the Toolik River and above and below the point that this drainage entered the Toolik River. Note the log scale on the TSS concentration axis. The reported values are means and range of values for n = 3 thermokarst samples and n = 2 junction samples. From Bowden et al. (2008), used with permission from the American Geophysical Union.

km^{-2}. For comparison, an estimated 2000 m^3 of soil was displaced from the gully thermokarst on the Toolik River tributary in the first 2–3 years; assuming a density of 2 T m^{-3} (not unusual for a silty soil), it would have yielded 4000 tons of sediment (less minor volumes of organic matter and boulders). Thus, over a period of 2–3 years, this single thermokarst feature in a small (0.9 km^2) sub-watershed on the Toolik River delivered 18 times more sediment than would normally be delivered by the entire 143 km^2 upper Kuparuk River over the same time period.

Thermokarst features may also elevate nutrient concentrations in streams. Concentrations of ammonium and SRP measured above thermokarst features are typically low and consistent with long-term observations from other pristine streams and rivers in the area. However, ammonium and SRP concentrations downstream of thermokarsts (Figure 7.35) are usually high, sometimes orders of magnitude higher than the "reference" upstream concentrations (Bowden et al. 2008). Ammonium concentrations in these disturbed reaches are among the highest measured anywhere in the region, and phosphate concentrations are similar to the target enrichment levels used for the long-term experimental fertilization of the Kuparuk River. Thus, it is possible that if high levels of nutrients persist, there may be significant changes to the structure and function of thermokarst-impacted streams, such as has been documented in the experimentally fertilized reaches described earlier in this chapter.

In summary, we have observed that thermokarst hillslope failures in the foothills of the Brooks Range can substantially alter the loadings of sediments and nutrients to headwater stream ecosystems. These loadings can persist over many years, and even small thermokarst features can affect long reaches of headwater streams. Thus, even a low density of thermokarst features might have widespread impacts on arctic headwater stream ecosystems. As permafrost continues to warm and begins to thaw under the influences of continued warming in the Arctic, thermokarst terrain will likely become more extensive, and hillslope failures in foothills of the Brooks Range will become more numerous. The expectation is that between now and 2100 a considerable portion of the permafrost in the Arctic will thaw (Schuur et al. 2011); when it does, thermokarst terrain will likely increase.

Changes in Arctic Streams: Past and Future

General Context

Chapter 2 provides an overview of the past and probable future climate changes in the vicinity of Toolik Lake. At a wider regional level, mean annual temperatures in the Arctic increased 2°C–3°C over the last half century, primarily due to winter warming. It has become cloudier in the spring and summer but less cloudy in the winter. Albedo has decreased during all seasons, but most in fall and winter. Trends in precipitation are equivocal. However, the patterns near Toolik have been somewhat different. There has been no significant increase in the mean annual temperature. Similarly, there does not seem to be a significant trend in precipitation over time, though there is a strong increasing trend of the differences in precipitation from north to south (closer to the Brooks Range) that may be due to orographic effects. Significant trends in mean annual temperature and precipitation are potentially obscured by high interannual variation and in the case of snow (which affects runoff) by high measurement error. Nevertheless, there is other evidence of changing environmental conditions near Toolik over the last 20–30 years, including

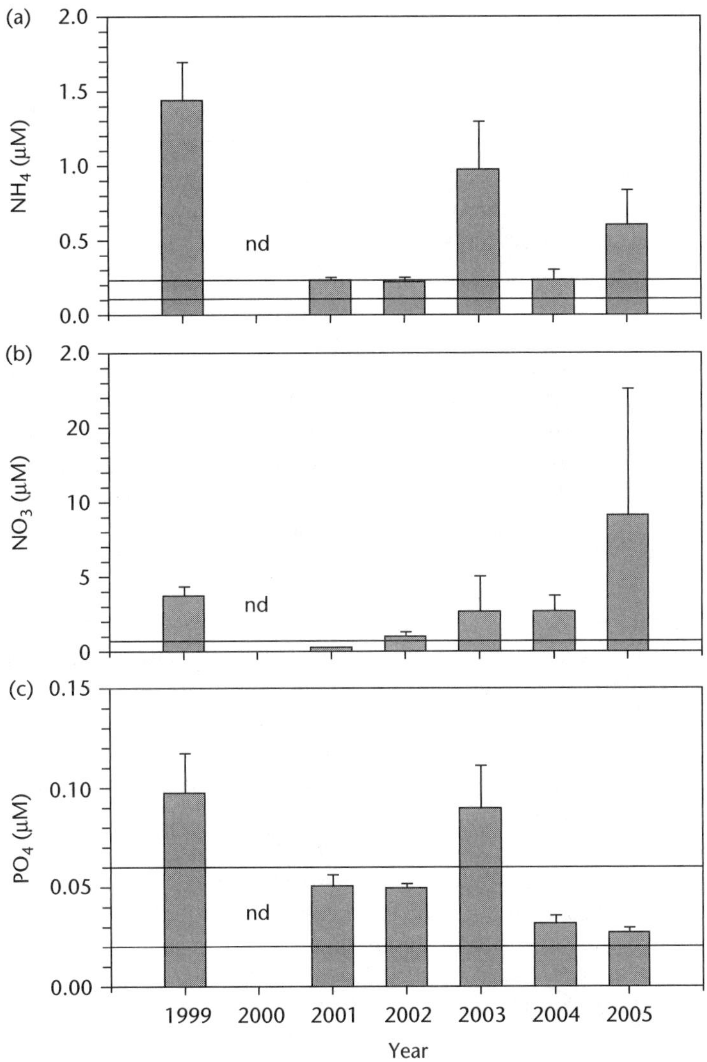

Figure 7.35 Concentrations of (A) ammonium, (B) nitrate, and (C) phosphate within the stream reaches most impacted by the thermokarst on Imnavait Creek. The bars are the mean values for the stations between 25 m and 350 m, by year. No samples were collected in 2000 (nd). The horizontal lines indicate the 95% confidence intervals for the mean reference values measured in Imnavait Creek. The 95% confidence interval for reference nitrate values included zero so there is only one (upper) boundary line. The error bars are ± 1 SE for n = 7, except for 2003, 2005 (n = 5) and 2004 (n = 4). From Bowden et al. (2008), used with permission from the American Geophysical Union.

warming trends in nearby permafrost boreholes (e.g., Romanovsky et al. 2010) and increased formation of thermokarst terrain (e.g., Bowden et al. 2008).

The descriptions in chapter 2 used a suite of five general circulation models to explore scenarios for the climate of the Toolik region in 2100. The results of these models suggest that it will be slightly warmer and considerably wetter than now, especially in the winter, with much more snow. These observations and projections provide a context for what has been observed in streams over the last 30 years and what might be expected in the future.

Discharge

The best record available for the long-term trends in daily river discharge is from the USGS gauge at the mouth of the Kuparuk west of Prudhoe Bay. The data strongly suggest that the seasonal hydrologic regime of the basin is shifting systematically in response to the regional warming of the North Slope, although this trend is less evident at the scale of the Toolik Lake region. Total annual runoff has not changed significantly over the period 1972–2005. The spring runoff typically occurred during the months of May and June (Figure 7.36), but prior to 1980 runoff in May was uncommon. From 1980 to 1990, there was a shift to significant runoff during May in about half the years. After 1990, runoff in May was common and dominated the spring runoff in a number of years. Overall the May runoff has increased while June runoff has declined, so that the sum of June plus May runoff has remained about the same. This shift in seasonality is expected to continue.

Interannual variability in average monthly runoff at the mouth of the Kuparuk River during the months of July and August was high and did not exhibit any particular trend. However in fall months there was strong evidence of an increasing trend in discharge (Figure 7.36). In September and October average monthly discharge nearly doubled over the period of record. Prior to 1990, discharge in November was typically low, but has increased after 1990. Runoff in December was nearly unheard of prior to 1990, but since 1990 there has been at least some flow in December in all but four years. Total runoff during the fall season was still a small part of the total annual runoff, but it is undeniable that fall discharge has been increasing in recent years.

Since the runoff total during the summer and fall was still small relative to that in the spring, the small, nonsignificant decline in spring runoff coupled with the substantial increase in summer and fall runoff resulted in no significant change in total annual runoff. However, the shift in runoff to earlier and later months has lengthened the flow season, and there has been a marked increase in summer high-flow events. Despite the increase in mean summer flow, the longer flow season with no increase in spring runoff, and potentially higher temperatures, leaves the river susceptible to droughts. These droughts have become more frequent in recent years at least in the upper Kuparuk, and have the potential to negatively impact the biota by drying the river channel, reducing available habitat areas, and impeding migration of the graying population to and from their overwintering sites.

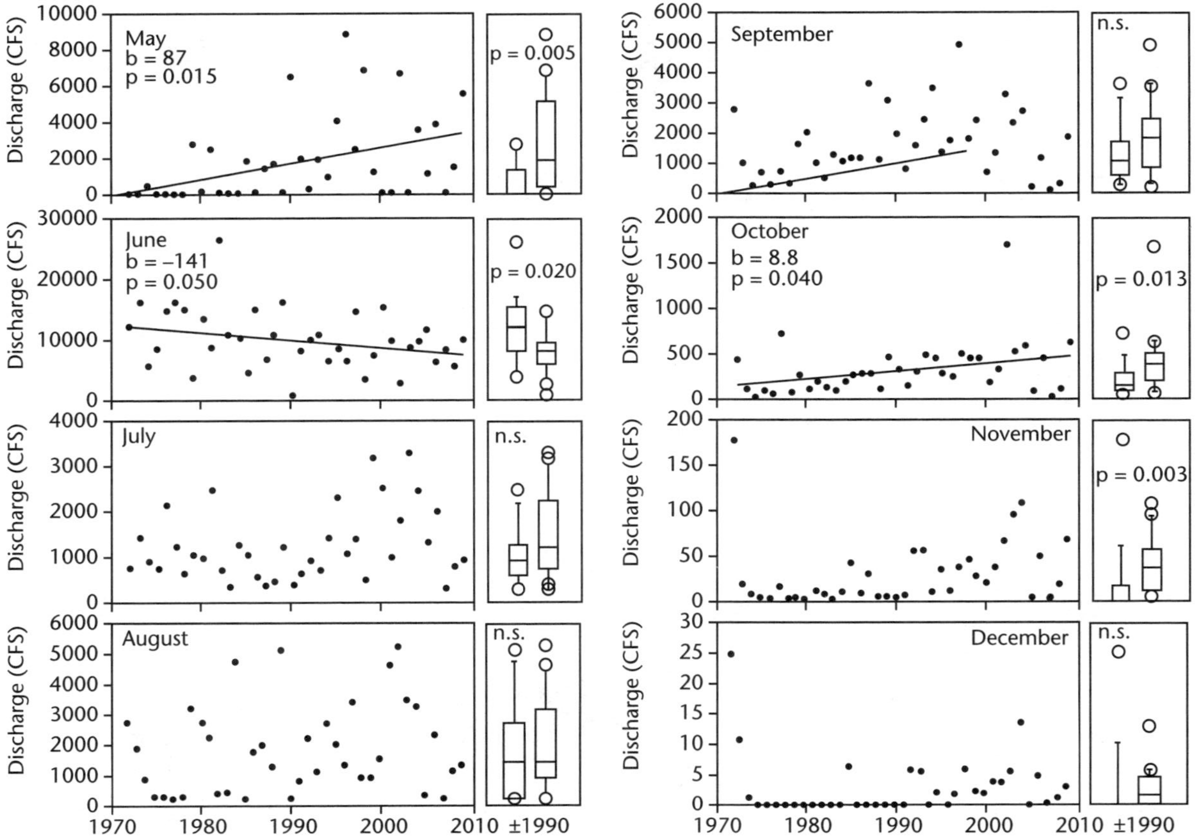

Figure 7.36 Mean monthly discharge from the USGS gauge (15896000) at the mouth of the Kuparuk River from 1972 to 2010. Significant trends are indicated with a regression line and statistics in the left panel of each monthly pair. The right panel of each pair shows a box and whisker comparison of discharges between the early period of the record (1972–1989) and the late period of the record (1990–2010).

Nutrient Fluxes

Warmer and wetter summer seasons may have increased the concentrations and export of inorganic nitrogen in the form of NO_3 from the upper Kuparuk catchment, especially in the last 15 years (Figure 7.37; McClelland et al. 2007). It is possible that phosphorus input to the river is increasing, but a measurable change in dissolved phosphorus is not expected because of strong limitations of and rapid uptake by the algal and bryophyte communities. However, if bryophytes become increasingly abundant under the new hydrologic and nutrient regimes, more phosphorus may be exported in particulate, detrital forms. Finally, as described in the previous section, there is the possibility that the development of thermokarst terrain will increase as climate continues to warm, which may increase sediment and nutrient loading to streams (chapter 1; Bowden et al. 2008).

Biota

It is clear from our long-term enrichment studies of the Kuparuk River that increased loading with even small amounts of phosphorus may have important impacts on the epilithic algal, bryophyte, and benthic macroinvertebrate community structure and production. Over the nearly three decades of study of the Kuparuk River, the substantial changes in the biota have been ones induced by the experimental enrichment rather than by other natural environmental changes. Although this may change in the future, the larger impacts on these stream ecosystems may be increases in variability of key driving variables such as discharge rather than increases in the intensity variables such as temperature or natural levels of nutrients.

As a migratory species with numerous habitat requirements, arctic grayling populations are especially susceptible to alterations in their environment. Thus, increased annual temperatures in the Arctic due to climate change could have multifaceted, direct and indirect consequences for grayling. More importantly, variations in the distribution of precipitation and resulting discharge throughout the year could alter or disturb migration and overwintering patterns for grayling and may be of particular importance to this species. Additionally, grayling may be affected by disturbances related to climate change that modify the nutrient, acidity, or sediment composition of the river.

Changes in climate expected for the Toolik region and the Arctic will have mostly negative repercussions for grayling. Increased water temperature within spawning, feeding and overwintering habitats will affect grayling directly by increasing metabolic costs and oxygen consumption. In spawning areas, higher temperatures may exceed the range in which eggs remain viable and could increase biological oxygen demand, which will stress eggs and fry during their under-gravel phase. Higher water temperatures in feeding areas may lead to an increased need for consumption by grayling to maintain adequate body condition. If the need for consumption is not coupled with an equivalent boost in prey production, the fish population will decline. The relationship between oxygen demand by fish and temperature may be of particular importance in overwintering habitats where crowded conditions exist in an oxygen-limited environment, and local extinctions may result. Unless arctic

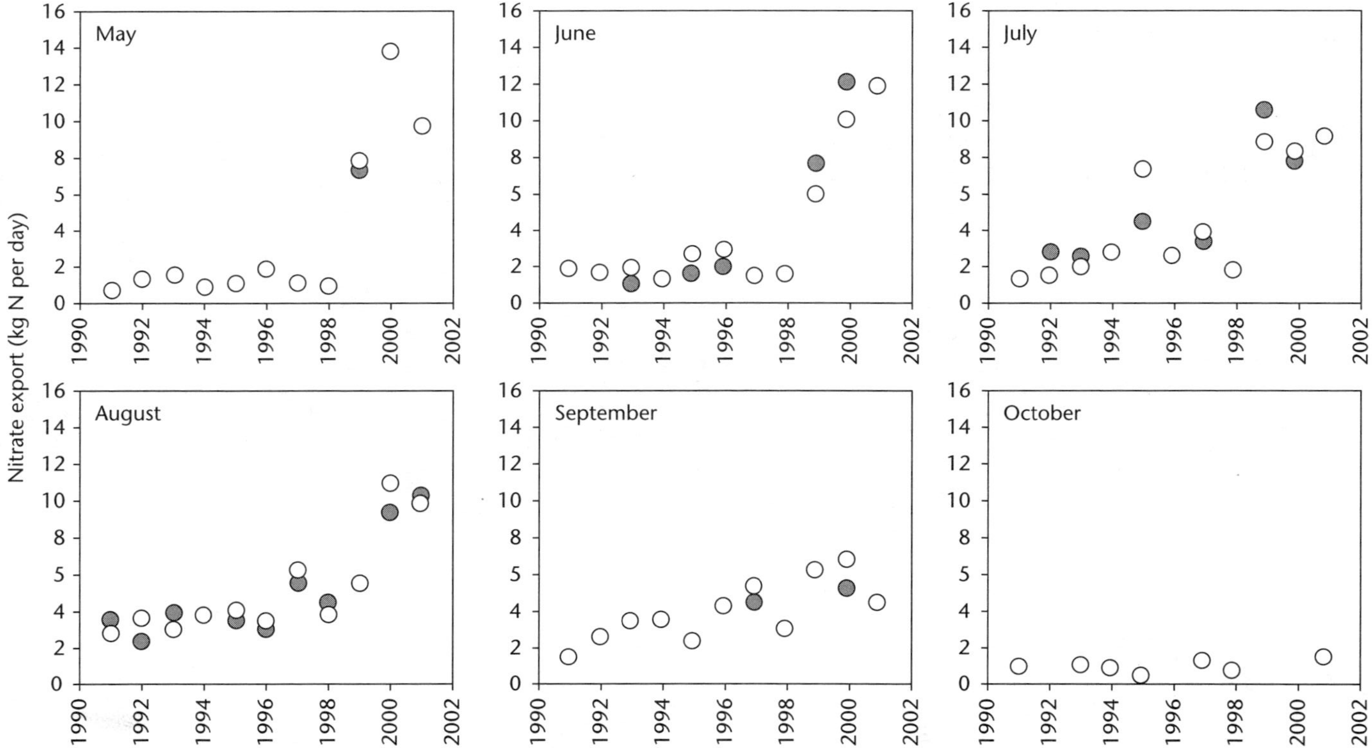

Figure 7.37 Nitrate export estimates in the upper Kuparuk River calculated using measured (filled circles) and simulated (open circles) river discharge to drive the program load estimator LOADEST from 1991 to 2001. Export estimates calculated using measured discharge are largely constrained to June, July, and August because of insufficient discharge data, whereas coupling of LOADEST with simulated discharge allows estimation of export over the complete flow period. From McClelland et al. (2007), used with permission from the American Geophysical Union.

grayling can evolve rapidly enough to compensate for the changes that are occurring, the rise in annual temperatures of the Arctic is not favorable for this species.

Variation in the severity of the arctic climate due to increased annual temperatures may cause alterations in species composition, biotic diversity, and food-web structure. Currently, due in part to the harshness of the climate, many tundra streams and rivers contain only a single species of fish, the arctic grayling. This may change as winter temperatures increase, becoming milder in response to global climate change. Milder winters may allow less-hardy fish species to increase their territory, forcing grayling to either shrink their range or compete with the invading species for food and space. Introductions of both fish and invertebrate species may modify prey species composition, affecting prey density, diversity, and overall food-web structure. Additionally, introduction of new species may establish undesirable parasites and diseases, previously unseen in grayling habitats, to which grayling may be susceptible. Arctic grayling are well adapted to the severity of the arctic climate, and alterations toward milder winters could have broad and varying implications for grayling populations on the North Slope of Alaska.

Though the total annual amount of precipitation in the Toolik region does not appear to have changed, changes in seasonal patterns of precipitation and timing of freezing and thawing are major factors controlling river discharge in arctic tundra streams and therefore main factors affecting population dynamics of arctic grayling within those streams. The ability of this species to survive stems in part from the manner in which different age classes respond to alternating discharge regimes. The pattern of alternating good and poor growth between age classes (described above) has created a biological safety net for the species, guaranteeing that one or the other age class will have a successful growing season regardless of river discharge during any particular year. However, this safety net does not take into account repeated and extended periods of drought, causing portions of the river to go dry, restricting habitat availability, and impeding migratory patterns.

Since the early 2000s the precipitation trend in the Kuparuk River basin has been toward dryer dry periods and wetter wet periods. In addition to water loss through increased evaporation as temperatures warm, in the future water may percolate downward as permafrost thaws, further reducing water levels in critical grayling habitats. The population in this area could come under considerable strain if portions of the river were to dry repeatedly year after year, not only impeding migration, but physically taxing individuals by restricting their livable habitat and forcing them into less than optimal thermal conditions. Grayling are highly territorial and, as documented, can actually lose weight when forced into situations of high population density. A grayling population may be able to rebound from a few consecutive years of drought due to the resilience of the species, but eventually the population would be pushed beyond recovery.

Although winter snowfall might increase in the future, the spring melt period is occurring earlier, which alters the dynamics of the spring-runoff process. In the past, snow that might persist in upland areas could supplement discharge when summer precipitation was lacking. Without this input, the river is more susceptible to large swings in precipitation-driven discharge. In recent years seasonal discharge has become flashier, with high and low extremes that have negative consequences for the grayling population, as discussed above.

In general, rainfall in the Kuparuk river basin has increased in the fall, which would appear to benefit grayling as they migrate to the headwater lake. However, this increased rainfall may not benefit grayling if it is poorly timed with freeze-up. Under normal conditions the headwater lake would already be quite cold as the migrating graying arrive. The low temperatures near freeze-up are a major factor that triggers torpor during overwintering and may help explain how such large numbers of fish can congregate within the relatively small volume of the Kuparuk River's headwater lakes. Our data suggest that freeze-up has been occurring later than previously recorded, which means that grayling may remain active in the headwater lake for a longer period of time and at higher temperatures and densities. Under these conditions the fish will become stressed and may expend considerable energy in territorial behavior that is needed for overwinter survival. Should the grayling enter the headwater lakes significantly earlier than freeze-up, a large portion of the population might become so stressed that they do not survive the winter.

As noted previously, runoff from thermokarst failures contains high concentrations of ammonium, phosphate, and sediments, which eventually find their way to lakes and streams in the vicinity. An increase in suspended sediments to the Kuparuk River could have broad implications for the grayling population. Grayling growth, spawning, and development of eggs and fry could all be influenced by increased sediment load to the river. Grayling are highly visual feeders, well adapted to clear-water streams. Increased turbidity due to suspended sediments could impact the ability of fish to locate and capture prey, negatively affecting summer growth. Should growth of individuals dip below the threshold necessary to survive the winter, the population could suffer greatly. Additionally, grayling choose gravelly shoals in which to spawn. If sediments cover prime spawning sites, it could lead to a reduction in suitable spawning habitat, thus increasing territorial behavior and inhibiting spawning success for the population. Furthermore, grayling eggs and fry remain in the substrate for an extended time, relying on water circulation within and through the substrate for respiration. Settling sediment could easily smother both eggs and fry. Although the full repercussions of increased sediment load from thermokarsts are unclear, the implications are not positive for arctic grayling.

In contrast, nutrient additions from thermokarst terrain may be beneficial to grayling because growth can be enhanced through bottom-up processes when nitrogen or phosphorus are added to the stream ecosystem. Both adult and young-of-the-year respond favorably to nutrient addition, and would probably respond similarly to nutrient release from thermokarst features if the release persisted and was not coupled with increased sediment load.

Based on our observations in the Kuparuk River and other nearby streams, the rapid climate changes that are projected to occur in the Arctic pose a serious threat to the survival of arctic grayling in the area. It takes an interconnected and varied landscape including clear, cold rivers, streams, and lakes to sustain a grayling population. With increased annual temperatures, alterations in patterns of precipitation, and disruptions to the landscape, all aspects of grayling habitat are threatened. Even if immediate steps are taken to curb global climate change, it may not be enough to ensure the continued abundance of this freshwater member of the salmonid family.

Vignette 7.2. Forecasting the Future for Fish, by Joseph Vallino

A 12-year data set on fish, insects, and primary producers in both reference and phosphate-fertilized reaches of the Kuparuk River was used to summarize key ecosystem processes and food-web dynamics with a general inverse modeling framework developed by Wan and Vallino (2005). The model employs a flow analysis–type approach, but uses first-order approximations between annual mean stocks and environmental drivers of temperature, discharge, and solar radiation. Consequently, the model is more robust and requires fewer observations than standard process-oriented models and can use observations that are difficult to incorporate into process models. Unlike standard inverse models, this model is capable of prediction provided sufficient data are available for model calibration and the environmental drivers are known.

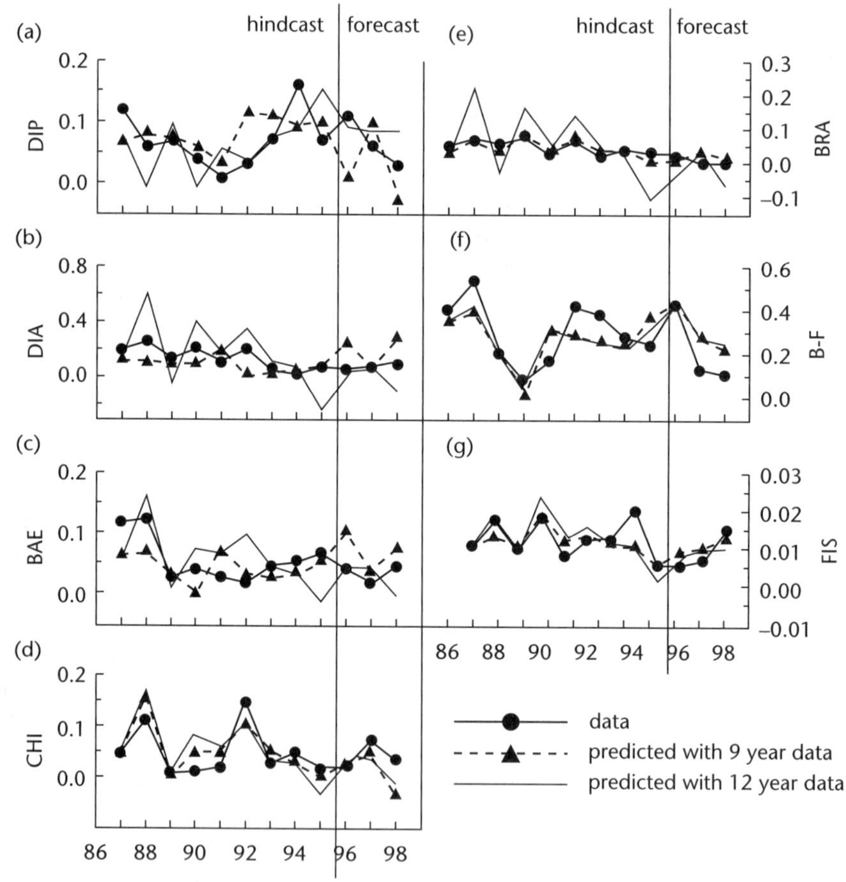

Figure 7.V2.1 Inverse model calibration, testing, and predictions. Comparison of model predictions (filled triangles) with observations (filled circles) in the Kuparuk River reference reach in which only the first nine years of data were used in model fit and the last three years are model forecasts. Solid line is the model result from a run calibrated using all data in a hindcast mode. Units are in mmol P m^{-2}. DIP = dissolved inorganic P, DIA = epilithic diatoms, BAE = *Baetis*, CHI = chironomids, BRA = *Brachycentrus*, B-F = black flies, and FIS = fish. From Wan et al. (2008), used with permission from *Ecological Modeling*.

When parameterized for the Kuparuk River data, the inverse model results show that the interannual variations of several components in the Kuparuk River ecosystem, including dissolved inorganic phosphate, chironomids, black flies, and arctic grayling, can be approximated accurately as a linear function of mean summer temperature, discharge, and solar radiation (Figure 7.V2.1; Wan et al. 2008). The model tracked the observations well for the fertilized reach, but had some difficulties fitting the control reach due to the low rates and concentrations observed there. In particular, the model indicates that changes in river habitat brought about by proliferation of the moss *Hygrohypnum* spp. in the P-fertilized reach caused a temporary shift in flowpaths supporting grayling from primary producer–based to detrital-based pathways. However, after moss establishment, primary producer flowpaths to grayling returned and detrital-based pathways weakened.

After validating the model against current conditions, it was used to explore how climate change in the Arctic might affect stream ecosystem food webs, and arctic grayling in particular. Precipitation, temperature, and solar flux were obtained from the Hadley GCM3 model for the North Slope (ACIA 2004; J. Walsh personal communication) for current (1987–1998) and future (2090–2099) conditions. Stream observations and GCM drivers from 1987 to 1997 were used to calibrate the model, and then model predictions were generated from 2090 to 2099 (Figure 7.V2.2). The predictions for solar radiation and for discharge were within the same range as for the contemporary (1987–1997) calibration period. However, temperatures averaged 1°C–2°C warmer for the 2090–2099 GCM projections. The model predictions indicate declines in epilithic diatoms, *Baetis* spp., and grayling production. On the other hand, chironomids are predicted to have high abundance and *Brachycentrus* sp. to maintain their abundance at levels similar to those in the calibration period.

There are, of course, uncertainties in the predictions of GCMs and the model developed for the Kuparuk River ecosystem. With these uncertainties in mind, the model was able to summarize the behavior of a complex food web and shows promise as a way to explore how changes in environmental drivers might affect food-web flows. The predictive power of the model could be improved by using data from more recent years to recalibrate the model and more rigorously test its predictive capability. In addition, the model does not yet include factors such the impacts of thermokarsts

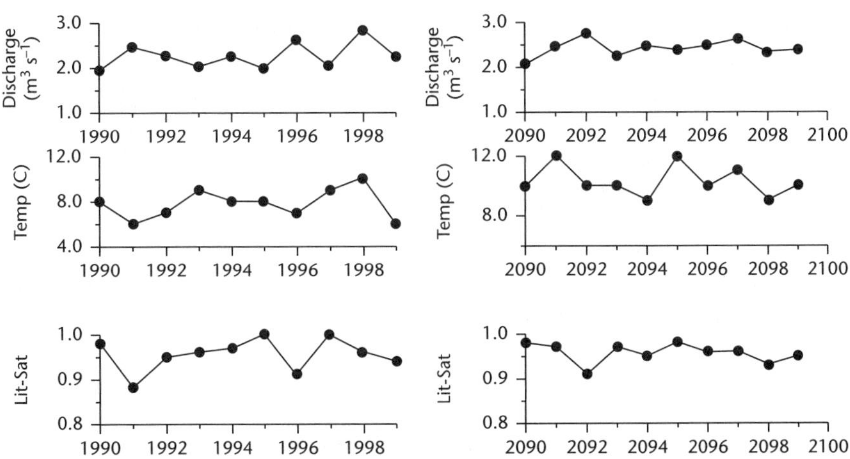

Figure 7.V2.2 Climate drivers during the period 1990–1999 and forecast for the period 2090–2100 according to an inverse linear model ("lit-sat" indicates PAR light saturation). From Wan et al. (2008), used with permission from *Ecological Modeling*.

on stream metabolism and the effects of hydrologic disruptions on grayling populations. These local and catchment-scale perturbations may have impacts on stream food webs that are as large or larger than the direct effects of flow, temperature, and light on stream reach processes and stream ecosystem structure.

Conclusions

The long-term research on arctic streams in the Toolik Lake region provides important insights into the structure and function of these ecosystems. These streams are physically diverse and temporally variable. However, they can be readily understood in a simple framework that includes easily measured variables such as discharge, freezing, and nutrient availability. Most of the stream types studied are oligotrophic and as a consequence have low rates of production that are strongly limited primarily by phosphorus and secondarily by nitrogen. Spring streams in the mountains and in the tundra environments are relatively rare but have higher levels of production and lower levels of physical disturbance that allow them to function as important refugia for some species.

Although rivers in the Toolik Lake region are relatively unproductive, the diversity of autotrophs and benthic macroinvertebrates in these streams is reasonably high. Diversity is lower than comparably sized temperate or tropical streams in near pristine condition. However, the biotic communities in arctic streams are nevertheless complex. The diversity is unevenly spread with many rare species and only a few dominant taxa. Although other fish species are important, especially in larger rivers, arctic grayling (*Thymallus arcticus*) is the dominant fish species in streams in the Toolik Lake region.

Climate change in the Toolik Lake region has been more subtle than in the Arctic as a whole. Nevertheless, there is evidence that the local climate is changing. If this persists, there will be a number of consequences for arctic streams (Figure 7.38). Changes in upland community structure and function will influence the structure and function of riparian and in-stream communities through changes in the fluxes of water, organic matter, and nutrients. These changes are likely to influence nutrient processing in streams and may especially affect the production of important greenhouse gases such as carbon dioxide, methane, and nitrous oxide. Alterations to riparian and in-stream ecosystems in the Arctic could have direct effects on humans, especially subsistence and recreational fishermen and hunters. The influences of climate change on fish communities were discussed at length above. Collectively these changes may strongly alter the total load and the nature of materials that are transported from the landscape to the Arctic Ocean. Increased loading of nutrients or usable carbon could enhance coastal productivity, especially in a warmer, future coastal environment that is ice-free. However, loading with less-usable materials or sediments from thermokarst disturbances might interfere with coastal production, with unforeseen and potentially negative impacts. Additional, integrated, land-river-coastal research will be needed to determine the magnitude and consequences of such impacts.

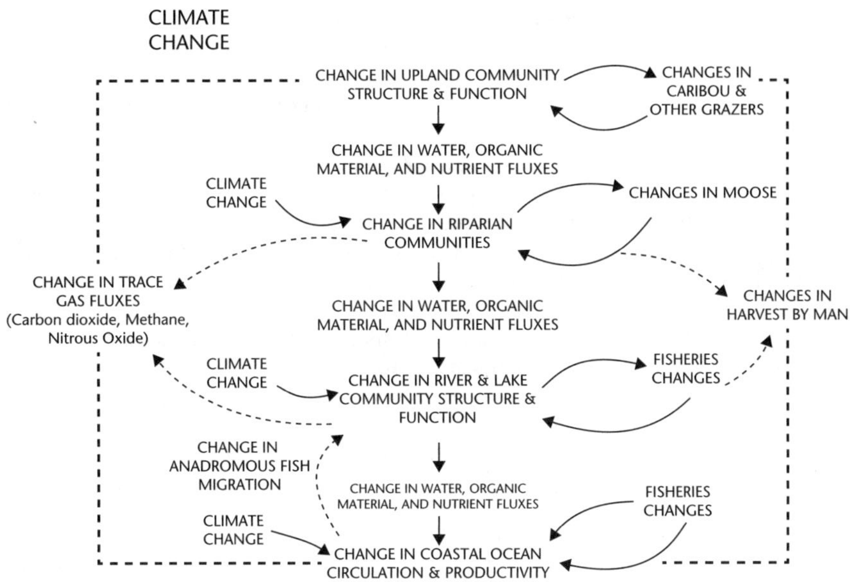

Figure 7.38 Interactions of arctic communities and biogeochemical cycles with global climate change.

References

ACIA. *Arctic Climate Impact Assessment*. 2004. Cambridge, Cambridge University Press.

Arscott, D. B., W. B. Bowden, and J. C. Finlay. 1998. Comparison of epilithic algal and bryophyte metabolism in an arctic tundra stream, Alaska. *Journal of the North American Benthological Society* 17:210–227.

Arscott, D. B., W. B. Bowden, and J. C. Finlay. 2000. Effects of desiccation and temperature/irradiance on the metabolism of two arctic stream bryophyte taxa. *Journal of the North American Benthological Society* 19:263–273.

Benke, A. C. 1993. Concepts and patterns of invertebrate production in running waters. *Verhandlungen Des Internationalen Verein Limnologie* 25:15–38.

Benstead, J. P., A. C. Green, L. A. Deegan, B. J. Peterson, K. Slavik, W. B. Bowden, and A. E. Hershey. 2007. Recovery of three arctic streams reaches from experimental nutrient enrichment. *Freshwater Biology* 52:1077–1089.

Benstead J. P., L. A. Deegan, B. J. Peterson, A. D. Huryn, W. B. Bowden, K. Suberkropp, K. M. Buzby, A. C. Green, and J. A. Vacca. 2005. Responses of a beaded Arctic stream to short-term N and P fertilisation. *Freshwater Biology* 50: 277–290.

Bott, T. L. 2007. Primary productivity and community respiration. Pages 663–690 in F. R. Hauer and G. A. Lamberti, editors. *Methods in stream ecology.* Burlington, MA: Academic Press, Inc.

Bowden, W. B., B. J. Peterson, J. C. Finlay, and J. Tucker. 1992. Epilithic chlorophyll a, photosynthesis, and respiration in control and fertilized reaches of a tundra stream. *Hydrobiologia* 240:121–131.

Bowden, W. B., J. C. Finlay, and P. E. Maloney. 1994. Long-term effects of PO_4 fertilization on the distribution of bryophytes in an arctic river. *Freshwater Biology* 32:445–454.

Bowden, W. B., M. N. Gooseff, A. Balser, A. Green, B. J. Peterson, and J. Bradford. 2008. Sediment and nutrient delivery from thermokarst features in the foothills of the

North Slope, Alaska: Potential impacts on headwater stream ecosystems. *Journal of Geophysical Research* 113:G02026. doi:10.1029/2007JG000470, 2008.

Bradford, J. H., J. P. McNamara, W. B. Bowden, and M. N. Gooseff. 2005. Imaging depth-of-thaw beneath arctic streams using ground-penetrating radar. *Hydrological Processes* 19:2689–2699.

Brosten, T., J. H. Bradford, J. P. McNamara, J. P. Zarnetske, M. N. Gooseff, and W. B. Bowden. 2006. Temporal thaw depth beneath two arctic stream types using ground-penetrating radar. *Permafrost and Periglacial Processes* 17:341–355.

Buzby, K. M., and L. A. Deegan. 2004. Long-term survival of adult Arctic grayling (Thymallus arcticus) in the Kuparuk River, Alaska. *Canadian Journal of Fisheries and Aquatic Sciences* 61:1954–1964.

Cappelletti, C. 2006. Photosynthesis and respiration in an arctic tundra river: modification and application of the whole-stream metabolism method and the influence of physical, biological, and chemical variables. M.S. thesis, University of Vermont, Burlington.

Craig, P. C., and P. J. McCart. 1975. Classification of stream types in Beaufort Sea drainages between Prudhoe Bay, Alaska and the Mackenzie Delta, NWT. *Arctic and Alpine Research* 7:183–198.

Crump, B. C., H. E. Adams, J. E. Hobbie, and G. W. Kling. 2007. Biogeography of bacterioplankton in lakes and streams of an arctic tundra catchment. *Ecology* 88:1365–1378.

Deegan, L. A., and B. J. Peterson. 1992. Whole-river fertilization stimulates fish production in an arctic tundra river. *Canadian Journal of Fisheries and Aquatic Sciences* 49:1890–1901.

Deegan, L. A., B. J. Peterson, H. Golden, C. C. MacIvor, and M. C. Miller. 1997. Effects of fish density and river fertilization on algal standing stocks, invertebrate communities and fish production in an arctic river. *Journal of the Canadian Fisheries and Aquatic Sciences* 54:269–283.

Deegan, L. A., H. E. Golden, C. J. Harvey, and B. J. Peterson. 1999. Influence of environmental variability on the growth of age-0 and adult arctic grayling. *Transactions of the American Fisheries Society* 128:1163–1175.

Edwardson, K. J., W. B. Bowden, C. Dahm, and J. Morrice. 2003. The hydraulic characteristics and geochemistry of hyporheic and parafluvial zones in arctic tundra streams, North Slope, Alaska. *Advances in Water Resources* 26:907–923.

Edwardson, K. J. 1997. Characterizations of Hyporheic Influences on the hydrology and biogeochemistry of arctic tundra streams. M.S. thesis, University of New Hampshire, Durham.

Fellows, C. S., H. M. Valett, and C. N. Dahm. 2001. Whole-stream metabolism in two montane streams: Contribution of the hyporheic zone. *Limnology and Oceanography* 46:523–531.

Gooseff, M. N., A. Balser, W. B. Bowden, and J. B. Jones. 2009. Effects of hillslope thermokarst in northern Alaska. *Eos Transactions: American Geophysical Union* 90:29–30.

Greenwald, M. J., W. B. Bowden, M. N. Gooseff, J. P. Zarnetske, J. P. McNamara, J. H. Bradford, and T. R. Brosten. 2008. Hyporheic exchange and water chemistry of two arctic tundra streams of contrasting geomorphology. *Journal of Geophysical Research* 113:G02029. doi:10.1029/2007JG000549.

Hamilton, T. D. 1978. *Surficial geologic map of the Philip Smith Mountains quadrangle, Alaska*. Miscellaneous field studies map MF-879-A, scale 1:250,000. U.S. Geological Survey.

Harvey, C. J., B. J. Peterson, W. B. Bowden, L. A. Deegan, J. C. Finlay, A. E. Hershey, and M. C. Miller. 1997. Organic matter dynamics in the Kuparuk River, a tundra river in Alaska, USA. *Journal of the North American Benthological Society* 16:18–22.

Harvey C. J., B. J. Peterson, W. B. Bowden, A. E. Hershey, M. C. Miller, L. A. Deegan, and J. C. Finlay. 1998. Biological responses to fertilization of Oksrukuyik Creek, a tundra stream. *Journal of the North American Benthological Society* 17:190–209.

Harvey, J. W., and B. J. Wagner. 2000. Quantifying hydrologic interactions between streams and their subsurface hyporheic zones. Pages 3–44 in J. B. Jones and P. J. Mulholland, editors. *Streams and ground waters.* Academic Press, San Diego, CA.

Hershey, A. E., W. B. Bowden, L. A. Deegan, J. E. Hobbie, B. J. Peterson, G. W. Kipphut, G. W. Kling, M. A. Lock, R. W. Merritt, M. C. Miller, J. R. Vestal, and J. A. Schuldt. 1997. The Kuparuk River: A long-term study of biological and chemical processes in an arctic river. Pages 107–129 in A. M. Milner and M. W. Oswood, editors, *Freshwaters of Alaska*. New York: Springer-Verlag.

Hershey, A. E., A. L. Hiltner, M. A. J. Hullar, M. C. Miller, J. R. Vestal, M. A. Lock, S. Rundle, and B. J. Peterson. 1988. Nutrient influence on a stream grazer: *Orthocladius* microcommunities respond to nutrient input. *Ecology* 69:1383–1392.

Hershey, A. E., R. W. Merritt, and M. C. Miller. 1995. Insect diversity, life history, and trophic dynamics in arctic streams, with particular emphasis on black flies (Diptera: Simuliidae). Pages 283–295 in F. S. Chapin, III, and C. Korner, editors, *Arctic and alpine biodiversity: Patterns, causes and ecosystems consequences.* Ecological studies 113. Berlin: Springer-Verlag.

Hiltner, A. L. 1985. Response of the black fly species (Diptera: Simuliidae) to phosphorus enrichment of an arctic tundra stream. M.S. thesis, University of Wisconsin, Madison.

Hopkins, D. M., T. N. V. Karlstrom, R. F. Black, J. R. Williams, T. L. Pewe, A. T. Fernald, and E. H. Muller. 1955. Permafrost and ground water in Alaska. *U.S. Geological Survey Professional Paper* 264F:113–146.

Huryn, A. D. 1998. Ecosystem-level evidence for top-down and bottom-up control of production in a grassland stream system. *Oecologia* 115:173–183.

Huryn, A. D., K. A. Slavik, R. L. Lowe, D. S. Anderson, S. M. Parker, and B. J. Peterson. 2005. Landscape heterogeneity and the biodiversity of arctic stream communities: A habitat template analysis. *Canadian Journal of Fisheries and Aquatic Sciences* 62:1905–1919.

Kane, D. L., L. D. Hinzman, C. S. Benson, and K. R. Everett. 1989. Hydrology of Imnavait Creek, an arctic watershed. *Holarctic Ecology* 12:262–269.

Kane, D. L., L. D. Hinzman, R. E. Gieck, J. P. McNamara, E. K. Youcha, and J. A. Oatley. 2008. Contrasting extreme runoff events in areas of continuous permafrost, arctic Alaska. *Hydrology Research* 39:287–298.

Kasahara, T., and S. M. Wondzell. 2003. Geomorphic controls on hyporheic exchange flow in mountain streams. *Water Resources Research* 39:1005. doi: 1010.1029/2002wr001386.

Kling, G. W., W. J. O'Brien, M. C. Miller, and A. E. Hershey. 1992. The biogeochemistry and zoogeography of lakes and rivers in arctic Alaska. *Hydrobiologia* 240:1–14.

Kriet, K., B. J. Peterson, and T. L. Corliss. 1992. Water and sediment export of the Upper Kuparuk River drainage of the North Slope of Alaska. *Hydrobiologia* 240:71–81.

Larouche, J. R., W. B. Bowden, R. Giordano, M. B. Flinn, and B. C. Crump. 2012. Microbial biogeography of streams in the Feniak Lake region, Noatak National Preserve, Alaska: Exploring influences of lithology and habitat. *Frontiers in Microbiology* 3:article 309. doi:10.3389/fmicb.2012.00309.

Lee, J. O., and A. E. Hershey. 2000. Effects of aquatic bryophytes and long-term fertilization on arctic stream insects. *Journal of the North American Benthological Society* 19:697–708.

Lock, M. A., T. E. Ford, M. A. J. Hullar, M. Kaufman, J. R. Vestal, G. S. Volk, and R. M. Ventullo. 1990. Phosphorous limitation in an arctic river biofilm: A whole ecosystem experiment. *Water Research* 24:1545–1549.

McClelland, J. W., M. Stieglitz, F. Pan, R. M. Holmes, and B. J. Peterson. 2007. Recent changes in nitrate and dissolved organic carbon export from the upper Kuparuk River, North Slope, Alaska. *Journal of Geophysical Research—Biogeosciences* 112:G04S60. doi:10.1029/2006JG000371.

McNamara, J. P., D. L. Kane, and L. D. Hinzman. 1998. An analysis of streamflow hydrology in the Kuparuk River Basin, Arctic Alaska: A nested watershed approach. *Journal of Hydrology* 206:39–57.

Niemi, G. J., P. DeVore, N. Detenbeck, D. Taylor, A. Lima, J. Pastor, J. D. Yount, and R. J. Naiman. 1990. Overview of case studies on recovery of aquatic systems from disturbance. *Environmental Management* 14:571–587.

O'Brien, W. J., and J. J. Showalter. 1993. Effects of current velocity and suspended debris on the drift feeding of arctic grayling. *Transactions of the American Fisheries Society* 122: 609–615.

Oswood, M. W. 1989. Community structure of benthic invertebrates in interior Alaskan (USA) streams and rivers. *Hydrobiologia* 172:97–110.

Oswood, M. W, J. G. Irons, III, and D. M. Schell. 1996. Dynamics of dissolved and particulate carbon in a tundra stream in arctic Alaska. Pages 275–287 in J. F. Reynolds and J. D. Tenhunen, editors, *Landscape function and disturbance in arctic tundra*. Ecological Studies 120. Berlin: Springer.

Parker, S. M. 2008. Effects of natural disturbance on benthic communities of Arctic headwater streams, North Slope, Alaska, U.S.A. Ph.D. dissertation, University of Alabama, Tuscaloosa.

Parker S. M., and A. D. Huryn. 2006. Food web structure and function in two artic streams with contrasting disturbance regimes. *Freshwater Biology* 51:1249–1263.

Parker S. M., and A. D. Huryn. 2011. Effects of natural disturbance on stream communities: A habitat template analysis of arctic headwater streams. *Freshwater Biology* 56:1342–1357.

Parkinson, D., J. C. Philippart, and E. Baras. 1999. A preliminary investigation of spawning migrations of grayling in a small stream as determined by radio-tracking. *Journal of Fish Biology* 55:172–182.

Peterson, B. J., J. E. Hobbie, and T. L. Corliss. 1986. Carbon flow in a tundra stream ecosystem. *Canadian Journal of Fisheries and Aquatic Sciences* 43:1259–1270.

Peterson, B. J., J. E. Hobbie, A. E. Hershey, M. A. Lock, T. E. Ford, J. R. Vestal, V. L. McKinley, M. A. Hullar, M. C. Miller, R. M. Ventullo, and S. S. Volk. 1985. Transformation of a tundra river from heterotrophy to autotrophy by addition of phosphorus. *Science* 229:1383–1386.

Peterson, B. J., L. Deegan, J. Helfrich, J. E. Hobbie, M. Hullar, B. Moller, T. E. Ford, A. E. Hershey, A. Hiltner, G. Kipphut, M. A. Lock, D. M. Fiebig, V. McKinley, M. C. Miller, R. Westal, R. Ventullo, and G. Volk. 1993. Biological responses of a tundra river to fertilization. *Ecology* 74:653–672.

Poff, N. L. 1997. Landscape filters and species traits: Towards mechanistic understanding and prediction in stream ecology. *Journal of the North American Benthological Society* 16:391–409.

Richardson, E. S., J. D. Reist, and C. K. Minns. 2001. Life history characteristics of freshwater fishes occurring in the Northwest Territories and Nunavut, with major emphasis on lake habitat requirements. Can. MS Rpt. Fish. Aquat. Sci. 2569: vii+146p.

Romanovsky, V. E., S. L. Smith, and H. H. Christiansen. 2010. Permafrost thermal state in the polar northern hemisphere during the International Polar Year 2007–2009: A synthesis. *Permafrost and Periglacial Processes* 21:106–116.

Scarsbrook M. R., and C. R Townsend. 1993. Stream community structure in relation to spatial and temporal variation: A habitat template study of two contrasting New Zealand streams. *Freshwater Biology* 29:395–410.

Schuur, E. A. G., B. Abbott, and the Permafrost Carbon Network. 2011. High risk of permafrost thaw. *Nature* 480:32–33.

Slavik, K., B. J. Peterson, L. A. Deegan, W. B Bowden, A. E. Hershey, and J. E. Hobbie. 2004. Long-term responses of the Kuparuk River ecosystem to phosphorus fertilization. *Ecology* 85:939–954.

Stream Bryophyte Group. 1999. Roles of bryophytes in stream ecosystems. *Journal of the North American Benthological Society* 18:151–184.

Suberkropp, K. 2001. Fungal growth, production, and sporulation during leaf decomposition in two streams. *Applied and Environmental Microbiology* 67:5063–5068.

van Everdingen, R. Editor 1998, revised May 2005. *Multi-language glossary of permafrost and related groundice terms.* Boulder, CO: National Snow and Ice Data Center/World Data Center for Glaciology. http://nsidc.org/fgdc/glossary/

Vannote, R. L., G. W. Minshall, K. W. Cummins, J. R. Sedell, and C. E. Cushing. 1980. River continuum concept. *Canadian Journal of Fisheries and Aquatic Sciences* 37:130–137.

Wan, Z., and J. Vallino. 2005. An inverse ecosystem model of year-to-year variations with first order approximation to the annual mean fluxes. *Ecological Modeling* 187:369–388.

Wan, Z., J. J. Vallino, and B. J. Peterson. 2008. Study of the inter-annual food web dynamics in the Kuparuk River with a first order approximation inverse model. *Ecological Modeling* 211:97–112.

West, R. L., M.W. Smith, W. E. Barber, J. B. Reynolds, and H. Hop. 1992. Autumn migration and overwintering of arctic grayling in coastal streams of the Arctic National Wildlife Refuge, Alaska. *Transactions of the American Fisheries Society* 121:709–715.

Wipfli, M. S. 1997. Terrestrial invertebrates as salmonid prey and nitrogen sources in streams: Contrasting old-growth and young-growth riparian forests in southeastern Alaska, USA. *Canadian Journal of Fisheries and Aquatic Sciences* 54:1259–1269.

Wollheim, W. M., B. J. Peterson, L. A. Deegan, M. Bahr, J. E. Hobbie, D. Jones, A. E. Hershey, G. W. Kling, and M. C. Miller. 1999. A coupled field and modeling approach for the analysis of nitrogen cycling in streams. *Journal of the North American Benthological Society* 18:199–221.

Yoshikawa, K., L. D. Hinzman, and D. L. Kane. 2007. Spring and aufeis (icing) hydrology in Brooks Range Alaska. *Journal of Geophysical Research—Biogeosciences* 11:G04S43.

Zarnetske, J. P., M. N. Gooseff, T. R. Brosten, J. H. Bradford, J. P. McNamara, and W. B. Bowden. 2007. Transient storage as a function of geomorphology, discharge, and permafrost active layer conditions in arctic tundra streams. *Water Resources Research.* 35:L02501 doi:10.1029/2005WR004816.

Zarnetske, J. P., M. N. Gooseff, W. B. Bowden, M. J. Greenwald, T. R. Brosten, J. H. Bradford, and J. P. McNamara. 2008. Influence of morphology and permafrost dynamics on hyporheic exchange in arctic headwater streams under warming climate conditions. *Geophysical Research Letters.* doi:10.1029/2007GL032049.

Ziemann, P. J. 1986. Energetics of arctic Alaskan fishes: Carbon isotope evidence. M.S. thesis, University of Alaska, Fairbanks.

8

The Response of Lakes Near the Arctic LTER to Environmental Change

Chris Luecke
Anne E. Giblin
Neil D. Bettez
Greta A. Burkart
Byron C. Crump
Mary Anne Evans
Gretchen Gettel
Sally MacIntyre
W. John O'Brien
Parke A. Rublee
George W. Kling

The Natural Condition of the Lakes

The density and variety of lakes near the Toolik Field Station (Toolik) is large, with a wide range of lake type, water chemistry, and physical mixing characteristics. Toolik Lake lies on the North Slope of Alaska in the foothills of the Brooks Range. Lakes in this region were mostly formed as kettle lakes from melting ice blocks or from moraines at the end of several different Pleistocene glaciations. One distinguishing feature of all arctic lakes is the long period of ice cover and complete darkness in winter months, coupled with several months of 24-hour light during the summer. These lakes tend to be very nutrient poor, to have low biological productivity, and to have limited numbers of species especially at the higher trophic levels (Hobbie 1984). It is important to recognize, however, that the lakes near Toolik fall at the "low-arctic" end of the spectrum of arctic lakes and are different in many ways from lakes at the "high-arctic" end. For example, high-arctic lakes lack cladocera and may have only one species of fish.

Studies of the physical, chemical, and biological conditions of Toolik Lake and nearby lakes and ponds began in 1975. This long time-series of limnological conditions provides a basis for assessing the effects of environmental change expected in northern Alaska (chapter 2). In addition, our understanding of the biogeochemical and ecological processes of the lakes has been greatly informed by an extensive series of manipulative experiments that provide insight into how these systems may respond to changing environmental conditions. These experiments include multi-year fertilization treatments conducted in large mesocosms in Toolik, whole-lake fertilization in lakes N2, N1, E5, and E6, and a lake trout removal experiment in Lake NE12.

Landscape Settings of Lakes in the Toolik Lake Region

North of the Brooks Range, thick permafrost effectively prevents deep groundwater flow and increases the surface hydrological connectivity between water tracks on the hillslopes and downstream water bodies. This permafrost barrier limits transient storage and reduces residence time of water within catchments. Underneath streams and lakes lies an unfrozen thaw bulb (talik) that varies in thickness but may extend for many meters beneath deep lakes such as Toolik. Due to thickening of this thaw bulb, or to thawing of permafrost dams, lakes may drain; such drainage has recently affected thousands of lakes in the Arctic (Smith et al. 2005). Chapters 2 and 7 document the impact of permafrost on water flow in these arctic landscapes in more detail.

Lakes and ponds near Toolik were formed by glaciers over the past ~780,000 years. Extensive ice-deposits from five glacial advances resulted in the formation of glacial kettle and moraine lakes embedded in a complex landscape (chapter 1; Figure 3.3). Lakes in this region lie on surfaces resulting from glaciations of approximately 12–25 ka, 60–100 ka, and 250–300 ka in age (Hamilton 2003; chapter 3). Effects of erosion and solifluction are particularly apparent on surfaces of older glaciations (>60 ka), where these processes have led to the disappearance of some stream channels and a decreased depth and number of glacial kettle lakes. In contrast, landscapes affected by more recent glacial advances (<12 ka) are characterized by well-defined stream channels connecting kettle lakes.

The glacial history affects the distribution and the size of lakes in the region. On the older glacial surfaces (watersheds for lakes E5 and E6, Figure 8.1), lakes are less abundant, have smaller surface areas, and are not as deep as lakes on more recently glaciated landscapes (watersheds for Toolik, the I-series lakes, and the NE lakes, Figure 8.1). The oldest glacial surfaces (>250 ka) have less than 0.2% of the landscape area covered by lakes, compared to greater than 3% of the surface area covered by water on the newly glaciated regions (Burkart 2007). Mean surface area of individual lakes (4.2 ha) is much greater on young surfaces compared to a mean lake surface area of 0.9 ha on the oldest landscapes.

Toolik Lake, and most of the other large lakes in the region, are multi-basin kettle lakes formed ~12 ka ago (Hamilton and Porter 1975) from several ice blocks left behind as the glaciers retreated. Consequently the lake has a series of shoals and a resulting complex morphometry (Figure 8.2). This morphometry strongly influences the pathways of water flow inside the lake from runoff during spring snowmelt and during summer rainstorms, and may influence the distribution of chemical and biological species.

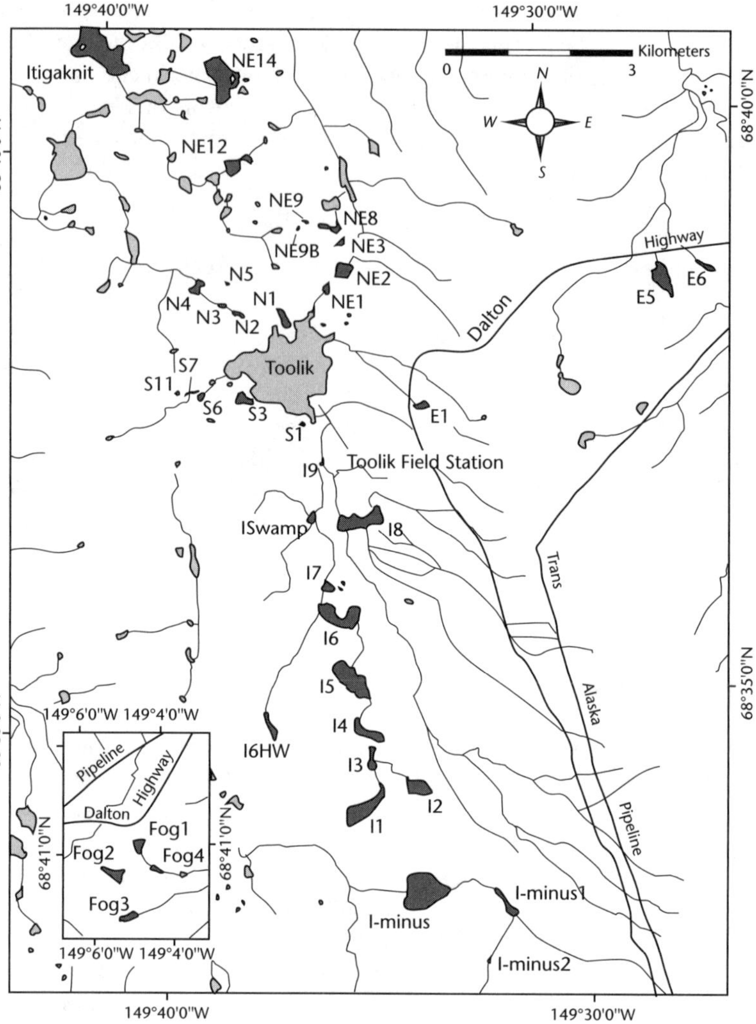

Figure 8.1 Map of the lakes commonly sampled in the Arctic LTER program. Lakes listed in Table 8.2 that are used for comparisons of effects of lake size and glacial history are shown. Lakes in the insert at the lower left (Fog1–4) are located about 20 km east of Toolik Lake along the Dalton Highway.

Physical Properties

Ice Cover and Runoff

During winter months air temperatures average less than –10°C near Toolik. The ice cover on lakes typically reaches 1.5 m in thickness but can reach 1.8 m when snow cover is sparse. When small ponds freeze to the bottom, the temperature in

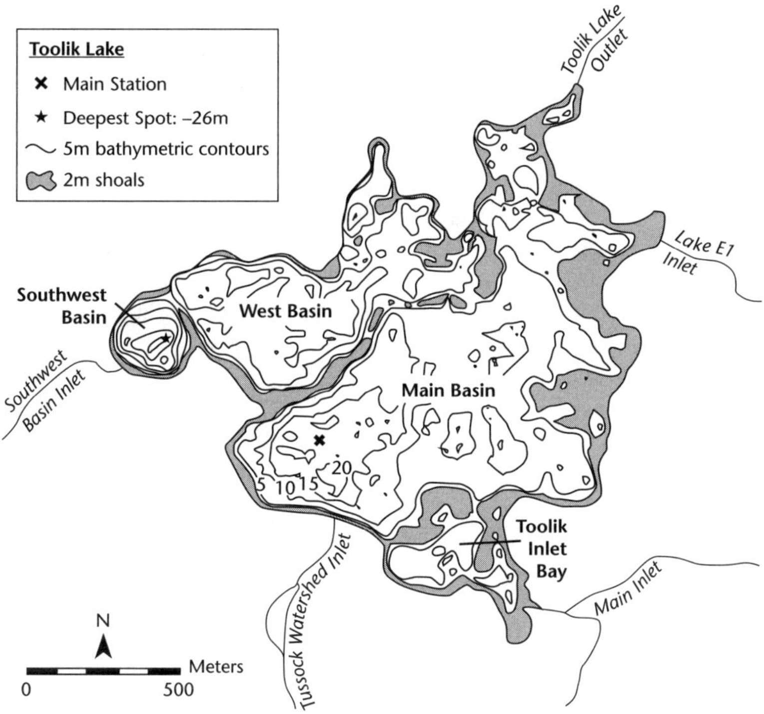

Figure 8.2 The bathymetry of Toolik Lake is shown with 5-m depth contours. The gray regions depict shoals that separate different kettle basins. The main sampling station is shown with an X and the deepest location with a star. The four dominant inlet streams and one major outlet stream are shown.

the sediments is rarely less than –5°C; when some water remains the temperature is usually 1°C–2°C.

During May, continuous sunlight and warming air temperatures begin to melt snow; as the snow becomes saturated with water, runoff begins into the still ice-covered lakes. The runoff peaks within days after initiation and soon penetrates the ice sheet at the inlet and flows into the lake and under the ice (Whalen and Cornwell 1985). At the same time, snow cover of lakes and ponds melts and sufficient light reaches the water beneath the ice to support photosynthesis of phytoplankton and benthic algae. During the previous decade (2000–2009), the date of the last presence of ice on Toolik Lake has ranged from June 8 to July 1. Ice completely melts from Toolik and other larger lakes in the region ~1–3 weeks after the ice is gone from medium- and small-sized lakes. Complete ice cover on Toolik Lake was reestablished during the fall between September 22 and October 10. The smaller lakes tend to become ice covered a week or two before Toolik Lake.

Light

Lakes in the foothills region tend to be highly stained with dissolved organic matter, which limits light penetration especially in the ultraviolet (UV) range (e.g., Scully and Lean 1994). Toolik Lake, for example, contains ~440 µM DOC and the range in lakes near Toolik is from ~290 µM (lake S6) to ~940 µM (lake N5). While there are a few relatively clear lakes, Secchi depths (depth of disappearance of a Secchi disc) in most lakes are less than 4 m (Table 8.1), and vertical light extinction coefficients for PAR (400–700 nm) are greater than 0.6 m^{-1}. The euphotic zone, where light is sufficient for net phytoplankton growth (~1% of surface irradiance), tends to be around 6–7 m in Toolik Lake. Light penetration tends to be higher in lakes on younger surfaces, but there is a great deal of variation. Extinction coefficients can increase to 1.4 m^{-1} in lakes undergoing nutrient addition. Due to the generally high color of the lakes, the penetration of UV light is relatively low. For example, in Toolik Lake in 2010 the extinction coefficients for light of wavelengths 395, 380, 340, 320, and 305 nm were 3.78 m^{-1} (± 0.01 S.D.), 5.02 (±0.18), 9.65 (±0.27), 13.2 (±0.5), and 17.2 (±1.7), respectively. These values are on the higher end of values reported by Morris et al. (1995) for 65 lakes in North and South America. The values these authors give for Toolik Lake from June 1993 are within 10% of measurements made in 2000 and again in 2010, indicating that no substantial changes in extinction coefficients in Toolik Lake have occurred during the last 18 years.

Thermal Stratification

The structure of thermal stratification of any lake or pond can vary considerably from year to year. Lakes normally contain a warmer, upper mixed layer (epilimnion) above the metalimnion (the middle depths where the most rapid change in temperature with depth defines the "thermocline"). The deepest portion of the water column (hypolimnion) is characterized by cold water temperatures that do not vary much with depth. The pattern of mixing and stratification varies consistently with the depth and morphometry of the lake basin. Deep lakes (>5 m) tend to stratify after ice leaves the lakes (i.e., ice-off), and usually remain stratified for the entire summer season. Shallow lakes and ponds (<5 m) may mix completely each day or night, or they can stratify on calm, sunny days and remain stratified for several days or weeks at a time before wind and cooling events mix the water column (see Vignette 8.1). Given these general features of nearly all lakes, it should be noted that the main difference between arctic lakes and those of other biomes is the long periods of ice cover and the extremes of solar radiation including continuous daylight and darkness.

The timing of ice-off, the onset of stratification, and the temperature of the water may vary considerably between years. Deeper lakes normally stratify 1–2 weeks after ice-off when solar radiation is near maximal values (O'Brien et al. 1997). Maximal epilimnetic water temperatures in Toolik Lake typically range from 14°C to 18°C, but hypolimnetic temperatures consistently have a narrower range from 4°C to 5°C. The structure of the water column also varies from year to year. For example, Figure 8.3 illustrates the differences between a cool and a warm summer at Toolik. In cooler summers (top panel of Figure 8.3) the epilimnion warms above 12°C only for brief periods and tends to be relatively thick, often extending to a depth of 8–10 m during mixing events. In contrast, in warmer summers epilimnetic temperatures are often well above 15°C and the epilimnion may only be from 3- to 5-m thick (bottom panel of Figure 8.3).

These differences in the structure of stratification between cool and warm summers have implications for the stability of the water column and for the response of the lake to external forcing such as wind or storm events. Such events can mix the lake deeply or cause the thermocline to tilt, which results in internal waves and possibly the upwelling of hypolimnetic waters (e.g., Imberger and Patterson 1989). When surface temperatures are cool and the density gradient from surface to deeper waters is small, such as in 2003 (Figure 8.3), lowered resistance to mixing and wind storms often cause deep mixing or upwelling of hypolimnetic water toward the surface. Internal waves may be formed even if the wind stress is insufficient to cause upwelling, and these waves may break and cause mixing of water across the thermocline. In warm summers, such as 2004, the stability of stratification is too great to allow upwelling to occur, but internal waves still form; if they break they result in mixing of deeper and shallower water layers around the thermocline. Essentially, the effect of these mixing phenomena is to move water, organisms, and nutrients from deeper waters to the surface waters, where they may stimulate algal productivity (see Vignette 8.2).

The physical structure and mixing of any lake can also be strongly influenced by storm runoff. The impact of storm runoff depends on the volume of inflow relative to the size of the lake, the inflowing water density, and the stability of stratification during and after the event (Fisher et al. 1979). Inputs of dilute, low-density water will form overflows and may deliver nutrients or mixing energy only to a relatively narrow layer of the lake surface. For example, this occurs in Toolik Lake during snowmelt when the runoff water is dilute and colder (0°C–1°C) than the lake water (3°C–4°C) and thus less dense, and the inflowing water is confined to a thin layer (~2–3 m) just under the ice. Intrusions of denser water at other times of the year will flow into the deeper strata. Organisms contained in these inflows may remain at the depth of intrusion, or be mixed into other layers if the stability of stratification is low as in Toolik Lake (MacIntyre et al. 2006; Evans 2007) or in other arctic lakes (Vincent et al. 1991). Thus the fate of the energy and materials that enter the lake during storm events will depend on the initial stratification, and whether the runoff is followed by cold fronts, which tend to mix the lake deeper, or warm fronts, which tend to restabilize the stratification.

Table 8.1 Mean (standard deviation) of selected physical and chemical characteristics of commonly sampled lakes in the Toolik region[a]

Lake		Light	Specific							
Name	Secchi	Extinction	Cond.	ALK	NH_4	NO_3	PO_4	TN	TP	Chl *a*
Toolik	4.4	0.63	55.3	0.48	0.46	0.16	0.05	16.77	0.16	1.40
	(0.47)	(0.13)	(30.5)	(0.09)	(0.82)	(0.30)	(0.13)	(4.07)	(0.07)	(0.84)
E1	4.2	0.82	69.7	0.71	0.24	0.12	0.03	7.57	0.20	1.78
	(0.34)	(0.26)	(36.0)	(0.31)	(0.28)	(0.24)	(0.04)	(0.86)	(0.02)	(1.29)
E5	3.6	1.09	11.8	0.09	0.25	0.61	0.06	18.35	0.31	1.22
	(0.40)	(0.18)	(1.1)	(0.01)	(0.33)	(1.03)	(0.14)	(4.49)	(0.13)	(0.31)
E6	2.2	1.42	11.6	0.08	0.23	0.23	0.05	20.57	0.42	1.83
	(0.33)	(0.56)	(2.7)	(0.01)	(0.24)	(0.32)	(0.06)	(3.51)	(0.20)	(0.65)
F2	8.4	0.66	136.3	1.42	0.36	0.09	0.08	11.72	0.14	0.71
	(1.3)	(0.22)	(11.3)	(0.09)	(0.76)	(0.09)	(0.15)	(2.91)	(0.04)	(0.47)
F4	2.2	0.79	90.7	0.85	0.63	0.18	0.06	24.29	0.44	3.59
	(0.61	(0.46)	(6.0)	(0.11)	(0.73)	(0.32)	(0.10)	(4.45)	(0.09)	(1.50)
I3	3.9	0.94	20.6	0.22	0.19	0.20	0.03	19.77	0.15	0.68
	(0.56)	(0.11)	(4.2)	(0.02)	(0.44)	(0.33)	(0.02)	(7.18)	(0.03)	(0.43)
I5	3.6	0.83	19.3	0.18	0.18	0.12	0.03	18.49	0.13	0.92
	(0.16)	(0.21)	(5.2)	(0.04)	(0.18)	(0.16)	(0.03)	(4.93)	(0.03)	(0.61)
I6	4.1	0.68	31.4	0.33	0.34	0.25	0.03	20.93	0.17	0.88
	(0.58)	(0.17)	(5.2)	(0.03)	(0.46)	(0.73)	(0.03)	(4.12)	(0.005)	(0.77)
I7	3.9	0.77	35.1	0.39	0.27	0.08	0.03	12.97	0.12	1.11
	(0.40)	(0.16)	(4.8)	(0.03)	(0.37)	(0.07)	(0.02)	(0.66)	(0.06)	(0.66)
I8	2.4	1.02	49.7	0.54	0.38	0.17	0.04	18.36	0.11	1.32
	(0.78)	(0.21)	(10.4)	(0.09)	(0.45)	(0.15)	(0.04)	(2.44)	(0.04)	(0.86)

N1	6.2	0.81	116.2	1.18	0.96	1.02	0.14	18.81	0.15	0.87
	(1.01)	(0.24)	(7.3)	(0.08)	(1.37)	(2.36)	(0.49)	(3.00)	(0.06)	(0.38)
N2R	6.4	0.80	109.9	1.08	0.62	0.13	0.11	15.21	0.17	1.14
	(0.23)	(0.21)	(32.3)	(0.26)	(1.10)	(0.20)	(0.28)	(0.49)	(0.11)	(1.10)
NE12	6.4	0.78	138.5	1.27	0.38	0.10	0.05	NA	NA	0.57
	(0.21)	(0.22)	(32.0)	(0.19)	(0.72)	(0.14)	(0.10)	—	—	(0.28)
NE9B	5.8	0.73	206.4	1.83	0.78	0.16	0.06	19.58	0.26	1.05
	(0.71)	(0.22)	(66.8)	(0.61)	(1.27)	(0.14)	(0.06)	(1.06)	(0.25)	(0.76)
S11	5.0	0.75	189.7	1.87	0.58	0.09	0.09	16.61	0.17	0.93
	(0.55)	(0.26)	(43.1)	(0.55)	(0.95)	(0.10)	(0.19)	(4.23)	(0.11)	(0.51)
S6	4.0	0.73	176.0	1.84	0.66	0.77	0.04	17.23	0.19	1.21
	(0.71)	(0.21)	(32.3)	(0.36)	(0.78)	(1.08)	(0.03)	(3.32)	(0.10)	(1.28)
S7	2.7	0.99	201.7	2.11	0.48	2.25	0.03	18.32	0.14	1.01
	(0.56)	(0.24)	(34.8)	(0.22)	(0.39)	(2.40)	(0.04)	(1.73)	(0.04)	(0.90)

[a] Means of chemical characteristics are for epilimnetic water (0–3 m) samples collected in July of each year. Means of specific conductivity (Specific Cond.), alkalinity (ALK), ammonium (NH_4), nitrate (NO_3), phosphate (PO_4), total nitrogen (TN), total phosphorus (TP) and chlorophyll (Chl *a*) are from the period 1990–2008. Means of Secchi depth, light extinction, total nitrogen, and total phosphorus are from the period 1999–2004. Units are as follows: Secchi depth (m); light extinction (m^{-1}); specific conductivity ($\mu S\ cm^{-1}$); alkalinity ($\mu Eq\ L^{-1}$); NH_4, NO_3, PO_4, TN, and TP (μM); and chlorophyll *a* ($mg\ m^{-3}$).

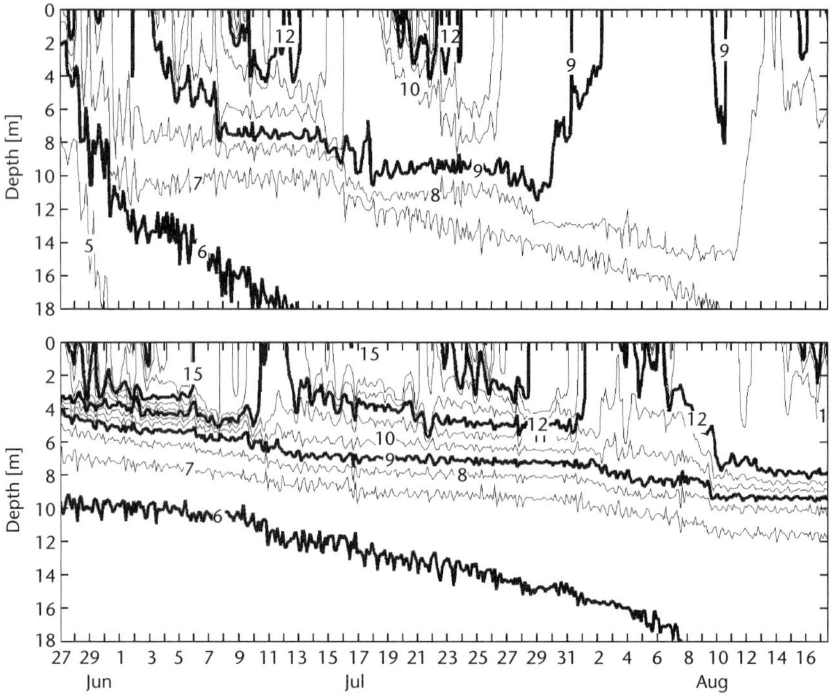

Figure 8.3 Thermal stratification of Toolik Lake during the cool summer of 2003 (top) and the warm summer of 2004 (bottom).

Vignette 8.1. Mixing Dynamics in Deep and Shallow Lakes, by S. MacIntrye

Patterns of thermal stratification are influenced by lake depth and morphology. Lakes that are deeper than 5 m stratify during summer, whereas shallower lakes are polymictic, mixing completely from the surface to the sediment boundary on time-scales from days to weeks.

The extent of stratification and mixing is determined by incoming solar radiation, air temperatures, wind, and the extent of absorption of light. Because the waters of Toolik and nearby lakes are stained with dissolved organic carbon (DOC), light is absorbed rapidly and the heat from the sun is trapped in a relatively shallow upper mixed layer. Whether the heat is mixed downward or held in the upper layer depends on how frequently warm air masses (with their accompanying sunny skies), warm air temperatures, and light winds are interrupted by cold fronts with increased cloud cover, air temperatures near freezing, and extended windy periods. During sunny periods, the lakes warm during the day, and shallow, diurnal thermoclines form. At night they cool and erase the diurnal thermocline, and over the course of the summer this causes the epilimnion to deepen slowly. During cold fronts, the mixed layer deepens, generally by a meter or two, and sometimes, as in 2002 (Figure 8.V1) and 2003 (Figure 8.3), much more extensively. Because the water temperatures are low, these lakes have low evaporation rates and thus reduced mixing due to convection driven by

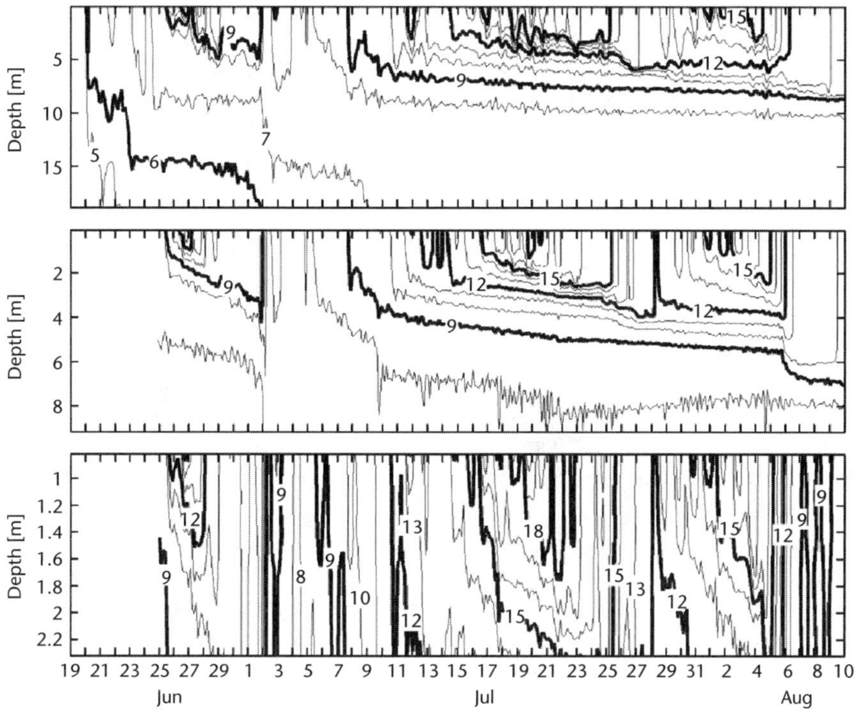

Figure 8.V1 Time series of temperature (°C) in Toolik Lake (top panel) Lake E5 (middle panel) and Lake E6 (bottom panel) for mid-June to mid-August of 2002. Data were collected using self-contained temperature loggers placed at approximately 1-m intervals in each of the lakes. Temperature contours are 1°C.

heat loss at the surface. Consequently, the epilimnia in arctic lakes tend to be shallow compared with those one would see in similarly sized tropical lakes.

The consequences of the alternation between warm and cold air masses can be seen in the thermal stratification of Toolik Lake (Figure 8.V1, upper panel). Thermal stratification was established slowly after ice-off in 2002 compared to other years (Figure 8.3). It was disrupted by a wind storm on July 2 with wind speeds over 20 m s^{-1}. This storm cooled the surface waters and caused the thermocline to rapidly tilt back and forth across the whole lake (not shown). The mixing from internal wave breaking was so intense that the 7°C isotherm abruptly descended to 15 m. After the large storm, stratification was reestablished. Surface temperatures were 15°C–16°C when warm air masses were present and 13°C and cooler when cold air masses returned. The epilimnion was only 5-m deep by the end of the study period, and the lake remained stratified. Because the mixing was so strong early in the summer, the hypolimnion was warmer (7°C) than in most summers (Figure 8.3).

Several aspects of the stratification dynamics vary with lake size (Figure 8.V1). The warming and cooling periods in the small stratified Lake E5 (Figure 8.V1, middle panel) and the polymictic lake E6 (lower panel) occur at the same time as in Toolik Lake, but the surface water temperatures are higher in the smaller lakes and are lower in E6 toward the end of summer. The depth of the diurnal thermocline is similar in all three lakes, but the seasonal thermocline is much thinner in E5 than in Toolik. The strong wind event on July 2 caused both E5 and E6 to mix to the bottom, but not Toolik. While such strong winds are infrequent in the region around Toolik, the shallow epilimnion combined with the thin thermocline in E5 sets the stage for moderate winds to induce mixing from internal wave breaking in this small lake just as in Toolik and to cause the movement of water between the upper and lower water column, or even all the way to the sediment–water interface in the shallowest lakes like E6.

Chemical Properties

Lakes in the Toolik region vary greatly in their chemistry even for lakes only a short distance apart. In the Toolik region the major factor controlling differences in the conductivity and major ion chemistry of lakes is glacial history. Chemical composition is only weakly affected by lake size, lake residence time, or lake-to-watershed area (Table 8.1).

Lakes surrounded entirely by the most recently glaciated landscape (<12 ka) have relatively high conductivities (>90 μS cm^{-1}), and have the highest concentrations of Ca (>500 μM), Mg (>100 μM), and bicarbonate (>1000 μM) (Figure 8.4). Sulfate concentrations are highly variable (5–250 μM), but are usually higher in lakes found on recently glaciated surfaces. These chemical characteristics of the lakes are influenced by the weathering of limestone and dolomite that dominate the watersheds of these younger surfaces (Brown and Berg 1980). Concentrations of Cl, Na, and K are all quite low and show little variation with landscape age (Kling et al. 1992a), suggesting low weathering rates of sedimentary rocks in the Brooks Range and low inputs from recent airborne sea salts to the region.

Lakes found on the oldest glacial surfaces (>250 ka) are dilute and have the lowest conductivities (<35 μS cm^{-1}) and major ion concentrations because soils on these surfaces are highly weathered. Sulfate concentrations in the lakes are extremely low and are generally less than 5 μM. Lakes on landscapes of intermediate glacial ages have chemistries that are intermediate to those located on the youngest and oldest surfaces. Many lakes, including Toolik Lake and the lakes in its catchment (I-series lakes), are influenced by landscapes of mixed age; their chemistries reflect the mix of the different geological surfaces, although in the I-series lakes the upland, older landscape appears to control the lake chemistry (Kling et al. 2000). Throughout the region, lakes have a circum-neutral pH, ranging from 6.7 to 8.1, that does not vary predictably with age of the landscape or lake size.

Nutrient concentrations in the lakes are very low and are dominated by organic forms (Table 8.1). The DOC concentrations in most lakes range from 200 to 1000 μM. Dissolved inorganic nitrogen and phosphorous concentrations in surface waters are nearly always below 0.2 and 0.1 μM, respectively, and frequently below analytical detection limits. Organic nitrogen and phosphorous concentrations are higher, but still very low. Dissolved organic nitrogen ranges from 5 to 20 μM in most lakes and dissolved organic phosphorus is seldom above 0.2 μM. Even in deeper lakes that stratify during the summer, there tends to be very little build-up

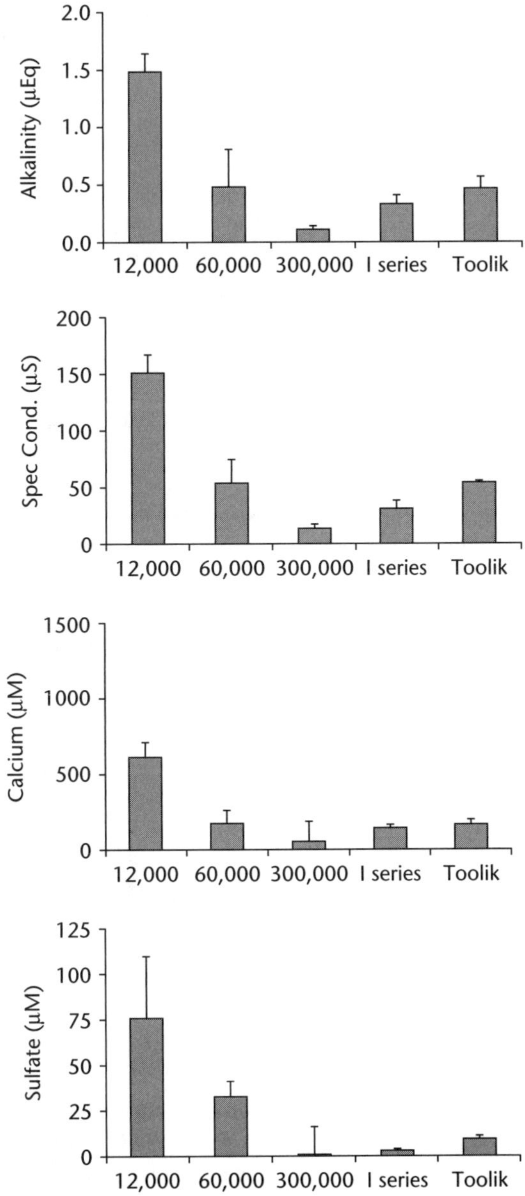

Figure 8.4 Mean (+SD) of specific conductivity (μS cm^{-1}), alkalinity (μEq L^{-1}), and concentrations of calcium and sulfate (μM) in surface waters of lakes near the Toolik. Numbers on the x-axis represent years before present of last glaciation. I-series and Toolik Lake have watersheds of mixed glacial ages as indicated in Table 8.1.

of nutrients in the bottom water. Ammonium concentrations at the end of the summer in the deeper portions of Toolik Lake may reach 0.8–1.0 µM, and there is no appreciable build up of phosphate.

Biogeochemical Budgets and Cycles

The main inlet to Toolik Lake accounts for ~90% of the discharge into the lake, and the remaining four lake inlets plus groundwater seepage account for ~10% of the total discharge on average (in 2004; Figure 8.5). Up to 30% of the N and P inputs occur during the first 10 days of stream flow when the lake is still ice covered (Whalen and Cornwell 1985). Nitrogen inputs from the two inlet streams are dominated by organic forms of dissolved N, which make up over 80% of the total N inputs. Dissolved P concentrations are usually higher than particulate P and dominated by organic forms, although not to as great an extent as for N (Table 8.1). The chemistry of seepage water has not been directly measured but has been assumed, when constructing nutrient budgets, to be similar to the volume-weighted inputs of the surface streams. Direct precipitation onto the lake makes up 2% of the total N input and 5% of the total P input.

Whalen and Cornwell (1985) constructed the most complete budget of N and P in Toolik Lake. Because early measurements of N_2 fixation in lakes suggested that

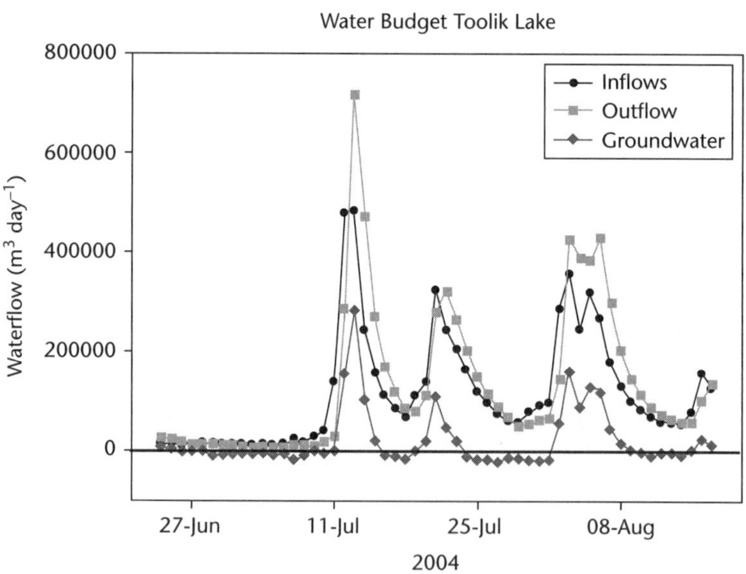

Figure 8.5. Water budget for Toolik Lake during the summer of 2004. Inflows include rain and discharge from the inlets, outflow is the discharge through the outlet and evaporation, and groundwater flows are assumed to be the difference between inflow and outflow. Negative values apply to groundwater only and indicate loss from the lake (data from E. Graham, unpublished).

water column rates were low, fixation was not considered in the budget (Alexander et al. 1989). Mass balance calculations using surface-water input and outputs for Na, a conservative tracer, were within 3%, suggesting that the water budget was well constrained. Using the same surface-water input and output data, it was calculated that about 18% of the N and 30% of the P was retained in the lake and its sediments.

A model of nutrient cycling in Toolik Lake (Whalen and Cornwell 1985) suggested that primary production was driven largely by pelagic recycling of nutrients. The unimportance of sediment-supplied nutrients was not due to lake depth, or low mixing, but rather to the very low release rate of nutrients from the sediments (Kipphut 1984). Nitrogen release from the sediments was estimated through the use of benthic chamber measurements. Chamber incubations indicated that shallow sediments only released 0.005–0.01 mmol N m^{-2} d^{-1}. This N release rate was similar to that needed to explain the small accumulation of dissolved N present under the ice at the end of the winter season. Benthic release of phosphorus from sediments of Toolik Lake was extremely low. Any P released during decomposition in the sediments was adsorbed onto iron oxides, which are abundant near the sediment surface (Cornwell 1987).

For nitrogen, research in other lakes in the vicinity of Toolik Lake shows that both benthic and pelagic habitats support rates of N_2 fixation that are significant to whole-lake function and to N budgets, especially in small lakes with low stream inputs (Gettel 2006). Summer-time epilimnetic rates of N_2 fixation ranged from 0.12 to 1.5 mg N m^{-2} day^{-1} for benthic habitats and from 0 to 2.6 mg N m^{-2} day^{-1} for pelagic habitats. Pelagic N_2 fixation rates integrated over the entire water column are higher than reported for many ecosystems and comprised up to 75% of N input to Lake Fog2. Benthic N_2 fixation is lower than has been reported for other oligotrophic systems, but is roughly equivalent to annual N inputs from atmospheric deposition when scaled over an entire year (~25 mg N m^{-2} $year^{-1}$). Benthic and pelagic rates of N_2 fixation when combined provide annual estimates that range from 22 to 696 mg N m^{-2} $year^{-1}$ in lakes. The N_2 fixation in the tundra near Toolik Lake is ~100 mg N m^{-2} $year^{-1}$ (Figure 5.6).

Studies in a wide variety of lakes show that nutrient release from sediments is extremely low (Kipphut, unpublished; Giblin, unpublished). Most lakes show essentially no N or P release during the summer under ambient conditions. The exceptions are in the few lakes in the region that experience low bottom-water dissolved oxygen conditions or prolonged anoxia due to infrequent mixing. In these cases some N release and, even more rarely, some P release may be observed. Lakes undergoing fertilization do begin to release nutrients, but it may take several years (see Vignette 8.2; O'Brien et al. 2005).

Zooplankton play an important role in the pelagic recycling of nutrients. Johnson et al. (2010) estimated that excretion by macrozooplankton in the epilimnion could account for 10%–80% of demand for nitrogen by phytoplankton. Rates of nutrient excretion were highest in lakes that were dominated by small-bodied copepods. Excretion by fish amounted to less than 5% of nutrient recycling rates estimated for macrozooplankton. Little information has been collected on excretion rates of microzooplankton, but our measured excretion rates of macrozooplankton are comparable to rates of macrozooplankton in temperate lakes (Vanni 2002).

At larger scales, the biological processing of nutrients in lakes is responsible for the spatial patterns of dissolved and particulate nutrients present in lakes within the I-series of lakes and streams (Kling et al. 2000). Here, in-lake processing reduced dissolved concentrations of C and N and resulted in higher concentrations of particulate organic carbon (POC) and nitrogen (PON) leaving the lakes. Accumulation of C and N by phytoplankton was responsible for this transformation. The temporal synchrony of changes in dissolved and particulate nutrients in this series of lakes highlights the importance of viewing lake processes within a landscape context.

Lakes near Toolik often have high concentrations of dissolved carbon dioxide and methane in the water column throughout the summer and function as a source of these greenhouse gases to the atmosphere (Kling et al. 1991, 1992b). The gases come from three sources: The first is soil water flowing directly into lakes, or into streams and then lakes, which contain high concentrations from plant and microbial respiration occurring on land. The second is the lake sediments, where microbial respiration produces both CO_2 and CH_4. The third is the water column itself, where mostly CO_2 is produced by respiration of DOC (and to a lesser extent POC), which enters the lakes from land. The deeper, stratified lakes like Toolik often have increasing CO_2 and CH_4 concentrations with depth, indicating that the sediments are an important source within the lake. Sediment concentrations can exceed local saturation, and the bubbles formed and released from the lakes may be an important source of methane to the atmosphere (Walter et al. 2006). The transport of carbon from terrestrial systems to the lakes is discussed in chapter 6.

Vignette 8.2. Changes in Sediment Chemistry Induced by Fertilization, by Anne Giblin

The addition of nutrients to lakes causes large changes in sediment chemistry due to increased deposition of organic matter and resulting changes in the oxygen content of bottom water. Long-term sediment data from Lake N2, which was fertilized for five years, provide insights into the biogeochemical changes that occurred during fertilization and suggest that the sediments may not fully return to pre-fertilization conditions.

Sediment cores have been taken from Lake N2 from a variety of depths several times since fertilization ceased in 1990 (Cornwell and Kipphut 1992). Below a water depth of about 3 m, there are dramatic differences between sediments from the control and fertilized side of the lake. Sediments taken from the fertilized side at 6 m, nine years after fertilization had stopped, had a higher percentage of carbon, nitrogen, sulfur, and phosphorus in the top several centimeters of the sediments than those taken from the control side (Figure 8.V2). These sediment samples also indicated that total sulfur had increased in sediments on both the fertilized and the control side since the experiment began (Figure 8.V2). The change on the control side may have been a result of the hypoxia at 6 m that reached into both sides of the lake (O'Brien et al. 2005), the slight increase in productivity on the control side during fertilization, or both hypoxia and productivity. No increase in total S was seen in sediments taken from the control side at 5 m, where the sediments did not experience hypoxia (data not shown), which suggests that oxygen changes in the bottom water were primarily responsible for the increase in S burial.

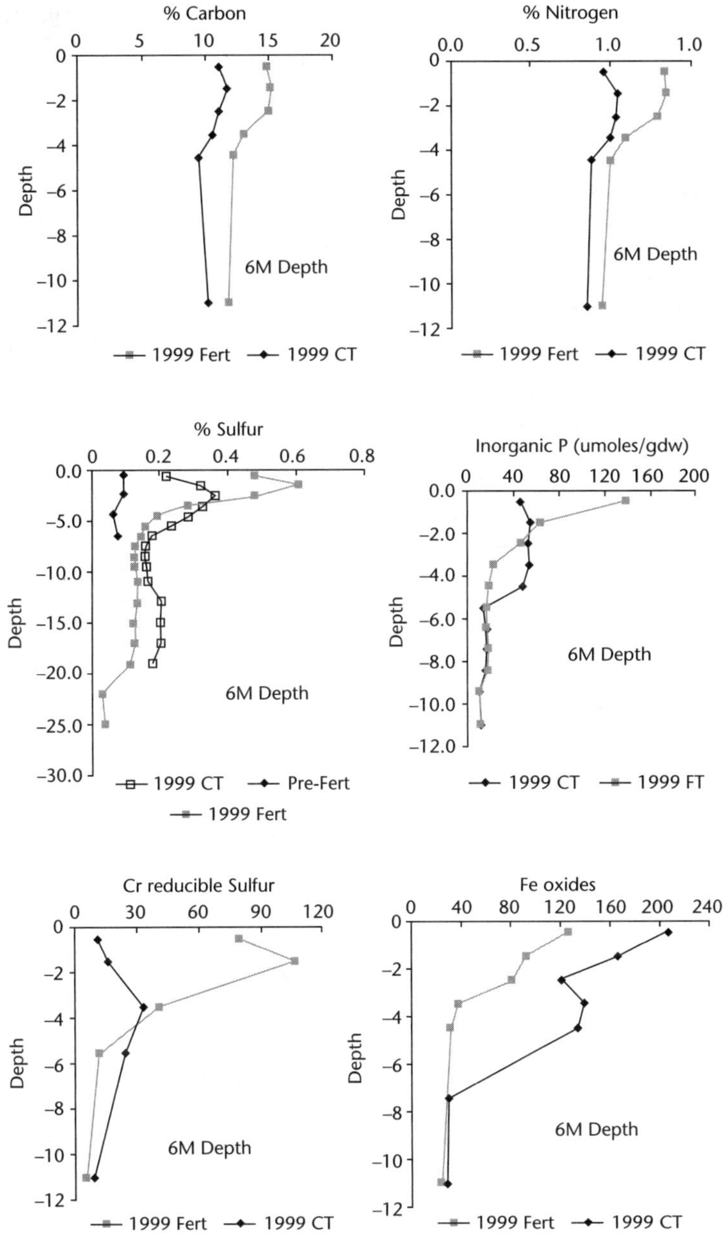

Figure 8.V2 Carbon, nitrogen, sulfur, phosphorus, and iron in sediments of Lake N2.

The changes in the total sulfur profiles in the sediments are due in large part to the incorporation of chromium reducible sulfur (Figure 8.V2). These iron-sulfide minerals form when sulfide produced via sulfate reduction reacts with iron in the sediments. The consequence of the formation on these iron-sulfide minerals is to decrease the pool of iron oxides in the sediments (Figure 8.V2). Sediments from the fertilized side of N2 contain substantially lower pools of iron oxides in the top 5 cm than do sediments from the control side.

What is the impact of this switch from iron oxides to iron sulfides in the sediments? Iron plays a major role in controlling phosphorus cycling in arctic lake sediments. Organic matter decomposition within the sediments releases P that diffuses toward the surface of the sediments. However, in most lakes in the Toolik region, this P is trapped by the adsorption onto iron oxides in surface sediments (Cornwell 1987). Fertilization has evidently depleted the iron oxide crust by trapping iron as sulfide minerals within the sediments. Because the total iron pools are similar on both sides of the lakes, we believe that iron-sulfide formation was the most important process leading to the loss of iron oxides. It took several years for P to be released from the sediments of N2 even after bottom waters had become anoxic (O'Brien et al. 2005). This was probably because there was a sufficient iron-oxide pool to adsorb the P being released. Sediments can retain P as long as the ratio of iron oxides to P is great enough.

These findings have implications for the ability of the lakes to recover from nutrient additions. Iron sulfides can be oxidized back to iron oxides; but with persistent hypoxia, it seems likely that the iron sulfides will be buried and the oxide pool on the fertilized side of the lake will remain lower than on the control side. The loss of this pool of iron oxides would accelerate the response time of sediment P fluxes to future nutrient additions.

Biological Properties

Variation in the composition of taxa can be explained by characteristics of lake size and glacial history. We group our most frequently sampled lakes according to both of these characteristics to facilitate discussion of the biology of these systems (Table 8.2). Small lakes are less than 5 m deep or have surface areas of less than 1 ha. Intermediate lakes have maximum depths from 5 to 10 m, and larger lakes are greater than 10 m deep or have surface areas exceeding 10 ha.

Food webs of arctic lakes are relatively simple, with low levels of species diversity at most trophic levels. Figure 8.6 depicts a simplified description of the food web for Toolik Lake. Phytoplankton species diversity can be substantial with over 140 taxa identified from Toolik Lake (O'Brien et al. 1997). This food web is similar to other medium-sized and large lakes near Toolik. Many of the small lakes do not support fish. The overall structure of the food web in these lakes is similar to those of temperate lakes, except that fish species richness is greatly reduced in the low Arctic (Hershey et al. 1999). The absence of fish in small shallow lakes allows for occupation of the highest trophic levels by invertebrate predators. Several other key aspects of the food webs of lakes vary with lake size (Table 8.3). Chlorophyll

Table 8.2 Morphometry of the most commonly sampled lakes near the Toolik Field Station

Name and category	Surface area (ha)	Maximum depth (m)	Landscape age (ka)[a]
Small lakes			
E6	1.9	3.2	>250
Fog4	1.9	4.4	Composite
S7	0.8	2.9	12
NE9B	0.4	6.0	12
S11	0.3	9.5	12
Medium lakes			
E1	2.6	11.0	60
N2	1.6	9.7	12
I8	16.7	9.5	Composite
I3	3.1	5.1	12
S6	1.1	5.2	12
Large lakes			
Fog2	5.9	20.3	Composite
E5	11.3	12.7	300
I5	18.7	8.5	Composite
I6	16.9	15.	Composite
I7	2.9	14.	Composite
N1	4.3	14.2	12
NE12	8.2	17.1	12
Toolik	149	25	Composite

[a] Landscape age refers to years since last glaciation. Composite refers to lakes draining watersheds with glacial ages of both ~12,000 and ~60,000 years.

concentration in surface waters is greater in the smaller lakes, as is the biomass of zooplankton (Table 8.3). The density of crustacean zooplankton is similar among different-sized lakes, but that is because these small, fishless lakes are frequently dominated by large-bodied *Daphnia middendorffiana.*

Bacterial Processing of Terrestrial Material in Lakes

Organic matter in particulate and dissolved form dominates the nutrient and carbon budgets of arctic surface waters. Decomposition through bacterial activity does not appear to be limited by protozoan grazing of bacteria (Rublee 1992). Some of this organic substrate originates in terrestrial environments, and some is produced within aquatic environments as exudates from algae and zooplankton. The relative importance of these two sources for bacterial activity varies, in part as a result of the amounts of dissolved organic matter (DOM) available from each source and in part as a result of the quality of the DOM (Judd et al. 2006).

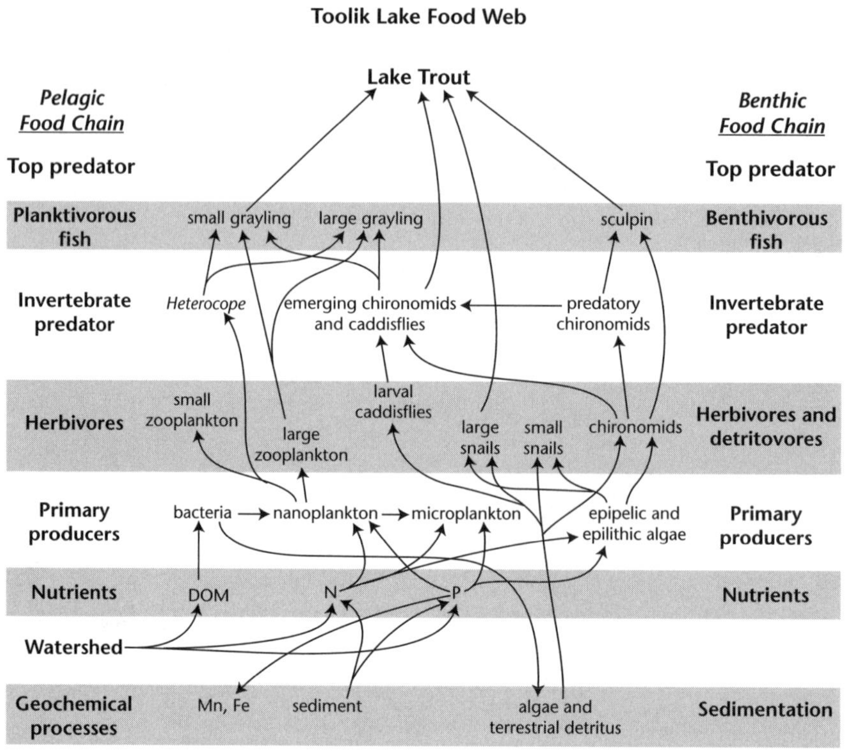

Figure 8.6 Simplified food web for Toolik Lake as modified from O'Brien et al. (1997), used with permission. Identification of pelagic and benthic aspects of trophic levels and associated processes are depicted on left and right sides of diagram respectively. Arrows indicate major pathways to material flow.

Table 8.3 Biological characteristics of small, medium-sized, and large lakes near Toolik. Mean summer values for years 2000–2005 are shown with standard error in parenthesis

	Lake size			$F_{2,15}$
Characteristic[a]	Small	Medium	Large	(p-value)[e]
Chlorophyll	2.74 (0.60)	1.06 (0.11)	1.22 (0.19)	7.17 (0.006)
Zoop density[b]	16.5 (7.43)	20.7 (9.82)	12.3 (3.00)	2.02 (0.16)
Zoop biomass[c]	384 (232)	156 (75.9)	48.0 (10.6)	3.85 (0.042)
Fish presence[d]	0 / 5	5 / 5	8 / 8	

[a] Lakes and units are those listed in Tables 8.1 and 8.2.

[b] Zoop density is crustacean zooplankton density in numbers per liter.

[c] Zoop biomass is biomass of crustacean zooplankton biomass based on length-mass regressions in Burkart (2007) and assuming individual zooplankton were of an average length for each lake.

[d] Fish presence is the number of lakes where fish were present out of the number of lakes included in this comparison.

[e] F-values and p-values in parentheses are based on one-way ANOVA of each biological characteristic among the three lake sizes using the mean value for each lake from the time-series.

Results of an enclosure study in Toolik Lake (Rublee and Bettez 1995) demonstrated that a fraction of terrestrial DOM leached from surface plant litter contains labile compounds and strongly stimulates bacterial activity in the lake. To assess the potential importance of this terrestrial DOM, we measured DOC inputs to Toolik Lake and corresponding bacterial activity through the seasons. During spring runoff, DOC concentrations increased and strongly stimulated bacterial activity in the lake (Crump et al. 2003). However, during storm events later in the summer, when DOC inputs to the lake also increased, there is no corresponding increase in bacterial activity. One possible reason for this pattern is changing quality of the DOC. The DOC leached from surface litter in spring runoff enters the lake directly, bypassing the frozen soils and associated microbes. In contrast, the DOC entering the lake during summer has been acted upon by microbes in the soil, which reduces its quality. Another possible reason for the pattern is that the bacterial composition in the lake changes as the season progresses (Crump et al. 2003), and may no longer be best adapted to consuming terrestrial organic matter. Terrestrial material that is incorporated into bacterial biomass in Toolik Lake is made available to higher trophic levels through trophic interactions in the microbial food web of arctic lakes (Rublee 1992; Kling 1994).

Bacterial Diversity and Biogeography

Bacterial abundance and productivity of Toolik Lake is similar to oligotrophic lakes in boreal and temperate regions. Bacterial densities of ~1×10^6 cells mL^{-1} and productivities of ~10 µg C L^{-1} d^{-1} are common in Toolik (O'Brien et al. 1997). The bacterioplankton of Toolik Lake is dominated by common, globally distributed freshwater phylotypes belonging to the *Proteobacteria*, *Bacteroidetes,* and *Actinobacteria* phyla (Crump et al. 2003), including an *Alphaproteobacteria* first discovered in Toolik Lake that is very closely related to the ubiquitous marine bacterium *Pelagibacter ubique* (Bahr et al. 1996). Nearly half of the phylotypes identified in one study persisted throughout the year, and the rest were specific to one of three seasons: winter, spring freshet, and summer. Bacterioplankton community composition, determined with a PCR-based community fingerprinting technique (Denaturing Gradient Gel Electrophoresis of 16S rRNA genes), showed a cyclical seasonal pattern in which a winter community shifted rapidly to a spring freshet community when the primary Toolik inlet stream started to swell with snowmelt water. At the same time below the ice, the bacterial production reached its annual peak, suggesting that this community shift was driven by rapid growth of certain phylotypes adapted to grow on labile terrestrial organic matter extracted by the melting snow and flushed into the lake. After the spring freshet slowed, ice left the lake and the bacterioplankton community shifted again to a summer community that changed little during the rest of the open-water season. Then, by the end of the following winter, a community formed that resembled the one from the previous winter. Similar, repeating annual cycles have now been seen in other lakes, rivers, estuaries, and the coastal ocean.

Bacterial community composition also varies spatially across the arctic landscape. One study of 10 lakes and 14 streams within the catchment of Toolik Lake showed that lake communities were on average 78% similar and did not appear to be influenced by lake morphology (Crump et al. 2007). Communities in hydraulically connected lakes changed gradually down-catchment, suggesting that dispersal via inoculation from upslope lakes influences the composition of these communities. In contrast, unconnected lakes contained somewhat different communities. Bacterial communities in streams arising from tundra were very different from those in lakes. Bacterial communities in streams that arise directly from lake outlets matched those in the lakes but shifted gradually downstream, apparently because of the addition of bacteria from soil water and hyporheic water. These results reveal large differences in lake-specific and stream-specific bacterial community composition over restricted spatial scales (<10 km), and suggest that geographic distance and hydrological connectivity influence the distribution of bacterial communities across landscapes.

Primary Producers

O'Brien et al. (1997) described the summer phytoplankton of Toolik Lake in some detail and concluded that the biomass was dominated by chrysophytes, with dinoflagellates and cryptophytes of second and third importance. However, unpublished data from E. A. MacIsaac for the summer of 1996 and U. Passow from the early years of the 21st century have shown that the tiny single-celled cyanophyta were missed in the early studies and are more numerous than other phytoplankton cells. Despite their numerical dominance, small cyanophyte cells such as *Synechococcus* do not dominate the biomass, which is dominated by flagellates in the divisions Cryptophyta, Chlorophyta, and Chrysophyta and the rarer, large-celled diatoms (Bacillariophyta) and dinoflagellates (Pyrrophyta).

Shallow lakes (E6 and Fog4) and deeper lakes (E5, Fog2, Toolik) had similar levels of phytoplankton diversity, but differed in the relative importance of taxonomic groups. The deeper lakes were dominated by chrysophytes and cryptophytes, whereas increased biovolume of cyanobacteria and the large chrysophyte *Dinobryon* occurred in shallow lakes (Figure 8.7). These shallow-lake taxa are generally resistant to grazing and are likely selected for in the shallow lakes with higher densities of cladoceran zooplankton. Burkart (2007) reported that grazing rates of macrozooplankton were 20%–50% higher when zooplankton were fed phytoplankton from deep lakes compared to trials when zooplankton were fed phytoplankton from shallow lakes. These differences were likely due to differences in phytoplankton species composition.

Primary Production

Overall primary productivity depends on the supply of both nutrients and light. Supply and availability of these resources is governed by several factors, including geological setting, lake morphometry, biological interactions, and hydrodynamic processes that are driven by climatic forcing. Primary production in pelagic regions of lakes near Toolik has been studied extensively since the 1970s. Recent studies of benthic primary production indicate that benthic algae are responsible for most of the primary production that occurs in shallow lakes, and a considerable portion of production in the larger lakes (Whalen et al. 2008).

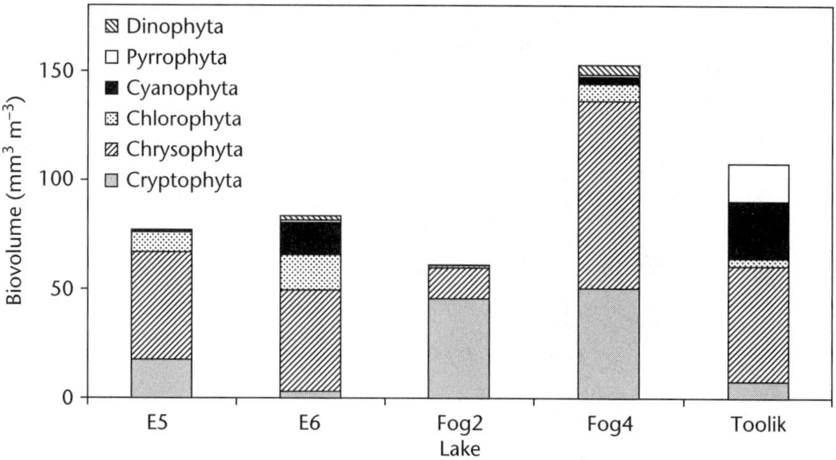

Figure 8.7 Biovolume of major divisions of phytoplankton present in lakes E5, E6, Fog2, and Fog4 on July 11, 2002, and in Toolik from August 1998. Lakes E6 and Fog4 are small, shallow lakes. Lakes E5, Fog2, and Toolik are deeper with maximum depths of 12, 20, and 23 m, respectively. Low-level fertilization of lakes E5 and E6 began on July 1 of that year.

Pelagic Primary Production

Rates of primary production in surface waters in lakes near Toolik range from approximately 5 to 50 mg C m^{-3} d^{-1} (Miller et al. 1986; Kling et al. 2000). These daily rates were estimated using ^{14}C techniques and are similar to rates measured in oligotrophic and ultra-oligotrophic lakes in the temperate zone but greater than those of larger lakes in the high arctic (Vincent and Hobbie 2000). Annual rates of primary production in arctic lakes are low because of the shortened season of open water and low rates of nutrient supply. Rates of primary production are highest shortly after ice-out and decrease due to nutrient limitation through the summer unless storm events provide nutrient additions through stream inputs or by upward mixing of nutrients from deeper strata (MacIntyre et al. 2006). More than half of the primary production typically occurs in the epilimnion (Figure 8.8). Although many of the lakes support chlorophyll maxima at metalimnetic depths, the high light extinction in these stained lakes limits rates of photosynthesis below the thermocline.

Integration of measured rates of ^{14}C uptake leads to areal primary production estimates of 20–200 mg C m^{-2} d^{-1}. This range mostly reflects seasonal variation within lakes rather than differences among lakes. Shallow lakes tend to have slightly greater rates of pelagic primary production per volume; however, areal production is similar for all lakes. Annual rates of pelagic primary production range from 2 to 10 g C m^{-2} for the lakes near Toolik (Kling et al. 2000; Evans 2007; Whalen et al. 2008).

Initial studies of nutrient limitation of phytoplankton growth in Toolik Lake were conducted using short-term batch bioassays (Whalen and Alexander 1986), followed by similar comparisons among 39 lakes near Toolik (Levine and Whalen 2001). Results indicated that the degree of nutrient limitation varied seasonally but

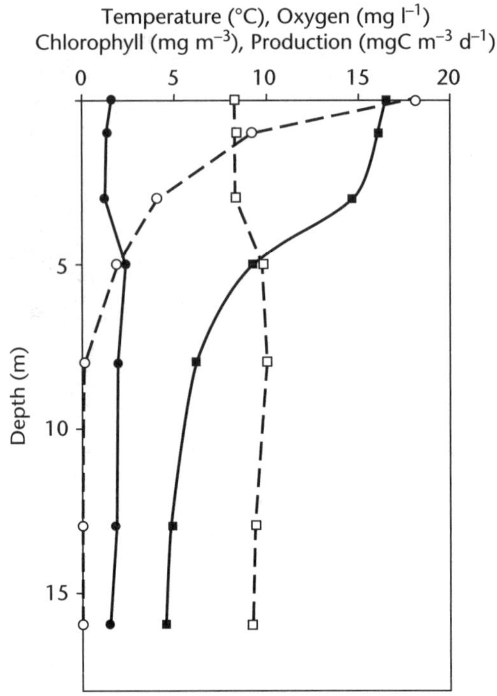

Figure 8.8 Depth profile of temperature (solid squares and solid line), dissolved oxygen (open squares and dashed line), chlorophyll *a* (solid circles and solid line), and primary production (open circles and dashed line) for Toolik Lake on July 9, 1999.

that phytoplankton were most often co-limited by both N and P. Nitrogen limitation was more common than phosphorus limitation in these experiments, and shallow lakes and lakes with higher alkalinities demonstrated a greater degree of P limitation. However, the greatest response in production (average increase of 60%) was seen in treatments with both N and P additions. Only six of the 54 bioassays failed to show significant effects of additions of N or P, but five of these six experiments occurred in headwater lakes. This is consistent with the idea that lake phytoplankton rapidly take up dissolved nutrients entering the lake from the catchment, and in turn export particulate nutrients to downstream lakes (Kling et al. 2000).

Changes in the chlorophyll concentrations during the nutrient enrichment bioassays provide further insights. Chlorophyll increased in many of the control treatments in which macrozooplankton were not present (Levine and Whalen 2001), suggesting that grazing by zooplankton has a measurable impact on phytoplankton biomass in bioassays. Chlorophyll declined in treatments where phosphorus was added alone, suggesting that bacterial populations may be P-limited and able to outcompete the algae for P in these treatments. These results support the findings of Vadstein et al. (1993), who concluded that planktonic bacteria outcompete phytoplankton for P in oligotrophic lakes in Norway. Overall, the controls observed on primary production by nutrient limitation are similar to those noted in the review by Lewis and Wurtsbaugh (2008) indicating that co-limitation of phytoplankton growth by N and P in lakes is the dominant condition.

The glacial history of the watersheds in the region had little effect on pelagic algal biomass as measured by chlorophyll *a* concentration (Table 8.3). Neither glacial history nor specific conductivity was significantly correlated with chlorophyll *a* concentration of epilimnetic water in these lakes. However, the only deep lake present on the oldest glacial surface (Lake E5) exhibited low specific conductivities and had some of the lowest concentrations of epilimnetic chlorophyll (Table 8.1). Shallow lakes on the older glacial surfaces had moderate chlorophyll concentrations and were similar to lakes in recently glaciated watersheds. While volumetric concentrations of chlorophyll tend to be higher in shallow lakes than in deeper lakes (Table 8.3), the tendency is not statistically significant; however, on an areal basis (per square meter) deeper lakes have more chlorophyll than shallow lakes (Whalen et al. 2008).

Lakes deeper than 5 m often developed deep chlorophyll maxima during the summer months when chlorophyll concentrations in the metalimnion or hypolimnion were frequently more than twice values measured in epilimnetic water (Figure 8.8). Despite the abundant phytoplankton in the deeper strata, these deeper layers did not contribute substantially to the primary production of the lake because of light limitation. However, internal waves can periodically move phytoplankton to depths with more light, resulting in increased rates of primary production (Vignette 8.3).

Vignette 8.3. Effect of Internal Waves on Phytoplankton Primary Production, by Mary Anne Evans

Just as surface waves form on lakes at the air–water interface, waves can form within a lake at boundaries between water layers of different densities. These internal waves can be much larger in amplitude than the surface waves on a given lake because the difference in density between the water layers is much smaller than that between water and air. Internal waves move water, chemicals, and any nonmotile or slow-moving organisms vertically through the water column. Because of this movement, internal waves can have important impacts on nutrient availability in the upper water column, spatial or temporal aggregation or patchiness of plankton, and light availability to deep phytoplankton.

Phytoplankton associated with internal waves are periodically moved closer to and farther away from the lake surface. Because light decreases exponentially with depth from the lake surface, this vertical movement can increase the average light exposure of the phytoplankton. The effects of this motion on the light climate for phytoplankton, and its implications for photosynthesis, have been previously described based on mathematical models for the situation where surface light is constant or varies very slowly compared to the period of the internal waves (Evans et al. 2008).

Research at Toolik Lake has shown that internal waves near the bottom of the euphotic zone, where light is sufficient for photosynthesis but not saturating, can increase photosynthesis. *In situ* experiments conducted at Toolik Lake during the summers of 2003 and 2004 tested the theory of internal wave effects. In each experiment a set of bottles containing lake water was incubated at depth in Toolik Lake for

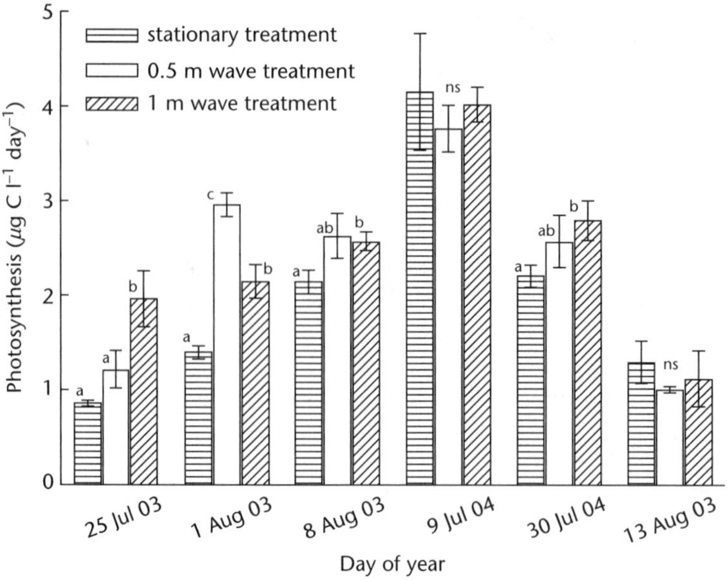

Figure 8.V3 Phytoplankton primary production (μg C L^{-1} d^{-1}) from a series of *in situ* internal wave experiments in 2003 and 2004. Stationary treatments (depth = 5 m) had generally lower primary production than treatments in which phytoplankton were moved sinusoidally through the water column with average depth of 5 m, period of 4 h, and amplitude in the vertical of 0.5 m or 1 m. Statistically significant differences were indicated by the letters a, b, and c, where "ns" indicates that there were no significant differences on that date.

24 hours, and the photosynthesis in each bottle was measured. During the 24 hours some of the bottles were held at a constant depth (5 m) and some were moved up and down with different amplitudes between 4 m and 6 m in a manner similar to the dynamics of natural internal waves in the lake. As predicted, enhancement of photosynthesis was observed in the moving bottles. However, on some days this enhancement was greater than expected, and on other days photosynthesis was unaffected by the vertical movement (Figure 8.V3).

To explore the mechanism causing these unexpected results, a simulation model was tested using the available surface light, light attenuation, and phytoplankton physiology measured on the day of each experiment. This model showed that variation around the photosynthetic enhancement predicted by earlier theory can be caused by changes in the amount of light entering the lake during the incubation. When surface light varies due to the passage of clouds with approximately the same period or time-scale as the internal wave motion, and the time of brightest light aligns with the crest of a wave, the enhancement of photosynthesis is increased. Conversely, when the time of brightest light aligns with a wave trough, the enhancement of photosynthesis is reduced or eliminated. These modeling results suggest that the interaction is robust and likely important in other systems with internal waves and moderate plus variable levels of cloud cover.

Benthic Primary Production

The extent of benthic primary production depends on lake morphometry and water clarity. Low chlorophyll concentrations in surface waters of most lakes leads to high water clarity and sufficient light penetration to support extensive benthic primary production (Vadeboncoeur et al. 2003). However, in lakes near Toolik, light penetration is limited due to high concentrations of colored dissolved organic matter. Benthic production comprises the majority of lake primary productivity in shallow lakes, but the contribution of benthic production to whole lake primary productivity is greatly reduced in lakes exceeding 8 m in depth.

Chlorophyll concentrations in shallow (<4 m water depths), soft sediments range from 100 to over 600 mg m^{-2}. These values typically exceed the total pelagic chlorophyll in the overlying water by 2 to 20 times. Chlorophyll concentrations are much lower on cobble and gravel substrates, but these substrates tend to make up a small portion of lake sediments in the lakes of the region (Cuker 1983). Sediment chlorophyll concentrations within lakes can show significant interannual variation (Gettel 2006) and are not well correlated with the age of the surrounding landscape.

Benthic primary production in shallow sediments ranges from 25 to 250 mg C m^{-2} d^{-1} (Figure 8.9). When integrated over the growing season, these rates are similar to or greater than those measured in the overlying water column. Rates of benthic primary production drop off rapidly with depth, and there is little net production below 6–7 m except in the clearest lakes (Gettel 2006; Whalen et al. 2008). At all but the very shallowest depths, gross primary productivity appears to be light limited; this is true even though benthic primary producers reach maximal photosynthetic rates at relatively low light intensities (Figure 8.10).

Benthic N_2 fixation is controlled most strongly by the availability of P and to a lesser extent light availability and grazing by the snail *Lymnaea elodes* (Gettel

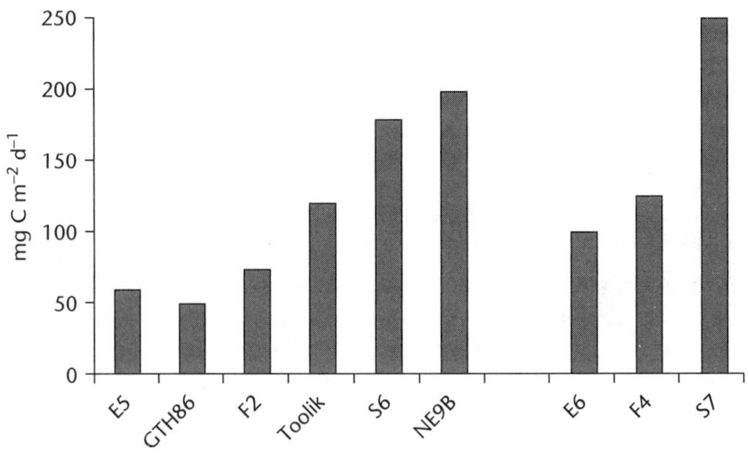

Figure 8.9 Gross primary production (GPP) estimated from changes in oxygen concentration in light and dark benthic chambers placed at 3-m depth in lakes near Toolik.

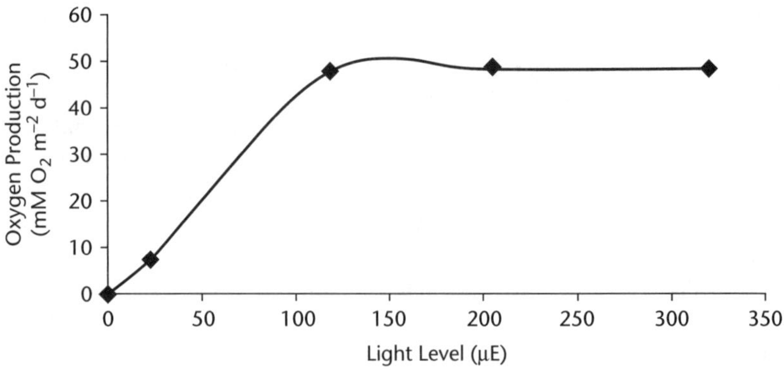

Figure 8.10 Gross primary production (GPP) estimated from oxygen production in sediment cores collected from Toolik Lake at 3-m depth and incubated under a range of light intensities (μE m^{-2} s^{-1}) in the laboratory.

et al. 2007). Results from nutrient and grazer manipulations suggest that P stimulates benthic N_2 fixation, while snail grazing slightly suppresses it. These controls play out at the landscape scale in the Toolik Lake area across patterns of glacial history. Benthic N_2 fixation is generally higher on younger-surface lakes (deglaciated ~12,000 yr BP) where P concentrations are higher than on older-surface lakes (deglaciated >25,000 yr BP). This pattern is evident despite the fact that *Lymnaea elodes*, which slightly suppresses benthic N_2 fixation, is largely absent on old-surface lakes due to a lack of calcium (Hincks and Mackie 1997).

Other controls on benthic production are beginning to be examined. While production on soft sediments does not seem to be greatly affected by snail grazing (Gettel et al. 2007), snails appear to play a larger role on cobbles, where benthic producer biomass is much less (Cuker 1983). While additional nutrients could stimulate benthic production, studies in fertilized lakes suggest that the stimulatory effect of nutrients on benthic production is countered by the decreased light reaching the bottom in all but very shallow lakes. In these fertilized lakes, light penetration is reduced by increased phytoplankton biomass (Gettel 2006).

Microzooplankton

The microzooplankton community in the lakes near Toolik is composed primarily of nanoflagellates, ciliated protozoans, rotifers, and copepod nauplii. Most of the small flagellates are mixotrophic, but are generally devoid of chlorophyll in Toolik Lake. Heterotrophic nanoflagellates can consume a substantial portion of bacterial productivity (Hobbie and Laybourn-Parry 2008). The ciliates, mainly oligotrichs <50 μm in length, are dominated by the three genera *Halteria, Strombidium,* and *Strobilidium,* while the rotifers are dominated by the four species *Keratella cochlearis, Kellicottia longispina, Polyarthra vulgaris,* and *Conochilus unicornis* (Rublee 1992).

In general, the biomass of the microzooplankton in these oligotrophic lakes is <10 μg C L^{-1} with the mean biomass of the protozoans usually <0.5 μg C L^{-1} (~150 individuals L^{-1}), the rotifers ~2-3 μg C L^{-1} (100 ind L^{-1}), and the remainder being nauplii at ~7–8 μg C L^{-1} (~7 ind L^{-1}). However, microzooplankton biomass can be higher in more productive lakes (up to 30 μg C L^{-1}, Rublee 1992), and can vary over time within a lake (Rublee and Bettez 2001).

The spatial and temporal patterns also vary by the different members of the microzooplankton community. Protozoan numbers are highest (up to ~1000 ind L^{-1} and 1-2 μg C L^{-1} biomass) in the epilimnion early in the summer and generally decrease with depth and over time. The rotifer population, however, does not correlate with depth and is low in the early summer (10–300 ind L^{-1}) and may increase five- to ten-fold by mid-July or early August. Naupliar biomass equals or exceeds rotifer biomass and shows a positive correlation with depth and time (Rublee 1992).

Macrozooplankton

Few species of crustacean zooplankton exist in this region with only seven common pelagic species present in over 100 lakes that have been sampled. Although the glacial history of the watersheds of this region has little impact on the distribution of most crustacean zooplankton, *Holopedium giberum* occurs more frequently in lakes on the older glacial surfaces (Burkart 2007). O'Brien et al. (2004) reported negative correlations between *H. gibberum* abundance and conductivity of surface waters in the lakes of the Toolik region. This correlation has been reported for *H. gibberum* from other regions of the world (Hessen et al. 1995). In a comparison of crustacean zooplankton biomass in seven lakes on recently glaciated landscapes compared to five lakes on older glacial surfaces, biomass of *H. gibberum* was ten-fold greater in lakes on older glacial surfaces (Burkart 2007). In contrast to *H. gibberum, Diaptomus pribilofensis* occurs more frequently in lakes with higher conductivity.

Lake size, depth, and the presence of fish have a pronounced effect on the abundance and distribution of crustacean zooplankton in the study lakes. *Daphnia middendorffianna* and *Heterocope septentrionalis* are more abundant in small than in medium and large lakes (Figure 8.11). These two large-bodied zooplankters are likely eliminated by fish predation in the medium-sized and large lakes. O'Brien et al. (2004) reported that densities of *D. middendorffiana* were twice as high in fishless lakes as in lakes with fish. In a more detailed study of 12 lakes, Burkart (2007) found that total crustacean zooplankton biomass was five-fold higher in shallow lakes without fish than in deeper lakes with arctic char, grayling, lake trout, or sculpin. These results from these low-arctic lakes are similar to those reported for alpine lakes (Knapp et al. 2001), and in low elevation lakes at lower latitudes (Carpenter et al. 1996).

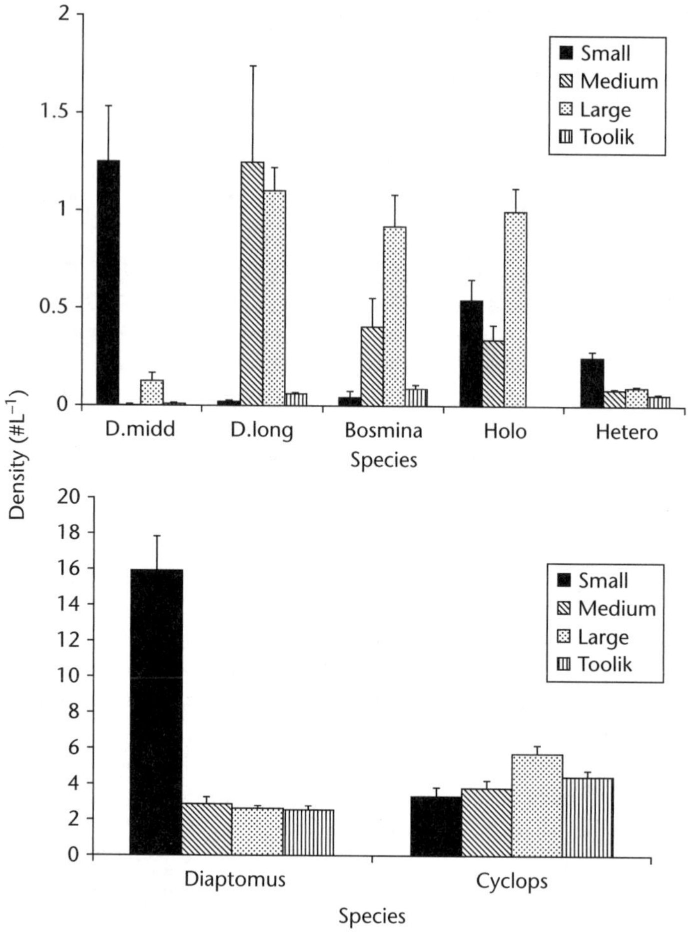

Figure 8.11 Density of cladocerans (top) and copepods (bottom) in upper portion of the water column in small, medium, and large lakes compared to densities in Toolik Lake. Mean + SE for lakes listed in Table 8.1 for the years 2000–2005. D. midd is *Daphnia middendorffiana*, D. long is *Daphnia longeremis*, Holo is *Holopedium*, and Hetero is *Heterocope septentrionalas*.

Vignette 8.4. Zooplankton Pigmentation: A Balance of Photo-protection and Fish Predation, by John O'Brien

If fish predation is so important in determining the abundance of zooplankton in alpine and arctic lakes, then we would expect that the zooplankton would be nearly transparent in order to reduce their visibility. Yet, the various populations of zooplankton in arctic lakes and ponds show a wide variety of pigmentation. For example, several copepod species range from bright red to blue to relatively clear. What causes this great variety in pigmentation and what are the advantages for survival?

One pattern is associated with the presence or absence of fish. The more colorful morphs are found in small ponds that lack fish, and the most transparent forms are found in deeper lakes where fish are present and predation is a risk. Another pattern is that pigmentation in zooplankton may be present in some species even when fish are present. This pattern seems to be tied to the necessity for photo-protection against sunlight. Particularly around the summer solstice, the solar radiation is intense, and exposure to ultraviolet light creates oxygen radicals that damage the cells of zooplankters. Experiments where individual pigmented and unpigmented zooplankton were exposed to surface light showed better survival by pigmented individuals. Further, zooplankton reduce photo-toxicity through production of a series of photo-protective pigments. Pigments like astaxanthin bind oxygen radicals and reduce tissue damage. The presence of these pigments reduces the transparency of pelagic zooplankton, thereby increasing their risk of being consumed by fish (Luecke and O'Brien 1981).

The hazards of photo-damage and fish predation interact to create a strong contrast between zooplankton assemblages in fish and fishless lakes (Luecke and O'Brien 1983a). Highly pigmented, large-bodied zooplankton dominate fishless ponds, whereas lakes with fish are occupied by relatively transparent or small-bodied zooplankton that avoid the top few meters of surface water. One ubiquitous species, *Diaptomus pribilofensis*, succeeds by combining small size at reproduction (to reduce predation from fish), with high concentrations of astaxanthin. This species thrives in almost all of the 100+ ponds and lakes in the vicinity of Toolik (O'Brien et al 2004).

Winter conditions in this low-arctic region create highly contrasting food webs through elimination of fish in shallow lakes. This occurs because lakes and ponds less than 1.5-m deep freeze completely, and oxygen concentrations are greatly reduced overwinter in slightly deeper lakes and ponds. In fishless habitats the pigmented invertebrates thrive, and invertebrate predators are important in structuring zooplankton assemblages. In lakes of the low Arctic, these predators are the copepods *Heterocope septentrionalis* and *Cyclops scutifer*, the phantom midge larva *Chaoborus,* and the amphipod *Hyalella azteca.* These organisms feed extensively on herbivorous zooplankton. Laboratory feeding trials and field enclosure experiments at Toolik indicated that the presence of *H. septentrionalis* determines the composition of herbivorous zooplankton in the ponds and lakes of the region (Luecke and O'Brien 1981). Yet these copepod predators have a range of feeding habits; stable isotope measurements showed some populations are mostly carnivorous and most are herbivorous, probably as a result of prey availability in different lakes (Kling et al. 1992c).

The short ice-free season and the cold temperatures present in most lakes restrict life-history options and patterns of energy allocation in arctic zooplankton. Yurista (1999) constructed an energy budget for *Daphnia middendorffiana* using populations from the ponds near Toolik Lake. His measurements of rates of ingestion, assimilation, and respiration at different temperatures indicated that *D. middendorffiana* can thrive at cold temperatures because ingestion rates at low temperature were higher than those reported for *Daphnia* species from temperate zone populations (McCauley et al. 1990). These higher ingestion rates of *D. middendorffiana* from the Arctic were coupled with lower assimilation rates than those measured for temperate-zone *Daphnia*.

Fish Distributions

Only six species of fish inhabit the lakes and streams near Toolik. These species are all native, a rare occurrence in North America where introduced fish are everywhere. Lake trout (*Salvelinus namaycush*), burbot (*Lota lota*), and arctic char (*Salvelinus arcticus*) occupy the highest trophic levels feeding extensively on benthic invertebrates, but some individuals will become piscivorous. Arctic grayling (*Thymallus arcticus*) and round whitefish (*Prosopium cylindraceum*) feed on benthic invertebrates and occasionally zooplankton, and slimy sculpin (*Cottus cognatus*) is a bottom-dweller also feeding on benthic invertebrates. Analyses of lake morphometry and landscapes indicated that lake trout and arctic char occupy only deep lakes that are part of well-established drainage systems (Hershey et al. 2006). Lake trout and round whitefish were most often found in lakes of the Itkillik drainage west of Toolik, and arctic char were most often present in lakes with drainage connections to the Sagavanirktok River. The distribution of grayling and sculpin indicated high dispersal abilities in that these two species were present in all watersheds sampled and were the only fish taxa found in lakes with weak drainage connections to other lakes.

Lake trout function as a top predator in these lakes, feeding on a mix of invertebrate and vertebrate prey. Investigations of spatial distributions and experimental manipulations indicated that the presence of lake trout not only has large impacts on the abundance and distribution of their prey but also exhibits significant effects on lower–trophic level organisms. In a detailed set of investigations, lake trout reduced numbers of slimy sculpin and restricted remaining sculpin to near-shore habitats (Hanson et al. 1992; Goyke and Hershey 1992). A variety of benthic invertebrates preyed on by sculpin respond positively to the presence of lake trout.

Results of Experiments and Process Studies

Experiments on the limnological processes in of these low-arctic lakes have been carried out to assess effects of land-use changes and climate on lake productivity and food-web structure. One effect of climate warming (ACIA 2004) and of increasing human populations is an increase in nutrients entering lakes. Various human activities have the potential to introduce new species into these lakes, and climate warming may allow species introduced from lower latitudes to thrive.

Over the past two decades a number of nutrient addition experiments were carried out on Toolik area lakes. These studies ranged from bottle assays to large mesocosm experiments to whole lake manipulations. The goals of these experiments were to assess the degree of nutrient limitation of algae as well as the response of the whole lake ecosystem to changes in nutrient inputs. The predominance of co-limitation of phytoplankton growth by nitrogen and phosphorus described above shaped the design of larger-scaled experiments to assess the impact of increased nutrient loading on lake ecosystems.

In addition to experiments designed to examine effects of nutrient additions to the base of the food web, a series of investigations to assess the importance of consumer effects on lake ecosystems has been conducted. These experiments ranged from small-scale observations of the behavior of predators and prey to whole lake additions or removals of individual species. Results of the experiments on food-web manipulations may be relevant to potential future effects of the loss or addition of species due to climate change or human introductions.

Mesocosm Experiments

From 1983 to 1985 a limnocorral experiment was established in one of the deep bays of Toolik Lake (O'Brien et al. 1992). Each of the corrals contained about 62 m^3 of lake water. Three nutrient treatments (no addition, low-level fertilization, and high-level fertilization) were coupled with two levels of planktivorous fish (grayling) abundance. The low-level nutrient addition was equivalent to the natural level of nutrient loading of N and P to Toolik Lake (Whalen and Cornwell 1985) and the high-level nutrient addition was either five or ten times that amount depending on the year. Fish were added to some limnocorrals as either free-ranging individuals or as caged individuals so that by moving the cage in and out of the mesocosm the rate of predation could be lower than that from a single fish. In all cases, the fish addition treatments likely exceeded natural abundance of fish in these lakes.

The major result of the limnocorral experiments was that phytoplankton responded dramatically to additions of nitrogen plus phosphorus. Rates of pelagic primary production were 10- to 20-fold higher in limnocorrals that received nutrients. These results indicated that not only were phytoplankton in these arctic lakes limited by N and P, but that sustained additions of these two nutrients had the effect of greatly increasing the biomass and primary productivity of Toolik Lake water. Chrysophytes and Chlorophytes increased in abundance in fertilized limnocorrals relative to other phytoplankton taxa (O'Brien et al. 1992).

The increase in phytoplankton had dramatic effects on physical and chemical parameters in the limnocorrals. The increased biomass of phytoplankton decreased light penetration into the corrals receiving nutrients. Secchi depth declined from 5 m in reference limnocorrals to less than 2 m in nutrient addition treatments. Both pH and oxygen concentrations increased greatly in the limnocorrals receiving nutrient additions due to the increases in rates of photosynthesis.

Response of heterotrophic organisms to the dramatic increases in phytoplankton abundance was mixed. Planktonic bacteria increased four- to eight-fold, but soon their abundance was drastically reduced by grazing of heterotrophic nanoflagellates in a classic predator–prey relationship (O'Brien et al. 1992). Microzooplankton grazers increased greatly in limnocorrals receiving nutrient additions (Rublee 1992). The response of macrozooplankton grazers, however, was mixed, likely due to predation by fish in the fish treatments. When fish predation was absent or present at low levels, larger cladocerans increased in abundance in response to higher biomass of phytoplankton in the fertilized treatments. The long and complex life histories of the pelagic copepods and many of the benthic invertebrates likely reduced their response to higher abundances of phytoplankton (Hershey 1992).

Whole-Lake Experiments

In order to more fully understand the effects of nutrient additions on low-arctic lake ecosystems, a series of whole-lake fertilization experiments was conducted near Toolik Lake. Results of these experiments allow inferences concerning response of fish and benthic invertebrates that could not be incorporated in the limnocorral experiments. In 1985, Lake N2 was divided in half by a polyethylene curtain that extended to the bottom of the lake. The downstream end of the lake was fertilized with phosphoric acid and ammonium nitrate such that annual loading of these two nutrients was five times natural values (O'Brien et al. 2005). This fertilization experiment ran for six years, and the lake was sampled in subsequent years to assess the time-scale of recovery from eutrophication.

Primary production in the fertilized side of the lake was three to five times higher than the reference side throughout the six years of nutrient additions. Nutrient addition also dramatically increased chlorophyll concentrations in the fertilized side of the lake (Figure 8.12). Within two years there was an increase in the ammonium flux from the sediments, but it took five years before phosphorus began to be released from the mineral-rich sediments (Figure 8.13). This P flux was most likely related to low oxygen concentrations in hypolimnetic water, which allowed P to be released as iron minerals were reduced (Figure 8.14). Oxygen concentration in the hypolimnion was lower with each year of fertilization, and by the fourth year of fertilization oxygen fell below 1 mg L^{-1} during July and August.

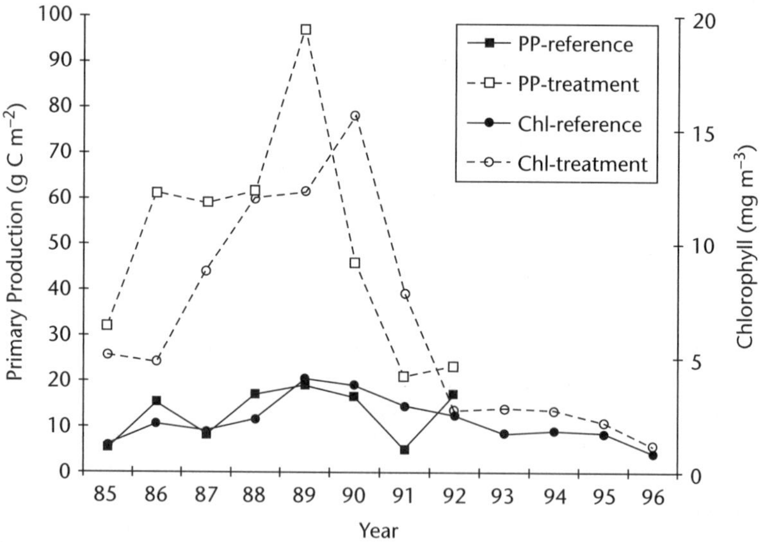

Figure 8.12 Mean summer concentration of pelagic chlorophyll *a* (circles) and total summer primary productivity (squares) in Lake N2. Filled symbols represent the reference section of the lake; open symbols represent the treated section of the lake. The treated section of this lake was fertilized with dissolved nitrogen and dissolved phosphorus during the summers of 1985–1990. From O'Brien et al. (2005), used with permission.

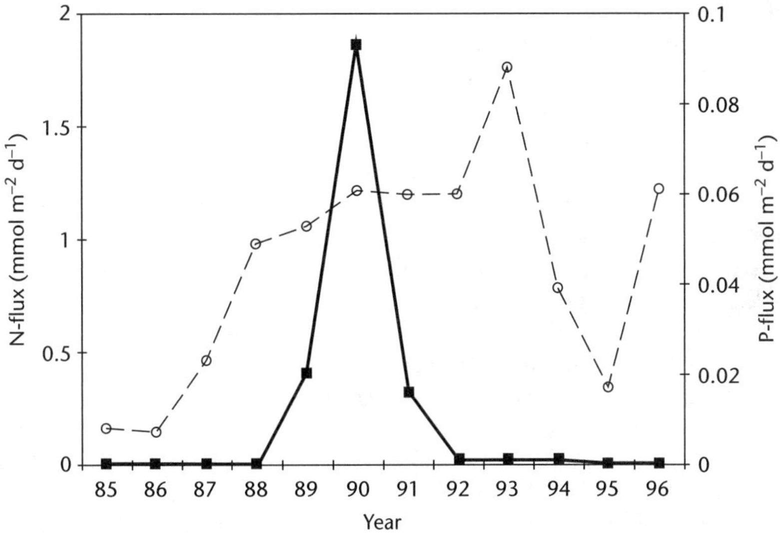

Figure 8.13 The average daily fluxes of inorganic nitrogen (open circles) and phosphorus (solid squares) from the sediments of the fertilized sector of Lake N2. The treatment fertilization was conducted each summer during from 1985 through 1990.

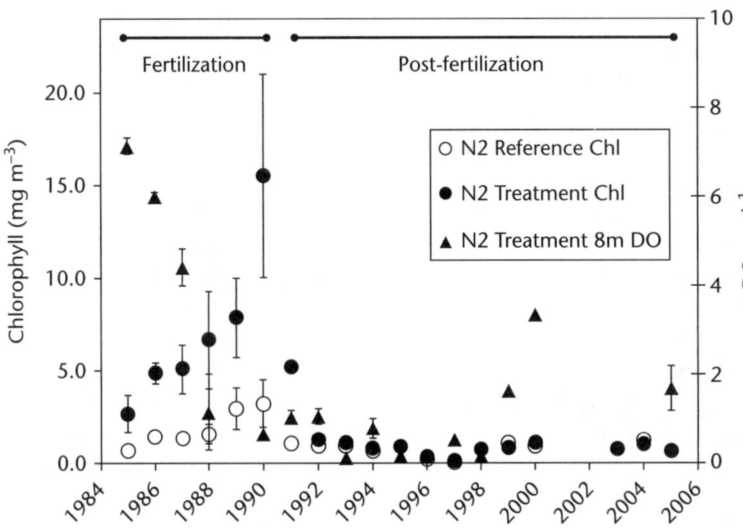

Figure 8.14 Oxygen concentration near the bottom (8 m) of the fertilized treatment in Lake N2 from 1985 to 2005 along with chlorophyll *a* concentrations in surface waters. Bottom-water oxygen concentrations in the reference side of Lake N2 remained high (>6 mg O_2 L^{-1}) during and after the fertilization.

Water-column levels of primary production in the fertilized side of Lake N2 were elevated over the reference side for two years after fertilization ceased (Figure 8.12); but by three years after the fertilization had ceased, chlorophyll *a* concentration and epilimnetic nutrients were similar to levels in the reference side. In contrast, oxygen concentrations remained below 1 mg L^{-1} until seven years after the cessation of fertilization. Late-summer oxygen conditions had still not returned to pre-fertilization levels 15 years after the cessation of fertilization. Ammonium fluxes from sediments were still elevated five years later. In spite of persistent low-oxygen conditions, no measurable P flux from the sediment was observed even two years after fertilization had ceased (Figure 8.13).

The response of the pelagic and benthic animal community to fertilization was mixed. One of the four macrozooplankton species (*Daphnia longiremis*) increased in number by about two-fold in the first five years. However, the copepod *Cyclops scutifer* showed no response during the treatment phase of the study and may be dependent more on the microbial food web than on primary production (Kling 1994). Benthic snails responded significantly, both *Valvata lewisi* and *Lymnaea elodes* increased in abundance on the fertilized side of this divided lake (Hershey 1992). In contrast there was no response of chironomid larvae. This suggests that chironomids may be limited by predatory sculpins rather than food in these arctic lakes. Because sculpins fed primarily on chironomid larvae in this lake, it was not surprising that sculpin growth rates were similar in the fertilized and control sections of the lake. We observed a similar lack of response of chironomid larvae and sculpin growth in the limnocorral experiment. This result from the divided lake, combined with isotope data (Kling 1994), suggests a lag in the response to fertilization of some benthic organisms.

A second whole-lake fertilization experiment was conducted on Lake N1 from 1990 to 1994. Nitrogen and phosphorus were added to this lake during summer months at a rate of approximately four times natural loadings. Lake N1 is deeper than Lake N2 and contains four of the six fish species present in the region: Lake trout, arctic grayling, burbot, and slimy sculpin.

As in previous nutrient addition experiments, phytoplankton biomass and primary production increased dramatically in response to additions of nitrogen and phosphorus (Figure 8.15). Chlorophyll concentrations and rates of primary production increased more than ten-fold over pre-fertilization conditions. These increased rates of primary production during the five years of the experiment eventually resulted in the development of an anoxic zone in the bottom of Lake N1 in 1994 (Lienesch et al. 2005).

As in the fertilization experiment in Lake N2, snail abundances in Lake N1 increased substantially during fertilization years. Densities of both *Valvata lewisi* and *Lymnaea elodes* were four times greater after the experimental fertilization. Densities of chironomid larvae did not show a change after fertilization, a result similar to the lack of response of chironomid larvae in the N2 fertilization experiment (Hershey 1992). Lienesch et al. (2005) report that growth rate of lake trout responded positively to this increase in snail abundance (Figure 8.16). The lack of a response of the numbers of lake trout to the fertilization was attributed to the effects of low oxygen conditions on the survival of lake trout eggs. In contrast to lake trout, individual slimy sculpin showed no increase in growth rate under fertilization treatments.

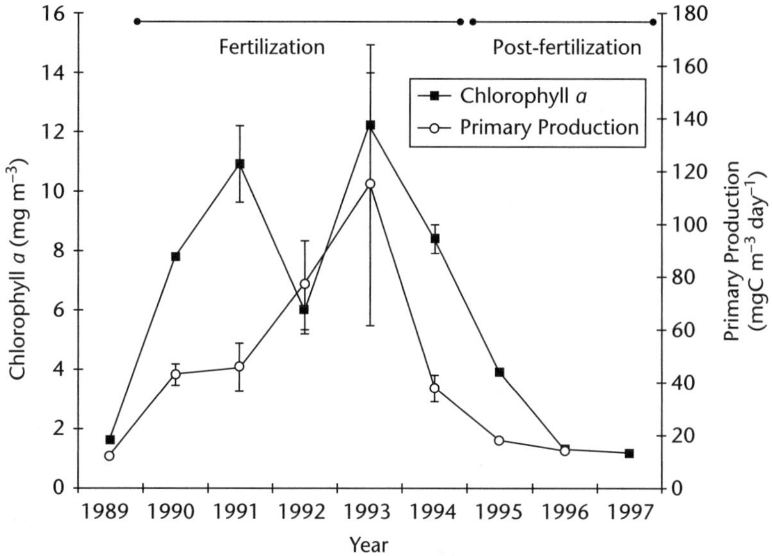

Figure 8.15 Response of chlorophyll *a* and rates of primary production in the water column of Lake N1 to fertilization. Mean summer values +1 SD are shown. Addition of phosphoric acid and ammonium nitrate occurred during summer months of 1990–1994. Redrawn from Lienesch et al. (2005), used with permission of Springer Publishing Company.

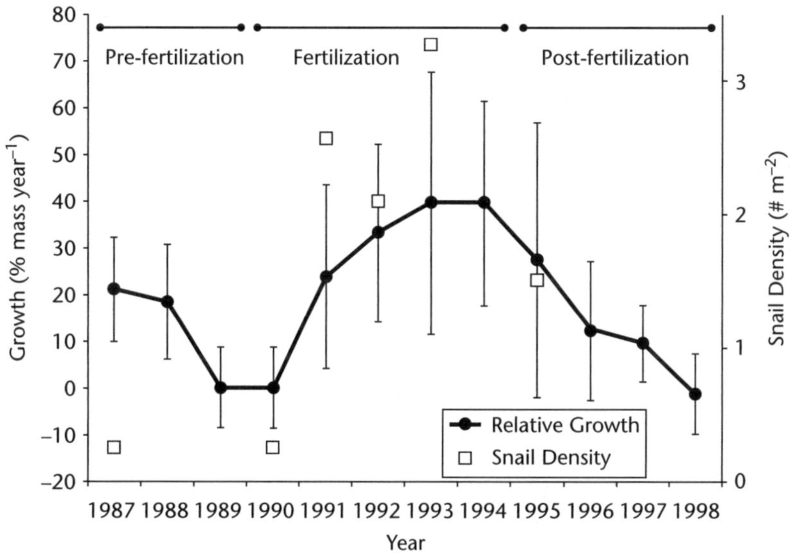

Figure 8.16 Relative growth of individually tagged lake trout in lake N1 before during and after fertilization (filled circles are mean values ±1 SE). Open squares represent density of snails observed by divers. From Lienesch et al. (2005), used with permission of Springer Publishing Company.

Chlorophyll concentrations as well as most limnological parameters returned to pre-fertilization values within one to two years after the nutrient additions ceased at the end of 1994, slightly more rapidly than in Lake N2. The slight difference in recovery times is likely due to the greater ratio of sediment area to lake volume in Lake N1; this allowed greater P binding by shallow oxidized sediments, limiting P return to the water column. An important exception to the rapid recovery in Lake N1 was that hypolimnetic oxygen concentration remained quite low through the summer of 1998 and had not completely recovered by 2010 (Lienesch et al. 2005; LTER database). The accumulation of sedimented phytoplankton during the fertilization years continued to provide a labile substrate for bacterial respiration long after the fertilization was ended. These results demonstrate the importance of maintaining long-term assessment of experiments to gain an understanding of processes that operate on very different time scales.

In the mesocosm and whole-lake experiments, changes in microzooplankton abundance were related to the degree of fertilization. In the limnocorral experiment additions of N and P at ten times natural loading resulted in a ten-fold increase in microzooplankton biomass while in lakes N1 and N2, which were fertilized at 3–5 times natural rates, the biomass of microzooplankton doubled (Rublee and Bettez 1995, 2001).

The response of microzooplankton taxa to fertilization treatments varied among lakes and among years (Rublee and Bettez 1995). These differences were likely due to differences in abundance of macrozooplankton present in these treatments. *Daphnia* is known to inhibit population growth of microzooplankton in whole-lake fertilization experiments in temperate lakes (Pace et al. 1998) and likely contributed to the variation in response of microzooplankton in fertilization experiments in arctic lakes.

In Lake N1, the only lake from which we have data on microzooplankton before, during, and after the experiment, there were shifts in both the number of individuals and species present throughout the experiment. The microzooplankton community shifted from one dominated by small particle-feeding rotifers (mainly *Conochilus unicornis)* during the first year of fertilization to a dominance of bacteriovorus peritrich protozoans (*Epistylus rotans*, and *Vorticella* sp.) during the second year, and finally to a dominance of the predatory rotifers *Synchaeta, Polyarthra,* and *Trichocerca* and zooplankton nauplii during the third, fourth, and fifth years of the fertilization (Rublee and Bettez 2001; Bettez et al. 2002).

After the end of fertilization, the biomass of the microzooplankton took one year in Lake N1 and two years in Lake N2 to return to the pretreatment level. In Lake N1 in the first year of recovery, the predatory rotifers decreased to their pre-fertilization levels and the composition of the microzooplankton returned to a dominance by oligotrichs and small particle-feeding rotifers. In Lake N2, the two years necessary for return of microzooplankton biomass to pre-fertilization levels was likely due to the longer time in which the lake maintained elevated chlorophyll and primary productivity.

Mesocosm experiments and whole-lake fertilization of lakes N1 and N2 indicated that phytoplankton and bacteria responded dramatically to nutrient increases of four to ten times natural loading rates. Results of food-web response to fertilization in mesocosms and whole-lake experiments in nearby lakes were similar to results of whole-lake fertilization experiments conducted in the temperate zone (Carpenter et al. 1996). In both the low-arctic and temperate lakes, phytoplankton populations increased dramatically with increased nutrient loading. The response of higher–trophic level organisms depended on the food-web configuration. When fish were present in nutrient-addition treatments, large-bodied zooplankton like *Daphnia* were not able to respond to increases in phytoplankton food resources. When large-bodied zooplankton densities remained low, microzooplankton increased dramatically in response to increased phytoplankton abundance, as also found by Pace et al. (1998). This similarity in response in large-scale fertilization experiments suggests that trophic cascades (Carpenter et al. 1996) function in a similar manner in arctic and temperate lake ecosystems.

Ongoing Whole-Lake Fertilization Experiments

The previous experiments were carried out with fairly large increases in nutrient levels. In 2001 a low-level fertilization experiment on a deep and a shallow lake was initiated to assess the impact of likely increases in nutrient loading due to climate warming and associated environmental changes. Fertilization at two times natural loading rates was begun in Lake E6 (maximum depth of 3 m, fishless) and Lake E5 (maximum depth 12 m, contains arctic char) to assess the degree to which any increases in primary production would result in increased biomass and productivity of zooplankton and fish. Lakes Fog2 (deep) and Fog4 (shallow) were chosen as reference systems. In these whole-lake fertilization experiments, N and P were added along with small amounts of ^{15}N to serve as a tracer of nutrient flow. This tracer allowed us to assess the time lags of when different food-web components are able to use the added nutrients.

This low-level fertilization of lakes E5 and E6 is ongoing. Preliminary results indicate that the fertilization treatments had immediate effects on phytoplankton populations. Chlorophyll concentration and rates of primary production of phytoplankton increased after the initiation of the fertilization treatments. These increased levels may have been enhanced each year indicating that the effects of fertilization during one summer carry over to subsequent years. The fact that this carryover occurs even though residence time of the water in these lakes is less than one year (Evans 2007) suggests storage and return of nutrients from the sediments.

As in the other whole-lake fertilization experiments, effects on higher trophic levels have been less obvious. Some taxa of crustacean zooplankton have increased in abundance in lakes E5 and E6 relative to the reference lakes, but only in years with higher than average epilimnetic temperatures. Lower temperatures increase the egg-development time of cladocerans and appear to limit their population response to added food during cooler summers.

In the fertilized lakes, the phytoplankton and zooplankton took up the ^{15}N tracer during the first summer. The concentration of ^{15}N in phytoplankton increased over the summer, and different zooplankton taxa accumulated the ^{15}N label at different rates with the faster uptake occurring in cladocerans and slowest uptake in cyclopoid copepods. Detectable increases in ^{15}N were not observed in benthic chironomids or snails until the second summer of the experiment. Increases in ^{15}N concentrations of arctic char were not observed until the third summer of the nutrient additions. This lagged response of the benthic components of the food web is similar to the lagged response we observed in the previous higher-level fertilization experiment in Lake N2 (Kling 1994).

Consumer Effects on Lake Processes

Both vertebrate and invertebrate planktivores in Toolik-area lakes have dramatic impacts on zooplankton assemblages. In laboratory experiments, invertebrate predators, particularly the large calanoid copepod *Heterocope septentrionalis*, fed selectively on smaller cladocerans (Luecke and O'Brien 1983b). In lakes where *H. septentrionalis* is abundant, the dominant zooplankton species possess a variety of defenses such as gelatinous coverings (*Holopedium gibberum*), extended helmets (*Daphnia longeremis*), or behaviors that prevent effective consumption as found in *Bosmina longirostris* and *Diaptomus pribilofensis*. In water bodies where *H. septentrionalis* is not present, the dominant zooplankter is *Daphnia pulex*, a highly vulnerable prey item. Rates of *H. septentrionalis* feeding on *D. pulex* in the laboratory indicated that predation rates could exceed reproduction of *D. pulex* under field conditions.

Studies on the feeding selectivity of planktivorous fishes indicate that large and heavily pigmented zooplankton were fed upon preferentially by both juvenile grayling and juvenile lake trout (O'Brien 1979). A simulation model of arctic grayling feeding (O'Brien and Evans 1992) indicated that predation rates on large-bodied zooplankton like *D. middendorffiana* would be sufficient to explain the differences in zooplankton assemblages present in fish and fishless lakes near Toolik. Results of the laboratory experiments and modeling simulations also explained why the limnocorral treatments that contained fish (O'Brien et al. 1992) had decreased abundances of *D. middendorffiana* and *H. septentrionalis*. These results help explain why the presence of fish has a dramatic impact on zooplankton assemblages in these lakes, even though diet composition and analyses of stable isotopes of carbon and nitrogen indicate that zooplankton comprise a very minor part of both grayling and lake trout diets. The high selectivity of the juvenile stages of these two fish species for large-bodied zooplankton greatly reduces these prey in lakes with high fish densities. After the large-bodied zooplankters are nearly eliminated, fish begin feeding on benthic invertebrates.

This conceptual model of the impact of fish predation is further supported by results of several whole-lake food-web manipulations. In 1988–1989 most of the lake trout greater than 300 mm were removed from lake NE12 by intensive gill-netting. These larger lake trout are potential piscivores and feed to some degree on sculpin and juvenile grayling. After the lake trout removal, juveniles of both grayling and lake trout began consuming more pelagic zooplankton prey, resulting in several very successful recruitment years (Keyes et al. 2007). As these juvenile fishes became more plentiful, their feeding rates on zooplankton resulted in a decrease in the abundance of *Daphnia*.

These interactions in lake food webs were further demonstrated by experimental reductions in lake trout density in a series of lakes. The removal of lake trout was accompanied by a large increase in recruitment of both juvenile lake trout and arctic grayling (Merrick et al. 1992). These successful year classes provided forage for burbot, formerly present at low densities, whose population increased as an indirect function of lake trout removal. These compensatory changes demonstrate the complexities inherent in food webs in arctic lakes.

Herbivorous macrozooplankton also exert top-down effects on lake communities (Carpenter et al. 1985). Natural densities of crustacean zooplankton exhibited grazing rates of between 12% and 22% of the water column per day in lakes E5, E6, Fog2, and Fog4 (Burkart 2007). A multiple regression model indicated that the variation in grazing rates was due to the species of zooplankton present and water temperature ($r^2 = 0.78$). These clearance rates accounted for approximately 20%–80% of the primary production of phytoplankton in the four lakes during the summers of 2000–2002, indicating that grazing greatly reduces the growth rate of phytoplankton populations during warm periods of high cladoceran biomass. The lack of large herbivorous cladocerans and relatively low densities of smaller copepods in Toolik Lake (Figure 8.11) reduces the ability of herbivorous zooplankton to exert strong grazing effects on phytoplankton (O'Brien et al. 1997).

Grazing by crustacean zooplankton appears to influence the species composition of phytoplankton assemblages in the lakes near Toolik. In fishless lakes where biomass of *Daphnia* often exceeds 100 μg L^{-1}, phytoplankton assemblages were dominated by taxa with cell sizes of greater than 80 μm in their longest dimension (Figure 8.7; Burkart 2007). Reciprocal grazing experiments indicated that zooplankton from two fishless lakes exhibited higher grazing rates when feeding on phytoplankton from lakes containing fish. These higher grazing rates only occurred during late summer when water temperatures exceeded 16°C and *Daphnia* populations were at maximal levels. The reported 30% increase in zooplankton feeding rates on phytoplankton from lakes with fish was likely due to the dominance of small, single-celled phytoplankton present in these lakes (Burkart 2007).

Evidence of Response to Recent Climate Change

The longest data sets on environmental change in lakes of this region come from Toolik Lake which has been monitored since 1975 and which serves as the primary long-term monitoring site for lakes in the region. Additional data come from more than 30 lakes (including those in Tables 8.1 and 8.2) that have been monitored less intensively but which have been sampled since the late 1980s or early 1990s.

Since 1975 there has been a very slight increase in average mid-summer epilimnetic water temperatures in Toolik Lake, but the increase is not statistically significant (Figure 8.17). This is in keeping with the data showing a slight but non-significant increase in mean annual air temperatures at Toolik Lake (chapter 2), and not unexpected when considering that most warming in the Arctic has occurred in the winter (chapter 2). In addition to the lack of trends in summer lake thermal conditions, no perceptible changes have been observed for many of the physical,

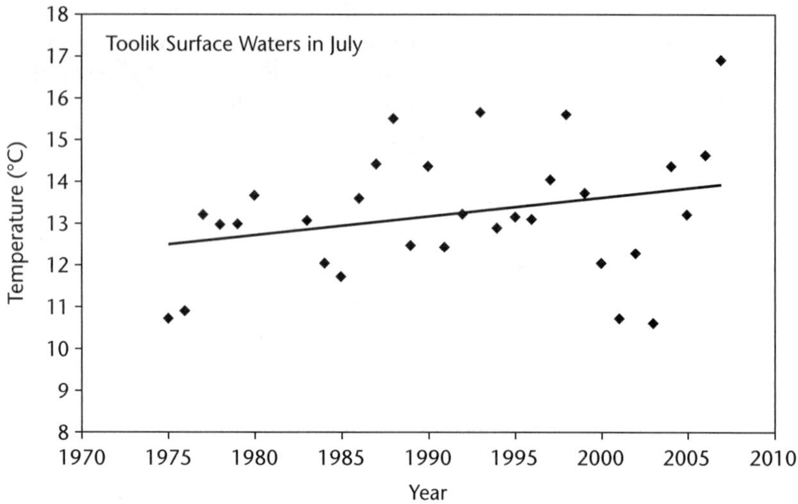

Figure 8.17 Mean temperature in the epilimnion of Toolik Lake each July 1975–2007. The slope of the regression line (y = 0.045x—76, R^2 = 0.082) is not statistically significantly different from zero.

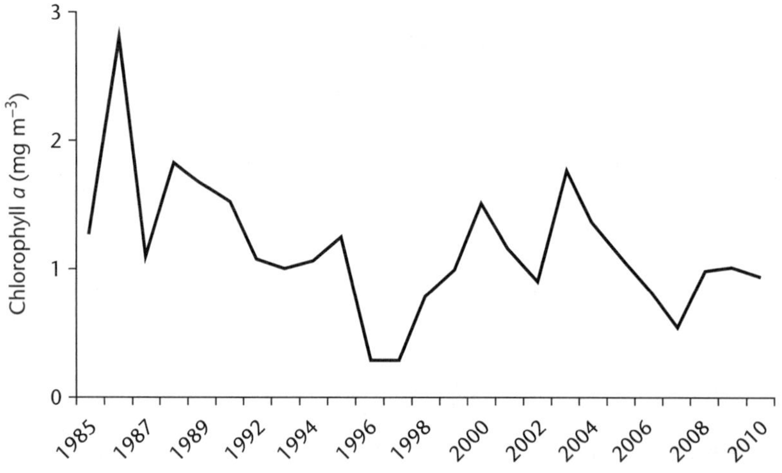

Figure 8.18 Concentration of chlorophyll *a* in the epilimnion (0–3 m) of Toolik Lake during July in period 1985–2010. No overall trend in this data series was present, but large interannual differences are apparent. Each data point represents the mean of 6–15 epilimnetic samples collected over the month.

chemical, and biological characteristics of Toolik Lake. For example, we have seen no substantial changes in chlorophyll concentrations (Figure 8.18), nor is there evidence of significant changes in nutrient concentrations. The apparent lack of changes in nutrient concentrations may be related to their extremely low concentrations and to the fact that our analytical methods for nutrients have become more sensitive during the course of the Arctic LTER program. In contrast, the most significant long-term trend we have observed is a relatively large increase in lake-water alkalinity (Figure 6.14), also observed in other lakes in the region. Although thaw probe measurements show no significant trend of increasing depth of the summer active layer (from 1990 to 2011; see chapter 6), measurements of strontium isotopes in the inlet stream to Toolik Lake suggest that the source of alkalinity is from weathering of previously frozen glacial till at the bottom of the active layer (Keller et al. 2007). One would expect that this increased contact with mineral soil would result in an increase in P delivery to lakes and streams as the mineral apatite (calcium phosphate) is present in the glacial till. While the P concentrations in streams and lakes have not increased, we do know that the Milky Way tributary to Toolik Inlet, which drains a gravel mine last active in the early 1970s, has the highest P concentrations of any tundra stream in the region. This area mined for gravel has subsided due to removal of several meters of material from the surface followed by thawing of permafrost and melting of contained ice; this is consistent with greater weathering of deeper soils that are apatite rich. The lack of a trend of increase in the lakes of P concentration over time could be a result of insufficient sensitivity in our chemical methods at such low P concentrations, or it could be that terrestrial plants and stream and lake organisms take up phosphate as soon as the mineral form is broken down.

Recent observations suggest that changes in lakes in this region are being driven more by an increase in major disturbances than by the direct or indirect impact of warming on lake or soil processes. One disturbance is an increase in wildfire. In 2007, a 1000 km^2 tundra fire occurred 35 km north of Toolik Lake (Rocha and Shaver 2011). In the fire-affected area it appears that land-surface failures due to thawing ground ice have greatly increased. Even in the absence of fire, there may be an increase in "thermokarst" formation in northern Alaska (chapter 7; Bowden et al. 2008). However, any increase in recent years is difficult to conclusively document as most of these thermokarst disturbances cannot be seen by satellites. Whether these thermokarsts are due to climate warming or fire increase, they lead to increased solute and nutrient transport to streams and lakes (Bowden et al. 2008; Gooseff et al. 2009), alter lake clarity, and represent a major disturbance to lake ecosystems that may persist for years.

Future Predictions of Ecosystem Processes in Arctic Alaskan Lakes

The general circulation climate models indicated that for the Arctic there will be increases in temperature of 4°C–6°C over the next century, but most of this change is expected to occur during winter periods (chapter 2). These higher air temperatures are coupled with a 10% increase in precipitation, although the multi-model predictions for just northern Alaska indicated higher precipitation (1.5- to 3-fold increases; see chapter 2), which would create a warmer and wetter environment for the Toolik region. The large interannual variability in climate measurements and forecasts for this region make it difficult to make precise predictions of how lake ecosystems will change in future years. Although indices of summer lake temperatures and lake productivity have shown little change at Toolik Lake since the mid-1970s, increases in air and soil temperatures have been recorded in nearby regions (Hinzman et al. 2005). The location of Toolik in the foothills near the Brooks Range increases local meteorological influences and makes identification of regional climatic trends more difficult. However, it is possible to use the mechanistic studies and results obtained under a variety of climate "extremes" at Toolik Lake in order to describe the potential future changes in ecosystem processes due to warming temperatures and altered moisture regimes.

Future increases in summer lake temperatures and increased thermal stratification of the lakes could greatly influence the diversity of biological communities and the rates of growth and other ecosystem processes of these lakes. For example, while phytoplankton productivity responds more strongly to light and nutrients than to temperature (e.g., Evans 2007), the response of zooplankton and fish to changes in temperature may be dramatic. In one study by Yurista (1999), the effects of temperature on the energy budget of *D. middendorffianna* indicated that the lakes of this region rarely exceed optimal temperatures for growth of this species. A warming of surface waters would reduce the ability of *D. middendorffianna* to thrive in the epilimnion of deeper lakes and may eliminate this species from shallow ponds. Yurista (1999) concluded that the combination of high ingestion rates at cold temperatures and low assimilation rates at warmer temperatures allows *D. middendorffiana* to thrive in the arctic lakes of today, but may pose difficulties if climate change increases lake temperatures and length of growing seasons such that temperate-zone *Daphnia* populations invade the Arctic.

An initial study on the effects of climate on fish using a bioenergetics simulation model showed that increased lake temperatures would greatly reduce the ability of lake trout to thrive in these low-arctic lakes (McDonald et al. 1996). Lake trout prefer water of 10°C with oxygen concentrations above 6 mg L^{-1} and show stress at temperatures above 15°C. The lack of abundant deep-water habitat in most of the lakes of the region means that usable summer habitat for lake trout would decrease by 30% if epilimnetic temperatures rise by 2°C (Hobbie et al. 1999). This reduction of habitat is exacerbated if nutrient loading to the lakes increases phytoplankton productivity resulting in decreases in hypolimnetic oxygen concentrations, or if groundwater inflows increase, leading to chemical stratification, which reduces mixing and also contributes to increased anoxia. In another region of northern North America, Sellers et al. (1998) suggest that loss of habitat for lake trout in lakes on the Canadian Shield may already be occurring.

Some insight into potential effects of climate warming on fish populations in the lakes near Toolik may be gained by studying warm summers. Analyses of the length-weight relationships of the common fishes of the Toolik area indicated that in summers when epilimnetic temperatures exceed 15°C for extended periods, a decrease in the slope of the length-weight relationship occurs for lake trout, arctic grayling, and arctic char (Figure 8.19). Additional increases in epilimnetic temperature would likely result in dramatic declines in fitness of the fishes in these lakes. The shallow nature of many of the lakes coupled with the lack of food resources below the epilimnion may make it difficult for these fish populations to persist.

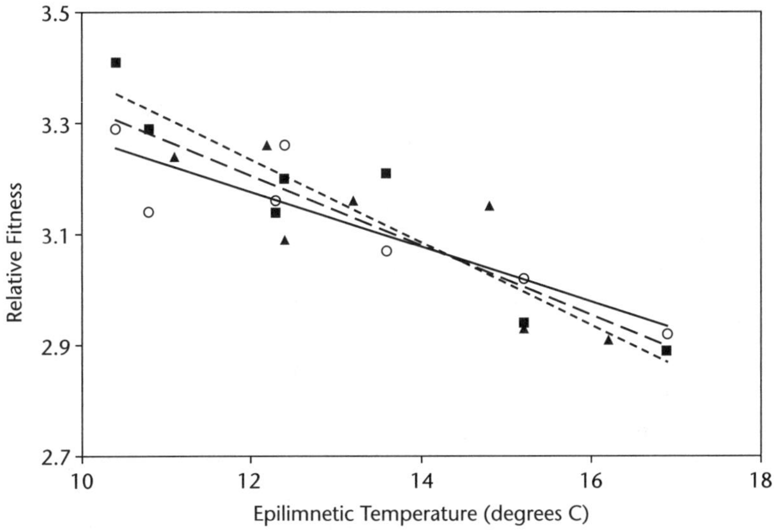

Figure 8.19 Relative fitness of three species of fish versus epilimnetic temperatures (°C) in lakes NE12 (grayling and lake trout) and Lake Fog2 (char). Fitness is defined as the slope of the length-weight relationship. Closed squares represent lake trout, closed triangles represent arctic grayling, open circles represent arctic char. The regression of relative fitness (y) versus temperature (x) for lake trout is $y = -0.077x + 4.14$ ($R^2 = 0.61$), for grayling is $y = -0.0497x + 3.77$ ($R^2 = 0.68$), and for char is $y = -0.0687x + 4.02$ ($R^2 = 0.84$).

In addition to future climate warming, it is likely that the patterns and amounts of precipitation will be changed. Although predictions of moisture change are inherently difficult and have a greater uncertainty than predictions of temperature change, we do know for example that more of the annual precipitation in a warmer world will fall as rain rather than as snow. As has been seen for much of the world so far, climate warming also tends to increase the extremes of moisture with longer droughts and more intense storms (chapter 2). The response of the lakes to increased stream discharge may be influenced by the magnitude of discharge as well as by changes in thermal stratification. During small storm events, stream temperatures remain fairly warm; if the lake is stratified, nutrient inputs enter the upper mixed layer and upper metalimnion (MacIntyre et al. 2006). In contrast, larger storm events discharge colder water, which would tend to deliver associated nutrients to deeper layers in the lake where light limitation reduces the potential for these nutrients to increase algal productivity. In either case, the fate of incoming solutes will depend on the strength of stratification and its effects on vertical mixing.

Increased solute loading to surface waters of lakes coupled with warmer summer temperatures has the potential to increase biotic production in the lakes of the region. Results from whole-lake fertilization studies indicate that phytoplankton abundance and productivity increase with nutrient loading even when relatively small amounts of nitrogen and phosphorus are added to the lakes (O'Brien et al. 2005). Unlike phytoplankton, zooplankton responded positively to lake fertilization only when temperature conditions allowed for more rapid egg development. Zooplankton showed no increase in response to the fertilization of lakes E5 and E6 when epilimnetic water temperatures remained below 15°C (C. Luecke, unpublished data). Under the climate warming predicted for the Toolik region, it is possible that increased water temperatures will be sufficient for zooplankton to respond to increases in primary production.

The net effect on the food web of enhanced rates of primary production from increased nutrient supply and increased water temperatures will depend on whether organisms at higher trophic levels will be able to use this extra food resource. The complex life histories of organisms that have evolved to thrive in low-arctic conditions may be disadvantageous in a warmer Arctic. The rapid increases in air temperatures forecast for much of the Arctic during the next 20 years (ACIA 2004) would provide little opportunity for organisms to evolve adaptations, but may allow for the invasion of arctic habitats by organisms currently present at lower latitudes. The combination of warming waters and invasive species may prove detrimental to the native aquatic fauna of low-arctic regions (Hellmann et al. 2008). The more moderate temperature changes predicted for the Toolik area may reduce the likelihood of invasions by exotic species.

References

Alexander, V., S. C. Whalen, and K. M. Klingensmith. 1989. Nitrogen cycling in arctic lakes and ponds. *Hydrobiologia* 172:165–172.

ACIA. 2004. *Arctic climate impact assessment.* Cambridge University Press, New York.

Bahr, M., J. E. Hobbie, and M. L. Sogin. 1996. Bacterial diversity in an arctic lake: A freshwater SAR11 cluster. *Aquatic Microbial Ecology* 11:271–277.

Bettez, N. D., P. A. Rublee, J. O'Brien, and M. C. Miller. 2002. Changes in abundance, composition and controls within the plankton of a fertilised arctic lake. *Freshwater Biology* 47:303–311.

Bowden, W. B., M. N. Gooseff, A. Balser, A. Green, B. J. Peterson, and J. Bradford. 2008. Sediment and nutrient delivery from thermokarst features in the foothills of the North Slope, Alaska: Potential impacts on headwater stream ecosystems. *Journal Geophysical Research* 113:1029–1041.

Brown, J., and R. L. Berg. 1980. Environmental engineering and ecological baseline investigations along the Yukon River—Prudhoe Bay haul road. *U.S. Army Cold Regions Research and Engineering Laboratory Report* 240, 18 pp.

Burkart, G. A. 2007. Energy flow in arctic lake food webs: The role of glacial history, fish predators, and benthic-pelagic linkages. M.S. thesis, Utah State University, Salt Lake City.

Carpenter, S. R., J. F. Kitchell, and J. R. Hodgson. 1985. Cascading trophic interactions and lake primary production. *BioScience* 35:634–649.

Carpenter, S. R., J. F. Kitchell, K. L. Cottingham, D. E. Schindler, D. L. Christensen, D. M. Post, and N. Voichick. 1996. Chlorophyll variability, nutrient input, and grazing: Evidence from whole-lake experiments. *Ecology* 77:725–735.

Cornwell, J. C. 1987. Phosphorus cycling in arctic lake sediments:adsorption and authigenic minerals. *Archiv Hydrobiologia* 109:161–179.

Cornwell, J. C., and G. W. Kipphut. 1992. Biogeochemistry of manganese and iron rich sediments in Toolik Lake, Alaska. *Hydrobiology* 240:45–59.

Crump, B. C., H. E. Adams, J. E. Hobbie, and G. W. Kling. 2007. Biogeography of bacterioplankton in lakes and streams of an arctic tundra catchment. *Ecology* 88:1365–1378.

Crump, B. C., G. W. Kling, M. Bahr, and J. E. Hobbie. 2003. Bacterioplankton community shifts in an arctic lake correlate with seasonal changes in organic matter source. *Applied Environmental Microbiology* 69:2253–2268.

Cuker, B. E. 1983. Grazing and nutrient interactions in controlling the activity and composition of the epilithic algal community of an arctic lake. *Limnology and Oceanography* 28:133–141.

Evans, M. A. 2007. The phytoplankton ecology of arctic lakes. Ph.D. dissertation, University of Michigan, Ann Arbor.

Evans, M. A., S. M. MacIntyre, and G. W. Kling. 2008. Internal wave effects on photosynthesis: Experiments, theory, and modeling. *Limnology and Oceanography* 53:339–353.

Fisher, H. B., E. J. List, R. C. Y. Koh, J. Imberger, and N. H. Brooks. 1979. *Mixing in inland and coastal waters.* San Diego, CA: Academic Press.

Gettel, G. 2006. Rates, importance, and controls of nitrogen fixation in oligotrophic arctic lakes, Toolik, Alaska. Ph.D. dissertation, Cornell University, NY.

Gettel, G. M., A. E. Giblin, and R. W. Howarth. 2007. The effects of grazing by the snail, *Lymnaea elodes*, on benthic N_2 fixation and primary production in oligotrophic, arctic lakes. *Limnology and Oceanography* 52:2398–2409.

Gooseff, M. N., A. Balser, W. B. Bowden, and J. B. Jones. 2009. Effects of hillslope thermokarst in northern Alaska. *Eos Transactions: American Geophysical Union* 90:29–30.

Goyke, A. P., and A. E. Hershey. 1992. Effects of fish predation on larval chironomid communities in an arctic ecosystem. *Hydrobiologia* 240:203–211.

Hamilton, T. D., and S. C. Porter. 1975. Itkillik glaciation in the Brooks Range, Northern Alaska. *Quaternary Research* 5:471–497.

Hamilton, T. D. 2003. *Glacial geology of Toolik Lake and the Upper Kuparuk River region.* Biological papers of the University of Alaska, no. 26. Fairbanks: Alaska Geobotany Center, Institute of Arctic Biology.

Hanson, K. L., A. E. Hershey, and M. E. McDonald. 1992. A comparison of slimy sculpin (*Cottus cognatus*) populations in arctic lakes with and without piscivorous predators. *Hydrobiology* 240:189–201.

Hellmann, J. J., J. E. Byers, B. G. Bierwagen, and J. S. Dukes. 2008. Five potential consequences of climate change for invasive species. *Conservation Biology* 22:534–543.

Hershey, A. E. 1992. Effects of experimental fertilization on the benthic macroinvertebrate community of an arctic lake. *Journal of the North American Benthological Society* 11:204–217.

Hershey, A. E., G. A. Gettels, M. E. McDonald, M. C. Miller, H. Mooers, W. J. O'Brien, J. Pastor, C. Richards, and S. A. Schuldt. 1999. A geomorphic-trophic model of landscape control of arctic lake food webs. *BioScience* 49:887–897.

Hershey, A. E., S. Beaty, K. Fortino, M. Keyse, P. P. Mou, W. J. O'Brien, A. J. Ulseth, G. A. Gettel, P. W. Lienesch, and C. Luecke. 2006. Effect of landscape factors on fish distribution in arctic Alaskan lakes. *Freshwater Biology* 51:39–55.

Hessen, D. O., B. A. Faafeng, and T. Andersen. 1995. Competition or niche segregation between *Holopedium and Daphnia*: Empirical light on abiotic key parameters. *Hydrobiologia* 307:253–261.

Hincks, S. S., and G. L. Mackie. 1997. Effects of pH, calcium, alkalinity, hardness and chlorophyll on the survival, growth and reproductive success of zebra mussel in Ontario lakes. *Canadian Journal of Aquatic Sciences* 54:2049–2057.

Hinzman, L. D., N. D. Bettez, W. R. Bolton, F. S. Chapin, M. B. Dyurgerov, C. L. Fastie, B. Griffith, R. D. Hollister, A. Hope, H. P. Huntington, A. M. Jensen, G. J. Jia, A. H. Lynch, A. H. Lloyd, A. D. McGuire, F. E. Nelson, W. C. Oechel, T. E. Osterkamp, C. H. Racine, V. E. Romanovsky, R. S. Stone, D. A. Stow, M. Sturm, C. E. Tweedie, G. L. Vourlitis, M. D. Walker, D. A. Walker, P. J. Bebber, J. M. Welker, and K. Yoshikawa. 2005. Evidence and implication of recent climate change in northern Alaska and other arctic regions. *Climate Change* 72:251–298.

Hobbie, J. E. 1984. Polar limnology. Pages 63–105 in F. B. Taub, editor, *Lakes and reservoirs.* Amsterdam: Elsevier.

Hobbie, J. E., B. J. Peterson, N. Bettez, L. Deegan, W. J. O'Brien, G. W. Kling, G. W. Kipphut, W. B. Bowden, and A. E. Hershey. 1999. Impact of global change on the biogeochemistry and ecology of an arctic freshwater system. *Polar Research* 18:207–214.

Hobbie, J. E., and J. Laybourn-Parry. 2008. Heterotrophic microbial processes in polar lakes. Pages 197–213 in W. F. Vincent and J. Laybourn-Parry, editors, *Polar lakes and rivers.* Oxford: Oxford University Press.

Imberger, J., and J. C. Patterson. 1989. Physical limnology. *Advances in Applied Mechanics* 27:303–455.

Johnson, C. R., C. Luecke, S. C. Whalen, and M. A. Evans. 2010. Direct and indirect effects of fish on pelagic nitrogen and phosphorus availability in oligotrophic Arctic Alaskan lakes. *Canadian Journal of Fisheries and Aquatic Sciences* 67:1635–1648.

Judd, K. E., B. C. Crump, and G. W. Kling. 2006. Variation in dissolved organic matter controls bacterial production and community composition. *Ecology* 87:2068–2079.

Keyes, M., K. Fortino, A. E. Hershey, and C. Luecke. 2007. Effects of large lake trout on the dietary habits of small lake trout: A comparison of stable isotopes and stomach content analyses. *Hydrobiology* 579:175–185.

Keller, K., J. Blum, and G. W. Kling. 2007. Geochemistry of soils and streams on surfaces of varying ages in arctic Alaska. *Arctic, Antarctic, and Alpine Research* 39:84–98.

Kipphut, G. 1984. Sediment nutrient and metal cycling in two arctic lakes. *Eos Transactions: American Geophysical Union* 66:1329.

Kling, G. W. 1994. Ecosystem-scale experiments in freshwaters: The use of stable isotopes. Pages 91–120 in L. A. Baker, editor, *Environmental chemistry of lakes and reservoirs,* Advances in Chemistry Series 237. Washington, DC: American Chemical Society.

Kling, G. W., G. W. Kipphut, and M. C. Miller. 1991. Arctic lakes and streams as gas conduits to the atmosphere: Implications for tundra carbon budgets. *Science* 251:298–301.

Kling, G. W., W. J. O'Brien, M. C. Miller, and A. E. Hershey. 1992a. The biogeochemistry and zoogeography of lakes and rivers in arctic Alaska. *Hydrobiology* 240:1–14.

Kling, G. W., G. W. Kipphut, and M. C. Miller. 1992b. The flux of carbon dioxide and methane from lakes and rivers in arctic Alaska. *Hydrobiologia* 240:23–36.

Kling, G. W., B. Fry, and W. J. O'Brien. 1992c. Stable isotopes and planktonic trophic structure in arctic lakes. *Ecology* 73:561–566.

Kling, G. W., G. W. Kipphut, M. C. Miller, and W. J. O'Brien. 2000. Integration of lakes and streams in a landscape perspective: The importance of material procession on spatial patterns and temporal coherence. *Freshwater Biology* 43:477–497.

Knapp, R. A., P. S. Corn, and D. E. Schindler. 2001. The introduction of nonnative fish into wilderness lakes: Good intentions, conflicting mandates, and unintended consequences. *Ecosystems* 4:275–278.

Levine, M. A., and S. C. Whalen. 2001. Nutrient limitation of phytoplankton production in Alaskan arctic foothill lakes. *Hydrobiologia* 455:189–201.

Lewis, W. M., and W. A. Wurtsbaugh. 2008. Control of lacustrine phytoplankton by nutrients: Erosion of the phosphorus paradigm. *International Review of Hydrobiologia* 93:446–465.

Lienesch, P. W., M. E. McDonald, A. E. Hershey, W. J. O'Brien, and N. D. Bettez. 2005. Effects of a whole-lake experimental fertilization on lake trout in a small oligotrophic arctic lake. *Hydrobiologia* 548:51–66.

Luecke, C., and W. J. O'Brien. 1981. Phototoxicity and fish predation: Selective factors in color morphs of *Heterocope*. *Limnology and Oceanography* 26:454–460.

Luecke, C., and W. J. O'Brien. 1983a. The effect of *Heterocope* predation on zooplankton communities in arctic ponds. *Limnology and Oceanography* 28:367–377.

Luecke, C., and W. J. O'Brien. 1983b. Photoprotective pigments in a pond morph of *Daphnia middendorffianna*. *Arctic* 36:365–368.

MacIntyre, S., J. O. Sickman, S. A. Goldthwait, and G. W. Kling. 2006. Physical pathways of nutrient supply in a small, ultra-oligotrophic lake during summer stratification. *Limnology and Oceanography* 51:1107–1124.

McCauley, E., W. W. Murdoch, R. M. Nisbet, and W. S. C. Gurney. 1990. The physiological ecology of *Daphnia*: Development of a model of growth and reproduction. *Ecology* 71:703–713.

McDonald, M. E., A. E. Hershey, and M. C. Miller. 1996. Global warming impacts on lake trout in arctic lakes. *Limnology and Oceanography* 41:1102–1108.

Merrick, G. W., A. E. Hershey, and M. E. McDonald. 1992. Salmonid diet and the size, distribution, and density of benthic invertebrates in an arctic lake. *Hydrobiologia* 240:225–233.

Miller, M. C., P. Spatt, P. Westlake, D. Yeakel, and G. R. Hater. 1986. Primary production and its control in Toolik Lake, Alaska. *Archiv Hydrobiologia* Supplement 74:97–131.

Morris, D. P., H. Zagarese, G. E. Williamson, E. G. Balseiro, B. R. Hargeaves, B. Modenutti, R. Moeller, and C. Queimalinos. 1995. The attenuation of solar UV radiation in lakes and the role of dissolved organic carbon. *Limnology and Oceanography* 40:1381–1391.

O'Brien, W. J. 1979. The predator-prey interaction of planktivorous fish and zooplankton. *American Scientist* 67:572–581.

O'Brien, W. J., M. Bahr, A. E. Hershey, J. E. Hobbie, G. W. Kipphut, G. W. Kling, H. Kling, M. McDonald, M. C. Miller, P. A. Rublee, and J. R. Vestal. 1997. The limnology of Toolik Lake. Pages 61–106 in A. M. Milner and M. W. Oswood, editors, *Freshwaters of Alaska: Ecological syntheses*, Ecological Studies, vol. 119. New York: Springer-Verlag.

O'Brien, W. J., M. Barfield, N. Bettez, A. E. Hershey, J. E. Hobbie, G. Kipphut, G. Kling, and M. C. Miller. 2005. Long-term response and recovery to nutrient addition of a partitioned arctic lake. *Freshwater Biology* 50:731–741.

O'Brien, W. J., and B. I. Evans. 1992. Simulation model of the planktivorous feeding of arctic grayling: Laboratory and field verification. Hydrobiologia 240:235–245.

O'Brien, W. J., A. Hershey, J. Hobbie, M. A. Hullar, G. W. Kipphut, M. C. Miller, B. Moller, and J. R. Vestal. 1992. Control mechanisms of arctic lake ecosystems: A limnocorral experiment. *Hydrobiologia* 240:143–188.

O'Brien, W. J., M. Barfield, N. D. Bettez, G. M. Gettel, A. E. Hershey, M. E. McDonald, M. C. Miller, H. Mooers, J. Pastor, C. Richards, and J. Schuldt. 2004. Physical, chemical, and biotic effects on arctic zooplankton community and diversity. *Limnology and Oceanography* 49:1250–1264.

Pace, M. L., J. J. Cole, and S. R. Carpenter. 1998. Trophic cascades and compensation: Differential responses of microzooplankton in whole-lake experiments. *Ecology* 79:138–152.

Rocha, A. V., and G. R. Shaver. 2011. Burn severity influences postfire CO_2 exchange in arctic tundra. *Ecological Applications* 21:477–489.

Rublee, P. A. 1992. Community structure and bottom-up regulation of heterotrophic microplankton in Arctic LTER lakes. *Hydrobiologia* 240:133–141.

Rublee, P. A., and N. Bettez. 1995. Change of microplankton community structure in response to fertilization of an arctic lake. *Hydrobiologia* 312:183–190.

Rublee, P. A., and N. D. Bettez. 2001. Lake characteristics influence recovery of microplankton in arctic LTER lakes following experimental fertilization. *Hydrobiologia* 446:229–232.

Scully, N. M., and D. R. S. Lean. 1994. The attenuation of ultraviolet light in temperate lakes. *Ergebnisse Limnologie* 43:135–144.

Sellers, T. J., B. R. Parker, D. W. Schindler, and W. M. Tonn. 1998. Pelagic distribution of lake trout (*Salvelinus namaycush*) in small Canadian Shield lakes with respect to temperature, dissolved oxygen, and light. *Canadian Journal of Fisheries and Aquatic Sciences* 55:170–179.

Smith, L. C., Y. Sheng, G. M. MacDonald, and L. D. Hinzman. 2005. Disappearing arctic lakes. *Science* 308:1429.

Vadeboncoeur, Y., E. Jeppesen, M. J. VanderZanden, H. Schierup, K. Christoffersen, and D. M. Lodge. 2003. From Greenland to green lakes: Cultural eutrophication and the loss of benthic pathways in lakes. *Limnology and Oceanography* 48:1408–1418.

Vadstein, O., Y. Olsen, H. Reinersten, and A. Jensen. 1993. The role of planktonic bacteria in phosphorus cycling in lakes. *Limnology and Oceanography* 38:1539–1544.

Vanni, M. J. 2002. Nutrient cycling by animals in freshwater ecosystems. *Annual Review of Ecology and Systematics* 33:341–370.

Vincent, W. F., M. M. Gibbs, and R. H. Spigel. 1991. Eutrophication processes regulated by a plunging river inflow. *Hydrobiologia* 226:51–63.

Vincent, W. F., and J. E. Hobbie. 2000. Ecology of arctic lakes and rivers. Pages 197–206 in M. Nuttall and T. V. Callaghan, editors, *The Arctic: Environment, people, policy.* The Netherlands: Overseas Publishers Association.

Walter, K. M., S. A. Zimov, J. P. Chanton, D. Verbal, and F. S. Chapin, III. 2006. Methane bubbling from Siberian thaw lakes as a positive feedback to climate warming. *Nature* 443:72–75.

Whalen, S. C., and V. Alexander. 1986. Seasonal inorganic carbon and nitrogen transport by phytoplankton in an arctic lake. *Canadian Journal of Fisheries and Aquatic Sciences* 43:1177–1186.

Whalen, S. C., B. A. Chalfant, and E. N. Fischer. 2008. Epipelic and pelagic primary production in Alaskan Arctic lakes of varying depth. *Hydrobiologia* 614:243–257.

Whalen, S. C., and J. C. Cornwell. 1985. Nitrogen, phosphorus, and organic carbon cycling in an arctic lake. *Canadian Journal of Fisheries and Aquatic Sciences* 42:797–808.

Yurista, P. M. 1999. Temperature-dependent energy budget of an arctic cladoceran, *Daphnia middendorffiana*. *Freshwater Biology* 42:21–34.

9

Mercury in the Alaskan Arctic

William F. Fitzgerald
Chad R. Hammerschmidt
Daniel R. Engstrom
Prentiss H. Balcom
Carl H. Lamborg
Chun-Mao Tseng

Introduction

The remotest regions of the globe are affected by natural processes and human-related activities occurring thousands of miles away. Atmospheric circulation is global, and the transport of heat is leading to warming arctic climates and increased melting of glaciers and sea ice. Atmospheric circulation and chemistry also play a pivotal role in affecting the transport and deposition of both inorganic and organic substrates on local and global scales. Results of the deposition now are found as a "chemical laundry list" in snow from various locations in the Arctic. For example, enhanced amounts of the neurologically toxic metals, lead (Pb) and mercury (Hg), are common at high latitudes (Vandal et al. 1995; Marteel et al. 2008). Other metals such as copper (Cu), zinc (Zn), and cadmium (Cd) are also elevated. Also evident are modern creations described as persistent organic pollutants (POPs), including chlorofluorocarbons (CFCs), polychlorinated and brominated biphenyls (PCBs and PBBs), flame retardants such as polybrominated diphenylethers (PBDEs), pesticides such as chlorinated hydrocarbons (DDT, DDE), and soot from combustion (e.g., Grannas et al. 2012; Halsall 2004; Garbarino et al. 2002).

The limited measurements of airborne metals and POPs reaching the North Slope of Alaska and the Toolik Lake region show the same list of chemicals as in other arctic regions. However, geography, far away from Europe across the Arctic Ocean and somewhat protected by the Brooks Range to the south, helps to keep concentrations low. Ford et al. (1995), for example, in a survey of the distribution of pollutants using concentrations in a species of moss as their indicator, found low concentrations of Pb, Cd, Cu, V (vanadium), and Zn in a series of samples across the North Slope. Two studies of sediments, lake trout, and arctic grayling in a deep lake 200 km east of Toolik Lake revealed organochlorine pesticides, PCBs, and lindane (Wilson et al. 1995), while Gubala et al. (1995) measured dichlorodiphenyl trichloroethane (DDT) and dichlorodiphenyldichloroethylene (DDE) in the same lake.

These authors also studied one core from the lake in detail to measure how Hg changed in concentration with depth.

A potential mechanism for the transport of these contaminants to the Arctic is presented in a series of reports by the Arctic Monitoring and Assessment Programme (AMAP). This mechanism, whereby POPs are transported from temperate-zone industrialized source regions of the northern hemisphere to the Arctic, involves multiple-stage distillation and condenser-like deposition processes (MacDonald et al. 2000; Simonich and Hites 1995). Such a mechanism would lead to seasonally and latitudinally dependent transport and accumulation. As suggested in Figure 9.1, similar processes may transport Hg from tropical to polar regions. Note that the atmosphere is the principal route by which Hg is transferred at the Earth's surface, and that the major species transported is the elemental form, Hg^0, which has an atmospheric residence time of one year or less (Selin et al. 2008; Lindberg et al. 2007). Along with this unusual volatility for a metal, Hg also exhibits sea-to-air evasion, which acts to extend atmospheric transport. Therefore, natural and pollutant Hg can be readily dispersed intra- and inter-hemispherically; as with Pb and POPs, the remote Arctic is not spared.

Recently, Helgason et al. (2008) reported that POP levels declined in seabird eggs from northern Norway between 1983 and 2003. Mercury, in contrast, remained

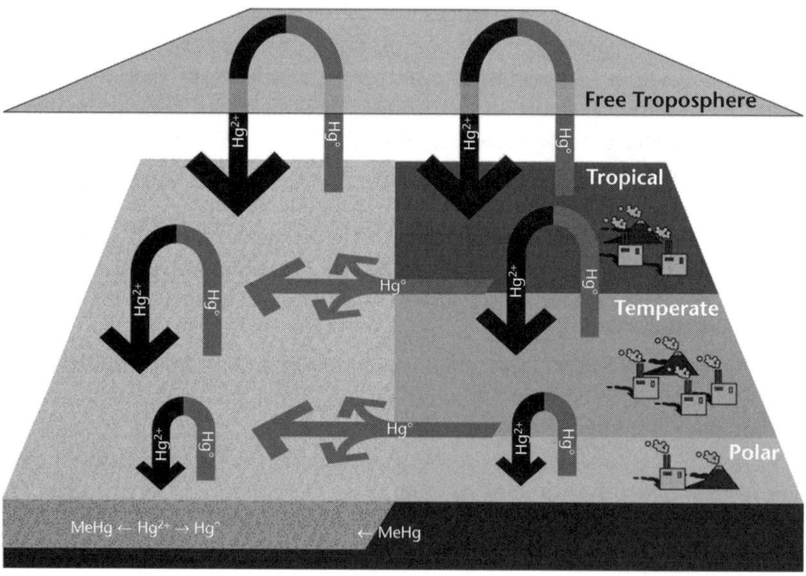

Figure 9.1 Atmospheric Hg is dispersed globally, largely in the volatile elemental form (Hg^o). Natural and anthropogenic emissions to the atmosphere are sufficiently long-lived to allow intra-hemispheric mixing. Re-emission of this material, both from land and ocean, is a hallmark of the Hg cycle. Similar to that of POPs, Hg re-emission is accelerated by warmer temperatures, but apparently so is the removal of Hg from the atmosphere by oxidation and convective mixing of the lower and free troposphere. This likely prevents Hg from being "globally distilled" in the same way as POPs.

relatively constant. The recent assessment on mercury in the Arctic (AMAP 2011) complements this chapter, which emphasizes results from one region around the Toolik Field Station, Alaska. Additional features of the global Hg cycle such as high temperature anthropogenic sources (e.g., coal burning) and natural Hg inputs associated with volcanoes are shown in Figure 9.1.

Contamination in Polar Regions: Lead and Mercury Connections

Studies on Pb distribution at high latitudes began in the 1960s with samples from recent and ancient snow in northern Greenland and continued with ice samples from the Rockefeller Plateau in West Antarctica. They provided unequivocal analytical and chronological evidence for widespread, modern environmental Pb contamination. Its undisputed cause was emissions from the combustion of leaded gasoline and the broad atmospheric dispersion of Pb on very fine aerosol particles (Murozumi et al. 1969). This work alerted the scientific community to global-scale pollution beyond that associated with radioactivity from nuclear explosions and carbon dioxide emissions from fossil fuel combustion. It also provided impetus for the use of dated glacial strata as a means of reconstructing historical patterns of contamination as well as temperature and other climate-related variables.

The Pb concentration profile versus sample age from the classic Greenland investigation shows that the Pb levels in 1965 are about 10 times greater than those in the mid-1700s and greatly enhanced (>200 times) compared to 800 BC (Figure 9.2). A dramatic increase in Pb deposition occurred following World War

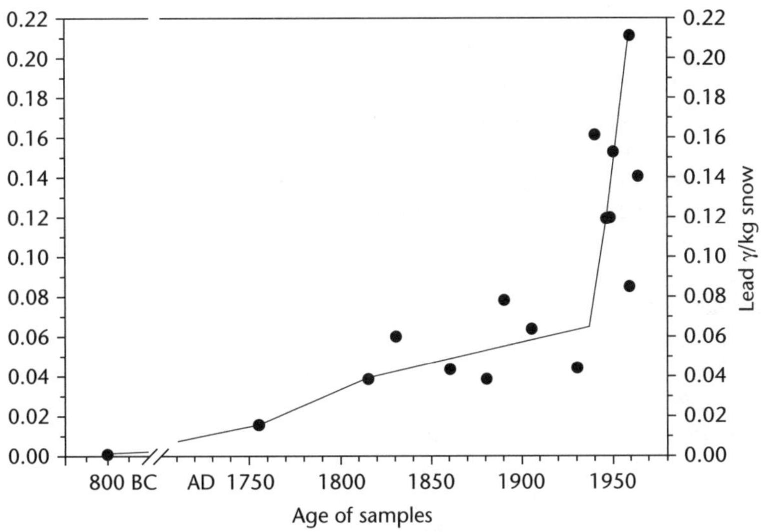

Figure 9.2 Increase of industrial lead pollution in snow (microgram/kg) at Camp Century (Greenland) since 800 BC. From Murozumi et al. (1969), used with permission.

II, which coincided with an increase in the use of gasoline containing tetra-ethyl Pb as an "anti-knock" additive. Lead-ore smelting was the major source of pollution before the advent of leaded gasoline, and its influence on Pb deposition to remote regions can be traced back to Roman times (Renberg et al. 2000; Brännvall et al. 2001). Lead contamination of the Earth continued effectively unabated until the phase-out of leaded gasoline began in the United States during the mid-1970s, and in Europe about a decade later. Although global-scale Pb pollution has been declining, contamination continues because leaded gasoline remains in use in many countries, especially in Africa and Asia (UNEP 2008).

The mosquito study summarized in Vignette 9.1, while pointing to the insects' potential as biomonitors of Hg deposition, also raises several related environmental questions. For example, is the Hg accumulating in mosquitoes primarily natural or enhanced by pollution? What is the origin of Hg observed in precipitation? Is it local or transported from some distance? Aspects of such questions can be addressed by careful reconstruction of atmospheric Hg deposition over time. For example, tree-ring records, glacial cores, and lake-sediment archives are potentially useful in determining temporal variations and spatial patterns for the deposition and accumulation of Hg and other contaminants. Ford et al. (1995) and Gubala et al. (1995) used moss monitoring and one lake core, respectively, to provide a semi-quantitative assessment for the local- and longer-range transport and deposition of contaminants (including Hg) to the Toolik environs. In detailed investigations reported here, multiple sediment cores from many lakes were analyzed and dated. Results from these dated lake sediment archives (given their broad geographic presence) have provided an experimental framework and reliable mass balance and modeling constraints for our studies of the biogeochemical cycling and contamination of Hg in the Arctic and in a variety of other remote environs.

Environmental Mercury Cycling

As suggested in Figure 9.1, Hg is not only a toxic trace metal, but it possesses one of the most complex and investigatively challenging biogeochemical cycles in nature (Fitzgerald and Lamborg 2003). Its three-phase character enables natural and contaminant Hg to be readily mobilized and dispersed worldwide, to interact in a myriad of complex reactions and transformations, to bioaccumulate as the very toxic monomethylmercury (MMHg) species in aquatic organisms (especially piscivorous fish), and to present a human-health risk and environmental concern (Fitzgerald and Clarkson 1991). A simple but useful view centered on the biogeochemical cycling of Hg in tundra aqueous systems and watersheds is presented in Figure 9.3.

There are a variety of pressing and challenging research issues relating to Hg contamination and its cycling and fate in nature. Health concerns associated with human exposure to MMHg are especially important international issues (Mergler et al. 2007). Indeed, fish consumption advisories have been issued by most countries and the World Health Organization (FAO/WHO 2006). In the United States, for example, the Environmental Protection Agency (USEPA 1997) and the Food

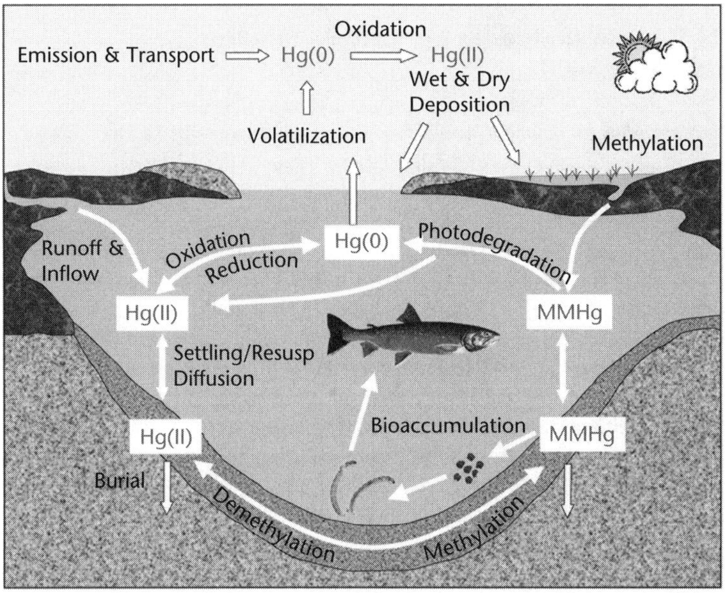

Figure 9.3 Biogeochemical cycling of Hg in a tundra lake and its watershed. Gaseous elemental mercury [Hg(0)] is oxidized in the atmosphere to reactive gaseous mercury [Hg(II)], and deposited in precipitation and by dry deposition. Bacteria convert a small portion of the incoming Hg(II) to monomethylmercury (MMHg), which is then bio-concentrated in the aquatic food chain. Various biotic and abiotic reactions interconvert the different forms of Hg, affecting uptake, burial, and evasion back to the atmosphere (data from Engstrom 2007).

and Drug Administration (USFDA/USEPA 2004) have each provided guidelines designed to reduce human exposure to MMHg through the limited and wise consumption of freshwater and marine fish. These warnings continue to raise a variety of questions as to the sources of Hg, the routes by which Hg enters aquatic systems, its chemically active forms (i.e., bioavailability), and the mechanisms and processes that lead to its bioaccumulation in fresh and salt water environments. The public debates associated with the USEPA Mercury and Air Toxics Standards (MATS) have raised awareness and worries related to linkages between Hg levels in aquatic organisms and anthropogenic Hg emissions from sources such as power plants. More broadly, interest in Hg is worldwide as reflected by the ten international conferences on "Mercury as a Global Pollutant" that have been held since 1990.

Although, this chapter is focused principally on mercury cycling and contamination in arctic Alaska, it provides a useful framework for investigating the behavior and fate of many contaminants in polar regions. Indeed, because the mobilization of Hg through industrial, medicinal, and mining applications is local to global in scale, because the interactions and reactions of Hg involve three phases (gas, liquid, and mineral), and because of well-documented environmental dispersion and ecosystem contamination, Hg provides a valuable analog for the biogeochemical cycling of many inorganic and organic pollutants, especially in the Arctic.

Vignette 9.1. Mosquitoes as Mercury Biomonitors, by William Fitzgerald

The summertime tundra near Toolik Lake is quiet except for the buzzing of mosquitoes attracted to scientists and other large animals. Their numbers are staggering; they transform the back of a jacket into a grey Harris tweed look-alike, with a packing density approaching 200 in the area of an average hand. Because mosquitoes could not be ignored and because they grow in water-saturated tundra, we wondered if they contained Hg—and not just inorganic forms, but the toxic monomethylmercury (MMHg) species.

Mosquitoes near Toolik Lake were hand-collected and analyzed (Figure 9.V1.1). They did indeed contain MMHg and at relatively high concentrations. The experiments were repeated and the results were strikingly reproducible, even from year to year. These initial tests led to the broader question as to whether MMHg would be present in mosquitoes from other locations. Samples were also analyzed from Michigan, California, and Florida. Not only did they contain MMHg, but even more importantly there was a direct relationship between the amount of MMHg in the mosquitoes and the amount of Hg deposited by way of precipitation (Figure 9.V1.2).

Notice the substantial difference between the delivery of Hg to the Arctic and to Florida—a consequence, in part, of Florida's much higher rainfall. There is almost ten times as much Hg deposited in rain from Florida as in rain from Alaska. The results are striking, and show a linear relationship between the concentration of MMHg in mosquitoes and the annual flux of Hg in precipitation. That is, as the amount of inorganic Hg increases with increased precipitation, there is an enhancement of MMHg in mosquitoes. Similarly, if Hg flux via precipitation were to diminish, then the MMHg content of the mosquitoes would be expected to decline. This was the first among recent studies lending strong support to models that had suggested that reducing the amount of Hg being delivered from the atmosphere (the principal source at the Earth's surface) would lead to the reduction of MMHg in ecosystems.

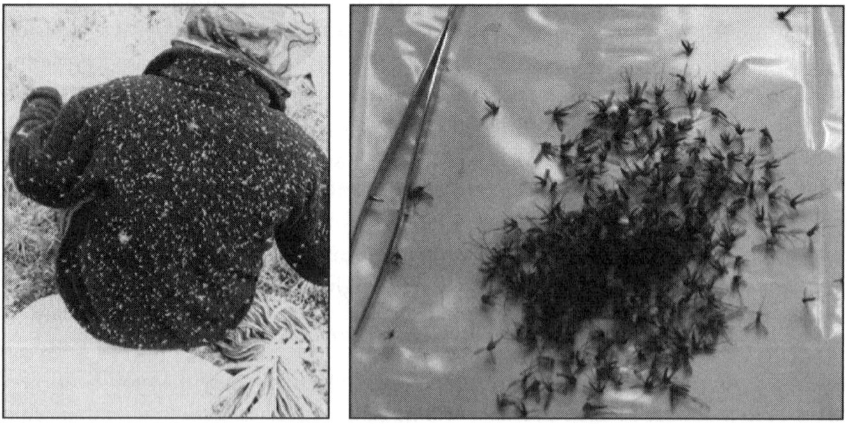

Figure 9.V1.1 To the left is a tweed-like mosquito-laden jacket, and on the right collected arctic mosquitoes for MMHg analysis.

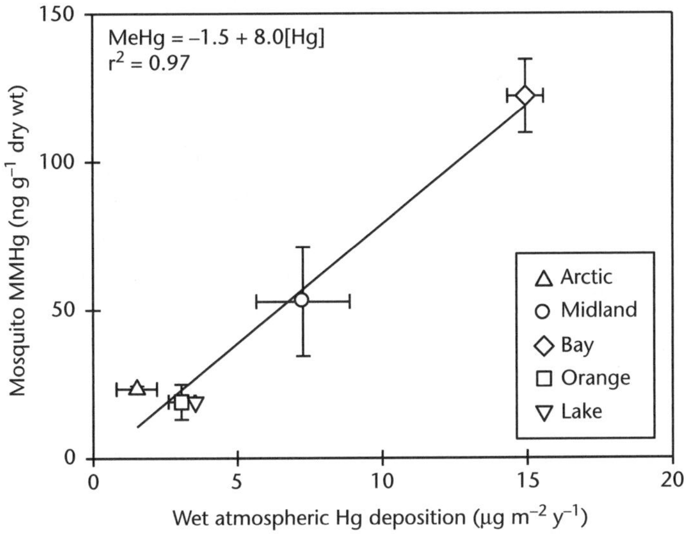

Figure 9.V1.2 The mean concentration of MMHg in adult mosquitoes and wet atmospheric deposition of total Hg (±1 standard error) at arctic tundra, AK; Midland County, MI; Bay County, FL; Orange County, CA; Lake County, CA (redrawn with data presented in Hammerschmidt and Fitzgerald 2005).

Arctic Mercury Depletion Events

A unique aspect of Hg cycling at high latitudes is atmospheric mercury depletion events (AMDEs). These phenomena, in which total gaseous mercury (TGM, >99% as Hg^0) is removed from the atmosphere and deposited to terrestrial and aquatic environs during polar sunrise, were first observed at Alert (Ellesmere Island, Canada), and have been confirmed in many other coastal polar locations, including Churchill (Canada; Kirk et al. 2006), Barrow (Alaska), Station Nord (northeast Greenland), Svalbard (Ny-Ålesund), and Neumayer (Antarctica). These extraordinary scavenging reactions are illustrated in Figure 9.4. Notice that over a period of about 8 to 12 weeks, there are significant drops in the atmospheric levels of TGM. Moreover, there is a corresponding increase in Hg associated with atmospheric particles (i.e., aerosols) and the amounts of Hg accumulating in snow (Lahoutifard et al. 2005). The removal of Hg from the troposphere during Arctic spring was immediately recognized to be connected with the general phenomenon of "Arctic Haze" because of the tight correlation between Hg and ozone. However, it was not until several more years of measurements had been conducted and at several locations in the Arctic that the possibility was brought forward that bromine chemistry in the atmosphere was behind much of the oxidation and removal of Hg from the Arctic atmosphere (Steffen et al. 2008). Indeed, it may very well be one of the primary drivers for the cycling of this toxic metal outside of the Arctic as well. Although research focused on AMDE has increased,

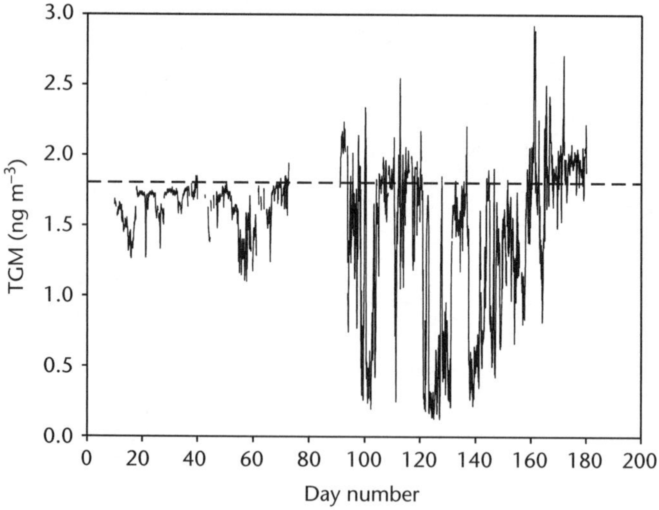

Figure 9.4 Total gaseous mercury (TGM, >99% as Hg^0) in the atmosphere at (a) Alert, Canada, in 1995 (adapted from Schroeder et al. 1998, with the original data provided by S. Steffen). Dashed lines indicate "normal" TGM concentrations (~1.8 ng m^{-3}). TGM is reduced significantly during atmospheric mercury depletion events.

knowledge is limited about the terrestrial, lacustrine, and marine cycling of Hg in polar regions and the ecosystem impact from pollution-derived Hg at high latitudes (AMAP 2011).

Accordingly, our studies in the U.S. Arctic are driven by the critical need for detailed scientific knowledge of Hg cycling in polar regions. These efforts have also been spurred by the associated human-health worries, social impacts, and ecological concerns that have been exacerbated by significant worldwide anthropogenic Hg emissions. We have focused on (a) atmospheric Hg deposition and contamination of lakes and watersheds, (b) in-lake cycling of Hg, especially the MMHg species, and (c) the behavior and fate of Hg^0, whose production and mobilization via water–air exchange is quite significant. The studies, which began in 2000, have been conducted in the terrestrial and lacustrine environs near the Arctic LTER site at the Toolik Field Station (68°38' N, 149°38' W). Although human impact affecting the transport and deposition (e.g., AMDEs) was evident, there was little knowledge about the reactions and mechanisms of the biogeochemical cycling of Hg in the Arctic. As detailed below, we found that North Slope lakes behave similarly to those in temperate latitudes, although the various reactions differed in their relative importance. Thus, research at Toolik has provided critical data regarding Hg cycling in this unique and changing landscape while also providing broad insights that will promote research elsewhere.

Atmospheric Mercury Fluxes in Northern Alaska

As illustrated in Vignette 9.1, Hg loading in precipitation on the tundra is much less than that in the Florida Everglades. Quantitatively this is good news but with Hg there is usually more to the biogeochemical story. For example, how do AMDEs affect total Hg loading, and how have the loadings changed with time? Is there significant evidence of anthropogenic impact, and how does it compare to other locations? Using sediment archives obtained from five carefully selected lakes with small watersheds, modest depths, and simple (single-basin) bathymetry, we reconstructed atmospheric Hg deposition to arctic Alaska over the last several centuries. The magnitude of this change is illustrated in Figure 9.5, where whole-lake Hg sedimentation based on 11 ^{210}Pb-dated cores from each of the five lakes demonstrates an average three-fold increase in atmospheric Hg deposition since the advent of the Industrial Revolution. In addition, the anthropogenic Hg impact in the Arctic is of similar proportion to that at other remote locations worldwide (Figure 9.6). Thus, and although the atmospheric loadings of Hg on the North Slope of Alaska are smaller than those delivered to Florida (in preindustrial times as well as today), the human-related impact is comparable. However, the Toolik core records show no evidence for recent declines in Hg deposition, which have been noted at sites closer to industrial Hg sources at lower latitudes (e.g., Drevnick et al. 2012). This signal of declining Hg emissions from eastern North America and Europe may not be evident in the Toolik region, either because the Arctic is recording the global trend in atmospheric Hg—which has not diminished owing to increased Hg emissions from Asia (Lindberg et al. 2007)—or because of low temporal resolution in the slowly accumulating sediments of Arctic lakes.

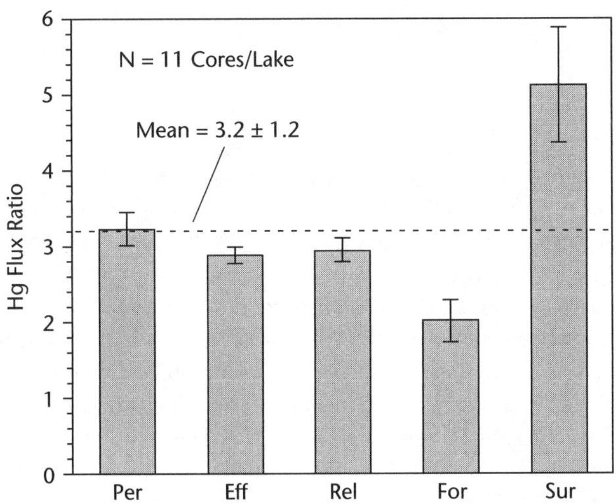

Figure 9.5 Ratio of Hg accumulation in recent (1990–2000) to preindustrial (pre-1850) sediments in five study lakes from the Toolik area. These lakes have been named provisionally as Perfect, Efficient, Relaxing, Forgetful, and Surprise. From Fitzgerald et al. (2005), used with permission.

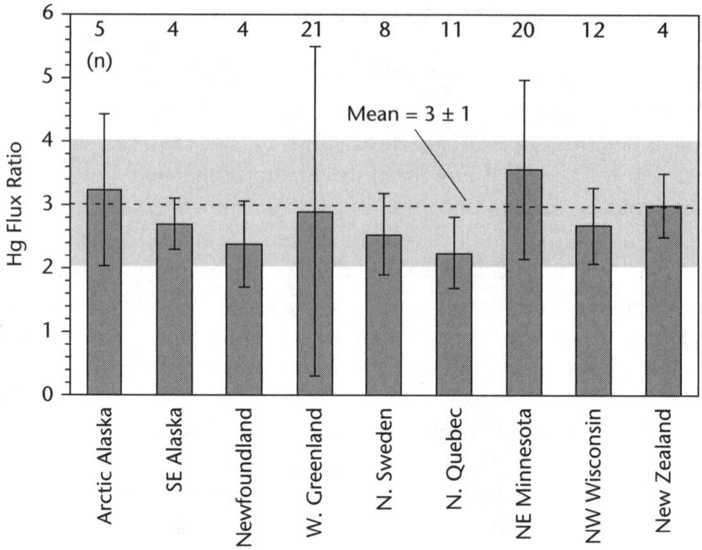

Figure 9.6 Ratio of Hg accumulation in recent (1990–2000) to preindustrial (pre-1850) sediments from remote lakes in various regions of the world (n = number of lakes in each study). Eleven cores were analyzed for each lake from arctic Alaska, and six cores per lake were analyzed from southeast Alaska and Newfoundland, one core per lake for all others. Data sources: arctic Alaska near Toolik (Fitzgerald et al. 2005); southeast Alaska and Newfoundland (the authors' unpublished results), northern Sweden (Bindler et al. 2001a); west Greenland (Bindler et al. 2001b); northern Quebec (Lucotte et al. 1995); northeastern Minnesota (Engstrom et al. 2007); north central Wisconsin (Rada et al. 1989); and New Zealand (Lamborg et al. 2002).

Whole-Lake Hg Budgets

The well-constrained mass balances for Hg and MMHg (Figure 9.7) provide much biogeochemical insight as well as the framework for our reaction and mechanistic investigations. Primary sources of Hg in arctic Alaskan lakes are direct atmospheric deposition to the lake surface as well as deposition to and runoff from watersheds. This atmospherically derived Hg is "environmentally active" and available for reactions that produce both Hg^0 (Tseng et al. 2004) and MMHg (Hammerschmidt et al. 2006). At the same time, there are significant erosional or mineral sources of Hg that do not appear to be labile and available for transformations. Another result is that an additional source term is needed to balance the evasion and sedimentation sinks for Hg (Figure 9.7, left; Fitzgerald et al. 2005). This additional flux (1.2 ± 0.7 µg m^{-2} yr^{-1}, or an input of 2.1 ± 1.3 µg m^{-2} yr^{-1} for a lake with a ratio of area-of-watershed to area-of-lake of 3) is likely due to some combination of springtime AMDEs or more generalized deposition of reactive gaseous mercury (RGM; Lindberg et al. 2002). In either case, it is likely that this is a source of Hg that is available for transformation to MMHg.

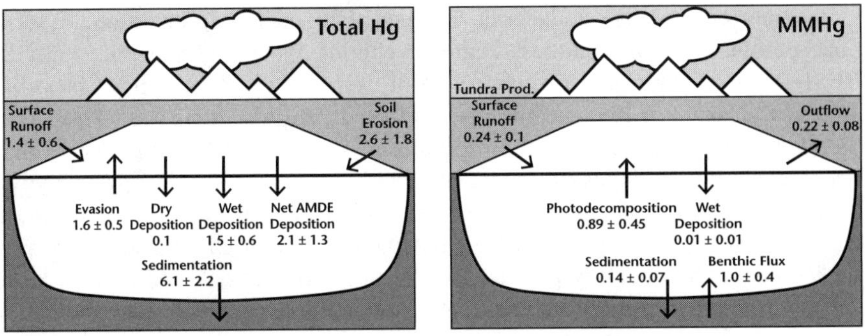

Figure 9.7 Typical mass balance for (a) total Hg (Fitzgerald et al. 2005) and (b) monomethylmercury (MMHg; from Hammerschmidt et al. 2006, used with permission) in small arctic lakes (fluxes are $\mu g\ m^{-2}\ yr^{-1}$). Total Hg cycling was examined in lakes without outflows.

Our subsequent investigations focused on the biogeochemistry of MMHg in four arctic Alaskan watersheds. We constrained the principal features of the MMHg cycling based on our summertime sampling and findings (Figure 9.7). These studies (Hammerschmidt et al. 2006; Hammerschmidt and Fitzgerald 2006a) have shown that: (1) the contribution of MMHg from tundra watersheds is modest relative to *in situ* benthic production with 80%–90% of total MMHg inputs provided by diffusion from sediments; (2) benthic MMHg production is related to loadings of environmentally active Hg from the atmosphere and is independent of sulfate; and (3) photodecomposition—a process confined to the short, ice-free season in the Arctic—is the primary control for MMHg in arctic Alaskan lakes (70%–90% of inputs) and attenuates bioaccumulation. These results provide a foundation for ongoing research that is testing critical reaction and process-oriented hypotheses dealing with the mechanisms affecting the behavior and fate of MMHg in arctic lakes and watersheds. In addition, this work raises important questions as to the spatial extent of AMDEs and their ecosystem impact—that is, what are the connections between the observed delivery of Hg to the ice and snow pack (Constant et al. 2007, Poulain et al. 2007), the net accumulation of Hg in a polar aquatic system (e.g., tundra lakes), and production and bioaccumulation of MMHg?

Summary and Conclusions

This research points to a strong connection between atmospheric Hg deposition and the production and bioaccumulation of MMHg. It is evident that loading of environmentally active Hg from the atmosphere, through deposition to lake surfaces and watersheds, is the major control on MMHg production in arctic Alaskan lake sediments, an observation supported by recent mesocosm and whole-watershed tests in northwest Ontario (Orihel et al. 2007; Harris et al. 2007). Moreover, and across broad gradients of atmospheric Hg deposition in

North America, there is a connection between atmospheric Hg fluxes and MMHg bioaccumulation in mosquitoes (Hammerschmidt and Fitzgerald 2005) and in fish (Hammerschmidt and Fitzgerald 2006b). This suggests that processes that enhance the atmospheric deposition or watershed transport of Hg to sites of methylation will increase the production and bioaccumulation of MMHg. Near Toolik the atmospheric Hg deposition associated with AMDEs (1.2 ± 0.7 µg m^{-2} yr^{-1}), or more generalized deposition of RGM, is equivalent to the wet atmospheric deposition (1.5 ± 0.6 µg m^{-2} yr^{-1}). This suggests that AMDEs and RGM may *double* the net sedimentary production and bioaccumulation of MMHg in lakes near Toolik and potentially have equal or greater impact on MMHg cycling in other regions of the Arctic.

Future mercury-related research should include an important seasonal (spring) and geographical component (e.g., the region between Toolik, Barrow, and Deadhorse, AK) that would complement our summer source–sink investigations. Given the likely significance of increased atmospheric Hg fluxes (e.g., AMDE) during spring in the Arctic and the paucity of information regarding Hg loadings and cycling of MMHg in aquatic systems, it would be appropriate to initiate a comprehensive suite of source-related, atmosphere–watershed–lake studies focused on the significance of springtime AMDE. While Hg deposition associated with AMDE increases near the coast, it is not clear if net deposition of Hg increases. Sedimentary archives will provide the answer. Moreover, if AMDE were to increase net Hg loadings to aquatic systems more proximate to the Arctic Ocean, as suggested by patterns of Hg in snow (Douglas and Sturm 2004), then one might expect MMHg to follow a similar pattern. That is, the bioaccumulation of MMHg in fish, zooplankton, and mosquitoes should reflect the influence of Hg loadings. Our initial survey of mosquitoes along a 200 km transect between Toolik Lake and the coast at Deadhorse revealed little variation in MMHg content, yet there is a 20-fold increase in gross Hg deposition. This surprising result suggests that net Hg deposition is comparable across the transect and caused by significant cycling and re-emission from the snow and ice pack (e.g., via photoreduction) of Hg deposited during AMDE (Hammerschmidt and Fitzgerald 2008). These hypotheses should be tested broadly in different arctic regions.

The social impacts of our Hg research spring from the ecological and public health threats of Hg accumulation in the pan-Arctic environment. For example, loadings of Hg, including AMDE, to a delicate and changing ecosystem and their lasting influence are of concern. Although relatively unpopulated, the Arctic is home to indigenous peoples, many of whom rely on wild fish and piscivorous animals for part of their diet. Furthermore, our process-based studies of Hg accumulation and cycling in arctic lakes enhance understanding of Hg dynamics in aquatic systems in general and in arctic lake environments in particular. This research also provides a foundation for other investigations and assessments (e.g., potential effects of climate change such as warming, permafrost melting, and enhanced weathering) of Hg and MMHg cycling in polar freshwater ecosystems. The results may be applicable to arctic marine systems that are home to fisheries of commercial, recreational, and cultural value to people outside of the Arctic.

References

AMAP (Arctic Mapping and Assessment Programme). 2011. *Mercury in the Arctic.* Oslo: Arctic Monitoring and Assessment Programme.

Bindler, R., C. Olafsson, I. Renberg, and W. Frech. 2001a. Temporal trends in mercury accumulation in lake sediments in Sweden. *Water Air Soil Pollution Focus* 1:343–355.

Bindler, R., I. Renberg, P. G. Appleby, N. J. Anderson, N. L. Rose. 2001b. Mercury accumulation rates and spatial patterns in lake sediments from west Greenland: A coast to ice margin transect. *Environmental Science and Technology* 35:1736–1741.

Brännvall, M.-L., R. Bindler, O. Emteryd, and I. Renberg. 2001. Four thousand years of atmospheric lead pollution in northern Europe: A summary from Swedish lake sediments. *Journal of Paleolimnology* 25:421–435.

Constant, P., L. Poissant, R. Villemur, E. Yumvihoze, and D. Lean. 2007. Fate of inorganic mercury and methyl mercury within the snow cover in the low arctic tundra on the shore of Hudson Bay (Quebec, Canada). *Journal of Geophysical Research—Atmospheres* 112:D08309. doi:10.1029/2006JD007961.

Douglas, T. A., and M. Sturm. 2004. Arctic haze, mercury and the chemical composition of snow across north-western Alaska. *Atmospheric Environment* 38:805–820.

Drevnick, P. E., D. R. Engstrom, C. T. Driscoll, E. B. Swain, S. J. Balogh, N. C. Kamman, D. T. Long, D. C. G. Muir, M. J. Parsons, K. R. Rolfhus, and R. Rossman. 2012. Spatial and temporal patterns of mercury accumulation in lacustrine sediments across the Laurentian Great Lakes region. *Environmental Pollution* 161:252–260.

Engstrom, D. R. 2007. Fish respond when the mercury rises. *Proceedings National Academy of Sciences* 104:16394–16395.

Engstrom, D. R., E. B. Swain, and S. J. Balogh. 2007. History of mercury inputs to Minnesota lakes: Influences of watershed disturbance and localized atmospheric deposition. *Limnology and Oceanography 52*:2467–2483.

FAO/WHO. 2006. *Summary and conclusions of the Sixty-First Meeting of the Joint FAO/WHO Expert Committee on Food Additives (JECFA).* Rome, Italy, June 20–29, 2006. 67/SC.

Fitzgerald, W. F., and T. W. Clarkson. 1991. Mercury and monomethylmercury: Present and future concerns. *Environmental Health Perspectives* 96:159–166.

Fitzgerald, W. F., and C. H. Lamborg. 2003. Geochemistry of mercury in the environment. Pages 107–148 in B. S. Lollar, editor, *Treatise on geochemistry*, vol. 9. Amsterdam: Elsevier.

Fitzgerald, W. F., D. R. Engstrom, C. H. Lamborg, C. -M. Tseng, P. H. Balcom, and C. R. Hammerschmidt. 2005. Modern and historic mercury fluxes in northern Alaska: Global sources and arctic depletion. *Environmental Science and Technology* 39:557–568.

Ford, J., D. Landers, D. Kugler, B. Lasorsa, S. Allen-Gil, E. Crecelius, and J. Martinson. 1995. Inorganic contaminants in arctic Alaskan ecosystems: Long-range atmospheric transport or local point sources? *Science of the Total Environment* 160/161:323–335.

Garbarino, J. R., E. Snyder-Conn, T. J. Leiker, and G. L. Hoffman. 2002. Contaminants in arctic snow collected over northwest Alaskan sea ice. *Water, Air and Soil Pollution* 39:183–214.

Gubala, C. P., D. H. Landers, M. Monetti, M. Heit, T. Wade, B. Lasorsa, and S. Allen-Gil. 1995. The rates of accumulation and chronologies of atmospherically derived pollutants in arctic Alaska, USA. *Science of the Total Environment* 160/161:347–361.

Grannas, A. M., R. M. Cory, P. L. Miller, Y. -P. Chin, and D. M. McKnight. 2012. The role of dissolved organic matter in arctic surface waters in the photolysis of hexachlorobenzene and lindane. *Journal of Geophysical Research* 117: G01003. doi:10.1029/2010JG001518.

Halsall, C. J. 2004. Investigating the occurrence of persistent organic pollutants (POPs) in the Arctic: Their atmospheric behavior and interaction with the seasonal snow pack. *Environmental Pollution* 128:163–175.

Hammerschmidt, C. R., and W. F. Fitzgerald. 2005. Methylmercury in mosquitoes related to atmospheric mercury deposition and contamination. *Environmental Science and Technology* 39:3034–3039.

Hammerschmidt, C. R., and W. F. Fitzgerald. 2006a. Photodecomposition of methylmercury in an arctic Alaskan lake. *Environmental Science and Technology* 40:1212–1216.

Hammerschmidt, C. R., and W. F. Fitzgerald. 2006b. Methylmercury in freshwater fish linked to atmospheric mercury deposition. *Environmental Science and Technology* 40:7764–7770.

Hammerschmidt, C. R., and W. F. Fitzgerald. 2008. Methylmercury in arctic Alaskan mosquitoes: Implications for impact of atmospheric mercury depletion events. *Environmental Chemistry* 5:127–130.

Hammerschmidt, C. R., W. F. Fitzgerald, C. H. Lamborg, P. H. Balcom, and C. -M. Tseng. 2006. Biogeochemical cycling of methylmercury in lakes and tundra watersheds of arctic Alaska. *Environmental Science and Technology* 40:1204–1211.

Harris, R. C., J. W. M. Rudd, M. Amyot, C. L. Barbiarz, K. G. Beaty, P. J. Blanchfield, R. A. Bodaly, B. A. Branfireun, C. C. Gilmour, J. A. Graydon, A. Heyes, H. Hintelmann, J. P. Hurley, C.A. Kelly, D. P. Krabbenhoft, S. E. Lindberg, R. P. Mason, M. J. Paterson, C. L. Podemski, A. Robinson, K. A. Sandilands, G. R. Southworth, V. L. St. Louis, and T. Tate. 2007. Whole-ecosystem study shows rapid fish-mercury response to changes in mercury deposition. *Proceedings National Academy of Sciences* 104:16586–16591.

Helgason, L. B., R. Barrett, E. Lie, A. Polder, J. U. Skaare, and G. W. Gabrielsen. 2008. Levels and temporal trends (1983–2003) of persistent organic pollutants (POPs) and mercury (Hg) in seabird eggs from northern Norway. *Environmental Pollution* 155:190–198.

Kirk, J. L., V. L. St. Louis, and M. J. Sharp. 2006. Rapid reduction and reemission of mercury deposited into snowpacks during atmospheric mercury depletion events at Churchill, Manitoba, Canada. *Environmental Science and Technology* 40:7590–7596.

Lahoutifard, N., M. Sparling, and D. Lean. 2005. Total and methyl mercury patterns in arctic snow during springtime at Resolute, Nunavut, Canada. *Atmospheric Environment* 39:7597–7606.

Lamborg, C. H., W. F. Fitzgerald, A. W. H. Damman, J. M. Benoit, P. H. Balcom, and D. R. Engstrom. 2002. Modern and historic atmospheric mercury fluxes in both hemispheres: Global and regional mercury cycling implications. *Global Biogeochemical Cycles* 16:1104. doi:1110.1029/2001GB1847.

Lindberg, S. E., S. Brooks, C. -J. Lin, K. J. Scott, M. S. Landis, R. K. Stevens, M. Goodsite, and A. Richter. 2002. Dynamic oxidation of gaseous mercury in the Arctic troposphere at polar sunrise. *Environmental Science and Technology* 36:1245–1256.

Lindberg, S., R. Bullock, R. Ebinghaus, D. Engstrom, X. Feng, W. Fitzgerald, N. Pirrone, E. Prestbo, and R. Seigneur. 2007. A synthesis of progress and uncertainties in attributing the sources of mercury in deposition. *Ambio* 36:19–32.

Lucotte, M., A. Mucci, C. Hillaire-Marcel, P. Pichet, and A. Grondin. 1995. Anthropogenic mercury enrichment in remote lakes of northern Québec (Canada). *Water Air Soil Pollution* 80:467–476.

MacDonald, R. W., L. A. Barrie, T. F. Bidleman, M. L. Diamond, D. J. Gregor, R. G. Semkin, W. M. J. Strachan, Y. F. Li, F. Wania, M. Alaee, L. B. Alexeeva, S. M. Backus, R. Bailey, J. M. Bewers, C. Gobeil, C. J. Halsall, T. Harner, J. T. Hoff, L. M. M. Jantunen, W. L. Lockhart, D. Mackay, D. C. G. Muir, J. Pudykiewicz, K. J. Reimer, J. N. Smith, G. A.

Stern, W. H. Schroeder, R. Wagemann, and M. B. Yunker. 2000. Contaminants in the Canadian Arctic: 5 years of progress in understanding sources, occurrence and pathways. *Science of the Total Environment* 254:93–234.

Marteel, A., C. F. Boutron, C. Barbante, P. Gabrielli, G. Cozzi, V. Gaspari, P. Cescon, C. R. Ferrari, A. Dommergue, K. Rosman, S. M. Hong, and S. Do Hur. 2008. Changes in atmospheric heavy metals and metalloids in Dome C (East Antarctica) ice back to 672.0 kyr BP (Marine Isotopic Stages 16.2). *Earth and Planetary Science Letters* 272:579–590.

Mergler, D., H. A. Anderson, L. H. M. Chan, K. R. Mahaffey, M. Murray, M. Sakamoto, and A. H. Stern. 2007. Methylmercury exposure and health effects in humans: A worldwide concern. *Ambio* 36:3–11.

Murozumi, M., T. J. Chow, and C. Patterson. 1969. Chemical concentrations of pollutant lead aerosols, terrestrial dusts and sea salts in Greenland and Antarctic snow strata. *Geochimica Cosmochimica Acta* 33:1247–1294.

Orihel, D. M., M. J. Paterson, P. J. Blanchfield, R. A. Bodaly, and H. Hintelmann. 2007. Experimental evidence of a linear relationship between inorganic mercury loading and methylmercury accumulation by aquatic biota. *Environmental Science and Technology* 41:4952–4958.

Poulain, A. J., E. Garcia, M. Amyot, P. G. C. Campbell, and P. A. Arlya. 2007. Mercury distribution, partitioning and speciation in coastal vs. inland High Arctic snow. *Geochimica Cosmochimica Acta* 71:3419–3431.

Rada, R. G., J. Wiener, M. R. Winfrey, and D. E. Powell. 1989. Recent increases in atmospheric deposition of mercury to north-central Wisconsin lakes inferred from sediment analysis. *Archives of Environmental Contamination and Toxicology* 18:175–181.

Renberg, I., M. -L. Brännvall, R. Bindler, and O. Emteryd. 2000. Atmospheric lead pollution history during four millennia (2000 BC to 2000 AD) in Sweden. *Ambio* 29:150–156.

Schroeder, W. H., K. G. Anlauf, L. A. Barrie, J. Y. Lu, A. Steffen, D. R. Schneeberger, and T. Berg. 1998. Arctic springtime depletion of mercury. *Nature* 394:31–32.

Selin, N. E., D. J. Jacob, R. M. Yantosca, S. Strode, L. Jaegle, and E. M. Sunderland. 2008. Global 3-D land-ocean-atmosphere model for mercury: Present-day versus preindustrial cycles and anthropogenic enrichment factors for deposition. *Global Biogeochemical Cycles* 22:GB2011. doi:10.1029/2007GB003040.

Simonich, S. L., and R. A. Hites. 1995. Global distribution of persistent organochlorine compounds. *Science* 269:1851–1854.

Steffen A., T. Douglas, M. Amyot, P. Ariya, K. Aspmo, T. Berg, J. Bottenheim, S. Brooks, F. Cobbett, A. Dastoor, A. Dommergue, R. Ebinghaus, C. Ferrari, K. Gardfeldt, M. E. Goodsite, D. Lean, A. J. Poulain, C. Scherz, H. Skov, J. Sommar, and T. Cemme. 2008. A synthesis of atmospheric mercury depletion event chemistry in the atmosphere and snow. *Atmospheric Chemistry and Physics* 8:1445–1482.

Tseng, C.-M., C. H. Lamborg, W. F. Fitzgerald, and D. R. Engstrom. 2004. Cycling of dissolved elemental mercury in arctic Alaskan lakes. *Geochimica Cosmochimica Acta* 68:1173–1184.

UNEP (United Nations Environmental Program). 2008. http://www.unep.org.

USEPA (U.S. Environmental Protection Agency). 1997. *Mercury study report to Congress Vol. VII: Characterization of human health and wildlife risks from mercury exposure in the United States*. EPA-452/R-97-009. Washington, DC: Office of Air Quality Planning and Standards and Office of Research and Development, U.S. Environmental Protection Agency.

USFDA/USEPA (U.S. Food and Drug Administration and U.S. Environmental Protection Agency). 2004. *What you need to know about mercury in fish and shellfish.* EPA-823-R-04-005. http://www.cfsan.fda.gov/~dms/admehg3.

Vandal, G. M., W. F. Fitzgerald, C. F. Boutron, and J. P. Candelone. 1995. Mercury in ancient ice and recent snow from the Antarctic. Pages 401–415 in R. Delmas, editor, *Ice core studies of global biogeochemical cycles,* NATO Advanced Study Institute Publications. New York: Springer-Verlag.

Wilson, R., S. Allen-Gil, D. Griffin, and D. Landers. 1995. Organochlorine contaminants in fish from an arctic lake in Alaska, USA. *Science of the Total Environment* 160/161:511–519.

10

Ecological Consequences of Present and Future Changes in Arctic Alaska

John E. Hobbie
George W. Kling

Introduction

In this chapter we bring together the evidence for changes in the ecology of an arctic Alaska site caused by climate changes, and predict the ecological changes by 2100. Some of these changes are the direct effects of a changing climate on physical, chemical, and biological processes, while others are indirect effects resulting from a number of ecological processes and interactions. One example of an indirect effect is the argument that an increase in plant biomass was likely not the direct effect of warmer soils on plant growth but was the result of a longer period of unfrozen soils on the total microbial activity. The microbial activity converts organically bound soil nutrients to forms usable by plants. Most of the evidence summarized here has been presented in the previous chapters of this book.

The Arctic LTER site at Toolik (the Toolik Field Station of the University of Alaska, Fairbanks) was chosen to represent a well-vegetated, low-arctic landscape including tundra, lakes, and streams. Data have been collected there since 1975 when the road supporting the Trans-Alaska Pipeline was completed. The rivers and lakes near Toolik drain into the Kuparuk River, which flows north to the Arctic Ocean.

In interpreting this long record of research, and in comparing predictions of change in this northern site with other landscapes in different climates, several key factors must be kept in mind: First, the climate at latitude 69° N is cold; summers are short and cool, with continuous daylight for most of the summer season and with continuous darkness in winter. The mean annual air temperature at Toolik is cold enough (–8.5°C) that the entire region is underlain by continuous permafrost. Second, there is a strong north-south climate gradient of ~3.5°C in the mean annual temperature of the Kuparuk River Basin (KRB) from the cool coastal plain bordering the Arctic Ocean to the warmer northern foothills some 170 km to the south where Toolik is located. Third, the site is quite isolated and in many ways pristine, with minimal direct impact of human activities both in the past and at present. Effects of human activities are limited to road dust and recreational and subsistence fishing and hunting. All wastes from the Toolik Field Station itself are trucked to Prudhoe Bay or Fairbanks. Streams and lakes in the area receive no nutrient enrichment, and even atmospheric nitrogen deposition is minimal. The original animal communities of caribou, bear, wolves, foxes, arctic grayling, arctic char, and lake trout are largely intact.

Alterations in Climate Drivers of Change: Effects in Northern Alaska

Air Temperatures

Based on the 22-year record (1989–2010), Toolik annual air temperature averages –8.5°C with a range from –10.6 to –5.8 °C (see Table 2.2). Over this period there is no statistically significant warming trend. One reason for this lack of a trend, despite significant warming at other areas in northern Alaska (Hinzman et al. 2005), is the particular period of record at Toolik. As noted in chapter 2, this lack of a warming trend is largely an artifact of this period of record and is entirely consistent with the lack of a warming trend at Barrow on the coast during exactly the same period. Barrow does, however, show a long-term warming trend over its 110-year record. For the 70 years beginning in 1940, the annual average air temperature at Barrow increased by a total of 2.0°C (Hinzman et al. 2005). At the scale of the entire North Slope, the near-surface air temperatures have warmed more than 3°C over the past 60 years (Shulski and Wendler 2007).

The lack of a significant warming trend in mean air temperatures at Toolik during the past 22 years contrasts with evidence of cumulative environmental changes near Toolik. One change linked directly to warming of the air is an increase in the ground temperatures at a depth of 20 m in a deep permafrost borehole located about 20 km south of Toolik (Romanovsky et al. 2011). At this depth the annual changes are damped out and most of the temperature increases (0.8°C over 20 years) are likely caused by warming air temperatures and perhaps changing snow cover (Figure 2.15). Another indicator of change at the North Slope scale is the continued shrinkage of all the Brooks Range glaciers, including one only 30 km from Toolik (Hinzman et al. 2005). Other cumulative effects, such as an increase in soil weathering linked to a deepening of the active layer and an increase in height and canopy density of vascular plants, are also linked to warming, but indirectly (see details later in this chapter).

Precipitation, Water Balance, and River Discharge

A review of the Arctic as an integrated system (Hinzman et al. in press) reports that pan-Arctic precipitation has increased by about 5% since the 1950s. Despite this increase, which also occurred in northern Alaska, Oechel et al. (2000) found that the summer water balance (precipitation minus potential evapotranspiration) has decreased since 1960 in North Slope villages because of longer and warmer summers. In addition, Muskett and Romanovsky (2011) report that runoff has also decreased overall on the Alaskan arctic coastal plain and foothills. The complete water balance is more complicated, however, and recent analyses indicate that groundwater storage on the coastal plain and foothills, probably in the unfrozen "talik" zones beneath lakes and streams, has increased from 1999 to 2009 (Muskett and Romanovsky 2011). At the same time, there has been a reduction in surface water as thousands of shallow lakes have dried (Smith et al. 2005). These apparently contrasting observations may in fact be consistent, considering that the surface active-layer depth of unfrozen ground has been increasing in many areas of the Arctic (Brown et al. 2000; Liu et al. 2003), which would lead to a redistribution of surface water to groundwater.

The data from Toolik (chapter 2) show no statistically significant trends in either annual amounts of precipitation from 1989 to 2010 or in summer and winter amounts. The range of annual precipitation was 201–462 mm and the average annual precipitation was 312 mm. During this period, 60% of the precipitation fell during summer months (June through August). Over the KRB, mean rainfall ranged from <50 mm yr^{-1} along the arctic coast to >240 mm in the foothills and Brooks Range.

The best record available for the long-term trends in Kuparuk River discharge is from the USGS gauge at the river mouth (chapter 7). The total annual runoff has not changed significantly. However, there has been a shift in the timing of the spring runoff, which prior to 1980 mainly occurred in June (Figure 7.36). After 1990 the runoff in May became significant and now dominates the spring runoff, although the total spring-runoff volume has not changed. Despite the shift in timing of the entire Kuparuk River discharge, there has been no trend in the timing of the snowmelt at Toolik. The fall runoff (September and October) of the entire Kuparuk River has increased in recent years, but it is still a small proportion of the total. The shift in runoff to earlier and later months has lengthened the flow season, and there has been a marked increase in summer high-flow events. This longer flow season with no increase in spring runoff, and potentially higher temperatures, leaves the river susceptible to short-term droughts. These droughts have become more frequent in recent years and have the potential to negatively impact the biota by drying the river channel, reducing available habitat areas, and impeding migration of the grayling population to and from their overwintering sites (details in chapter 7).

Arctic Sea Ice

The September ice extent of the Arctic Ocean has decreased by 11% per decade since 1979 (Serreze et al. 2009). One estimate is that the summer ocean will be ice-free by 2050, although the newest models predict an even more rapid loss of ice (Wang and Overland 2012). This reduced summer ice allows additional solar heat to be absorbed into the top 20 m of the ocean. The resulting heat is slowly released to the atmosphere during the following autumn, which increases atmospheric temperatures. The September-to-November deviation in air temperature in the Toolik area (compared to the temperatures for the same months in the period 1979–2007 is 1-2°C (Serreze et al. 2009). It is not yet determined if the major disturbance of reduction of sea ice in the summer has increased the number of convective storms in the KRB and in the area around Toolik. There are unpublished data showing an increase in the number of lightning strikes in arctic Alaska over the past decade, but a connection with a change in number of storms and an apparent increase in number of wildfires is not yet established.

Future Climates

Based on modeling exercises using mean output from the five general circulation models described in chapter 2, it seems likely that the future climate throughout the KRB will be warmer and wetter, with more snow than presently occurs. By 2099 the annual mean temperature in the KRB, compared with the present, is projected to be ~0.5°C warmer in the south and up to 3°C warmer in the north. For the same period, the precipitation is projected to be ~1.5 times the amount of current precipitation in the southern end of the basin and ~3 times the amount in the northern end (Figures 2.18 and 2.19). This large change in the precipitation of the future would greatly affect both stream flow and soil moisture.

However, it is well known that accurate measurement of precipitation is difficult because of wind and blowing snow, to mention only two problems. It is also well known that the global climate models are not very good at modeling precipitation. Hinzman et al. (in press) also discuss the difficulties of projecting future precipitation. They give one scenario in which permafrost degradation could lead to surface drying, and this would result in less precipitation because ~25% of high-latitude summer rainfall comes from recycled evapotranspiration (according to Serreze and Etringer 2003).

Physical Responses to Change in Northern Alaska

Active-Layer Thickness

As the climate warms, we expect that the thickness of the active layer near Toolik (i.e., the annual depth of thaw of the ground surface) will increase. This is the principal layer where the cycling of C, N, and P controls availability of nutrients and sequestration of C. The expectation of change is based on correlations of the thickness of the active layer and the square root of the degree days of thawing (Shiklomanov et al. 2010), an indicator of the total heating over the summer. For example, in the drained-lake landscape at Barrow, the 1995–2009 data showed a mean of 36 cm of summer thawing (range 14–62 cm). At Toolik, Figure 6.14 shows that for the period 1990–2011 the mean active-layer thickness in the tussock tundra was 40 cm, with a range of 28 to 51 cm. There was no statistically significant change of the active-layer thickness at Toolik over this period. No comparison of degree days of thawing and the thickness of the active layer has been made for Toolik.

In the Toolik region there is, however, strong evidence for an increase in the thickness of the active layer in at least some portion of the catchment, based on changes in the chemistry of Toolik Lake and its inlet streams. The alkalinity of the lake has increased from ~350 to 600 μEq L^{-1} from 1975 to 2007 (Figure 6.14). The most reasonable explanation is that there has been an increase in weathering of previously frozen glacial till during this time. This material is either in a thin layer that has thawed at the bottom of the active layer, or in layers under the water tracks, rivers, and lakes where there is deepening of the thawed layer caused mainly by changes in soil-water temperature or flow. As described in chapter 6, the changing ratio of strontium isotopes in stream water supports this weathering hypothesis to explain the changes over time in the chemistry of Toolik Lake.

Permafrost Warming

The best evidence for warming in the region near Toolik is the continuous warming of the permafrost at a borehole near Galbraith Lake, some 20 km south of Toolik. This is one of five deep boreholes along the Dalton Highway from Prudhoe Bay (West Dock) south to Galbraith Lake at the edge of the Brooks Range (Romanovsky et al. 2011). The temperatures in these boreholes are measured annually at a depth of 20 m, the depth where annual fluctuations are damped out. At the Arctic Ocean at West Dock, the present temperature is –8°C; at Happy Valley, halfway between the coast and Toolik, it is –4.5°C; and at Galbraith Lake, next to the Brooks Range, it is –5°C. The temperature has risen in all five boreholes; at Galbraith, the rise was 0.8°C from 1992 to 2011 (Figure 2.15). Some of this rise was due to an increase in snow on the ground, which insulates the ground from the very cold air of winter (Stieglitz et al. 2003). Some of the rise was due to an increase in the total amount of heat reaching the deep soil (see the preceding discussion of active-layer thickness). At the present rates of warming, there will still be permafrost through 2100. However, because permafrost soils contain tremendous stores of carbon that, when thawed, may be released to the atmosphere as greenhouse gases, there is an increasing debate on whether permafrost warming will feedback to lower latitudes and accelerate global warming (Serreze et al. 2009).

Thermokarst Formation

Collapse of the land surface when permafrost thaws is termed "thermokarst formation." This is a normal landscape process wherever permafrost contains a significant amount of ice (chapter 7). When the ice melts, the ground may collapse, and water moving over the surface may form a gully; or the soil may become saturated with water, and whole hillsides fail or slide downhill. Sometimes a piece of the shore may slump into a lake. On the South Slope of the Brooks Range, where the air temperature is warmer, it is plain that many more fresh thermokarsts have formed than on the North Slope (Figure 7.31). Evidence from the past, a sediment core from Lake NE14 just northwest of Toolik, shows that at least for ~5,500 yr BP there has been a series of thermokarst failures in the basin that left a recognizable mineral signature in the sediments; this indicates that recent thermal erosion of permafrost resulting in thermokarst disturbances has happened in the past in the Toolik area (F. S. Hu, personal communication). However, for the region around Toolik Lake, it is difficult to be completely sure that the rate of thermokarst formation has increased in recent decades. The thermokarst features are difficult to see on satellite pictures, and aerial photos from the past are rare in this region. Evidence for a recent increase in thermokarst formation comes from Gooseff et al. (2009), who used remotely sensed imagery on a helicopter transect to conclude that there were a number of new thermokarst features near Toolik.

Thermokarst failures can have impacts on terrestrial and aquatic systems. For example, over a period of two to three years, a single thermokarst gully carried 18 times more sediment than would normally be delivered by the Kuparuk River basin south of the road crossing near Toolik (~143 km^2) over the same period. Ammonium and total P in the disturbed streams are sometimes ten times the normal stream concentrations (Figure 7.35). Downstream rivers and lakes will be affected. Given the warming predicted by the end of this century, permafrost degradation and thermokarst formation will undoubtedly increase in frequency and become a major disturbance of the tundra landscape.

Wildfire

Fire has not been a normal part of disturbances on the North Slope. Until recently the assumption was that the tundra was too moist to sustain fire. But in 2007 a lightning storm sparked an intense fire that burned for two months and covered an area of 1039 km^2 northwest of Toolik during a period of low rainfall (chapter 6; Jones et al. 2009). This was the largest wildfire known to have occurred on the North Slope of Alaska. At the fire site, the albedo was reduced and the net radiation increased. As a result, in 2008–2010 the soils at the fire site were warmer and the depth of thaw greater than in unburned areas (Rocha et al. 2011). The change in surface energy balance contributed to the formation of new thermokarst features in the burned area. The recovering vegetation is mostly the same as before the fire, except for the absence of mosses and lichens, and except in those areas where the fire burned down to mineral soil. Here there are many seedlings of *Eriophorum* and forbs, indicating a remarkable ability of the tundra to rebound from this novel and severe disturbance.

Even though the area affected by the fire was only a small part of the entire North Slope, the carbon balance of the entire region was affected when this fire released >2 Tg of carbon to the atmosphere, mostly due to combustion of the upper few centimeters of soil organic matter. The amount of C released was equal to about half the annual net C sequestration of the entire North Slope region (chapter 5). Future increases in wildfire frequency, severity, and area burned have the potential to dominate the slower, climate-driven rates of regional changes. Disturbance due to wildfire will likely become another major disturbance of the tundra landscape in the same way as wildfire dominates large-scale disturbance in the boreal forest.

Microbial Response to Warming in Northern Alaska

The makeup of the soil and water microbial communities of the Toolik region is strongly shaped by the vegetation (Wallenstein et al. 2007) in the soils and the source of the organic carbon of the flowing waters (chapter 6). Within one landscape, such as MAT, there are even different microbial communities associated with tussocks and with shrubs (Judd et al. 2006). Most of the microbial data is about bacteria, but measurements of the carbon stocks reveal that the biomass of fungi is 50–500 times greater than that of bacteria in the tussock tundra (Vignette 5.2).

Response of Microbial Processes to Warming

Microbes in the laboratory and in nature respond to warming by increases in metabolic activity, including respiration. In surface waters, the maximum activity was found at both 12°C and 20°C (Adams et al. 2010). Surprisingly, the maximum activity for bulk soils is at 20–30°C in the Arctic and sub-Arctic (Nadelhoffer et al. 1991; Rinnan et al. 2011). Almost all soil measurements are of the total microbial respiration, but Pietikäinen et al. (2005) were able to separately measure growth of bacteria and growth of fungi in temperate soils incubated at a wide range of temperatures. Optimum growth occurred at 25–30°C, but below ~10°C the fungal activity was much higher than bacterial activity. At Toolik, the fungal biomass in the soil is many-fold greater than the bacterial biomass, but it is well known that microbial biomass does not necessarily equate with microbial activity. Certainly more research is needed to answer this interesting question of whether fungal or bacterial activity dominates in the Toolik region.

Beneath the snow cover, bacteria in soils continue to respire even when temperatures reach –5°C or even –10°C, because a thin layer of unfrozen water still covers soil-particle surfaces (chapter 5; Mikan et al. 2002). This under-snow respiration accounts for ~20% of annual respiration at Toolik (Schimel et al. 2006).

At the LTER site and at other sites in northern Alaska where nitrogen strongly limits plant growth, there is a continuing slow increase in plant growth that indicates the microbial mineralization of organic nitrogen, the major pool in soil, to N forms available to plants. One type of evidence comes from satellite views (Jia et al. 2002; Verbyla 2008) that show an increase over decades in the annual maximum NDVI (normalized difference vegetation index), an indication of leaf biomass. Another type of evidence of increase comes from a detailed analysis of plant growth in a large number of plots sampled four times over 20 years at Toolik and Imnavait Creek (Vignette 5.5), and from warming and fertilization experiments discussed in chapter 5. The plot samples and the satellite views agree with the slow changes in plant growth in the warming experiments (Figure 5.6) and with the results of fertilization experiments; all indicate increases in plant growth that are likely to be primarily related to the increase in the availability of N to plants. One of the first publications to point this out was Chapin et al. (1995), reporting Toolik data.

The observations discussed above, the change in NDVI recorded by satellite and the change in plant growth at Toolik over time, indicate that plants are responding to an increase in microbial activity caused by warming. This increase is not reflected in the measures of the mean annual temperature (Figure 2.4) at Toolik. A better way is to examine the total amount of heating during a summer; for example, Shiklomanov et al. (2010) found a correlation between the thickness of the active layer at Barrow each summer and the square root of the degree days of thawing. Applying the same approach, Jia et al. (2003) found that on the North Slope the NDVI-based greenness index of maximum plant biomass derived from satellite images correlated well with a summer warming index. This index was the sum of monthly mean air temperatures greater than 0°C. This summer warming index of total heating has not been calculated for Toolik, but it may well be a better way than the average annual temperature to describe possible warming effects on microbial process in the soil.

Effect of Major Disturbances on Microbes: Warming, Thermokarst, Active-Layer Deepening, and Fire

There is little documentation of effects of major arctic disturbances on microbes. Certainly the direct effect of warming on increased microbial activity, already discussed, is by far the major effect. Another effect is the potential for high concentrations of dissolved organic matter (DOM) to appear in a stream or a lake downstream from a thermokarst event. This DOM is undoubtedly decomposed by stream and lake microbes, but there is little information on rates (see chapter 6). Further, Walter et al. (2006) described the movement of particulate carbon into Siberian lakes when lake edges collapsed; concentrations were so high that the lakes became anaerobic and methane was released. Schuur et al. (2008) point out that the organic material released from thawing permafrost and processed *in situ* in soils is only slowly decomposed because most of the organic matter is of low quality; it has already been partially decomposed before it was incorporated into permafrost. However, if the previously frozen soil C is brought to the surface during a thermokarst failure, the exposure to ultraviolet (UV) light can increase the lability to bacteria (see Vignette 6.1). Wildfire leads to warmer, drier soils and increased thermokarst disturbance, all of which could lead to higher rates of microbial activity.

Ecosystem Responses Acting Through Microbes

When microbes make more nutrients available as a result of warming, the effects may extend beyond increased plant growth. As explained in chapter 5, long-term fertilizer and warming experiments consistently lead to higher terrestrial biomass, primary production, and changes in species composition. In MAT, the most common type of tundra in northern Alaska, the experiments indicate that deciduous shrubs should become more dominant in a warmer climate with higher nutrient availability. Experiments in other plant communities (e.g., moist nonacidic tundra, heath, and wet sedge) do not show this proliferation of shrubs (Hobbie et al. 2005). Yet, pairs of photographs taken 30–40 years apart do support the prediction of an increase in shrubbiness, although the careful monitoring at Toolik Lake and Imnavait Creek (Vignette 5.5) indicate a general increase in plant height and canopy density with little change in relative abundance of shrubs. Modeling calculations indicate that a regional shift to shrubs could have a greater effect on regional climate than would global warming. The whole system response would be a positive feedback (Figure 5.15) as shrub canopies reduce the springtime albedo of the snow cover, absorb solar radiation, and warm the air significantly. An increase in shrubs would also trap more blowing snow in the winter and in this way insulate the soil from very cold winter temperatures more effectively. Microbial activity would increase as warmer soils in the wintertime would allow nutrient mineralization to proceed through much or all of the year. However, a negative feedback also occurs in the summer when the denser canopies intercept more incoming solar radiation, thereby reducing the amount of solar radiation that reaches the soil surface. In this way the shrubs cause cooler soil temperatures during the summer. It is not yet known whether the positive or the negative feedback from increased arctic shrubs will be most important for microbes.

The LTER warming experiments produced an unexpected ecosystem effect on the ectomycorrhizal fungal community structure and function (Vignette 5.3; Deslippe et al. 2011). After 18 years of warming, the vegetation had shifted to the shrub *Betula nana*, and the shift was accompanied by changes in the symbiotic fungal species. These newly dominant species were able to mine the nitrogen in resistant organic matter, and their hyphae were able to explore far out from the roots. Evidently the nitrogen in labile organic matter of the soil lying close to the roots in the control plots had been depleted by microbial activity during the warming experiment. In the same warming experiments, Deslippe et al. (2012) found that the bacteria *Actinobacter* increased as did the fungi *Russula* ssp., *Cortinarius* ssp., and Helotiales.

The data on aquatic bacterial response to warmer temperatures also point to higher activity with warmer temperatures, but there are interesting nonlinearities in the response. There appear to be two major groups of aquatic bacteria in the Toolik area, adapted to two different temperature optima, one at ~12 °C and another at ~20 °C (Vignette 6.2). The result is that depending on the initial, average temperature of the system, warming may either increase or decrease microbial community activity. The concentrations and "quality" of DOC modify this response to temperature, and as described in chapter 6 the amounts and timing of terrestrial carbon inputs to surface waters exert a strong control on microbial activity and community structure. Although there are no arctic experiments on effects of warming on concentrations or quality of dissolved carbon, the results from Adams et al. (2010) indicate that both carbon limitation and temperature limitation of microbes occurs, apparently at different times of the summer season. Thus the response of aquatic microbes to future climate changes will depend on the matrix of interactions between landscape water balance and its effects on hydrological export of terrestrial DOC and also on inherent physiological responses of microbes to temperature.

Will the mineralization of organic N and P compounds in the soil result in increases in concentration of nutrients in soil water and eventually in streams and lakes? The experimental treatments of fertilization in streams and lakes have proven that both N and P nutrients limit primary production, but there is no evidence that this is already happening in undisturbed waters. For example, the concentration of water column chlorophyll in Toolik and other lakes, an indicator of algal growth, has not changed appreciably since 1985 (Figure 8.18).

Changes in Microbial Communities and Processes by 2100

In predicting changes by 2100, we begin with several assumptions: (1) that the types and rates of changes that have already occurred will continue; (2) that the predictions of changes in annual mean temperature from models for Toolik are correct; and (3) that the drastic changes of fire and thermokarst will increase in frequency but will not dominate the landscape. The slow increase in N mineralization, likely caused by an increase in the total degree days above 0°C, will continue. The predicted increase in the mean annual air temperatures will also add to the mineralization. However, as discussed in chapter 2, at Toolik the increase in mean annual temperature by 2100 is predicted to be less than 1°C. The net result will likely be a small but continual increase in the abundance of shrubs and their symbiotic mycorrhizal fungi. There are no long-term experiments that allow more than these general predictions. Changes in microbial communities will be linked to changes in vegetation communities through mutual feedbacks and interactions of both communities with climate. Presumably the microbial communities would become more similar to those existing in the boreal forest to the south of the Brooks Range. Chapter 6 suggests that terrestrial environments serve as reservoirs of microbial diversity for surface waters and that patterns of aquatic microbial diversity are structured by initial inoculation from upslope habitats. In this case, future environmental changes on land will likely impact the diversity and function of aquatic bacteria. Overall, there will be only slight changes in the landscape by 2100. Permafrost will still exist and the vegetation will be essentially the same as in 2012 unless, of course, the frequency of these now-rare events increases much more than expected.

Vegetation Responses to Change in Northern Alaska

Direct and Indirect Response of Vegetation to Change

All the evidence from the LTER research presented in chapter 5 confirms the conclusion by F. S. Chapin (1983), p. 47 that "temperature limits the rate at which resources become available but that temperature is not a strong direct limitation to plant growth in the Arctic." Thus, when plant photosynthesis is measured over the short term at different temperatures, there is a positive temperature effect. When plant growth was measured in experimental warming plots at Toolik, there was a positive response the first year, but it soon became evident that the plant growth had become limited by nutrients. Chapter 5 states that the limited responses to greenhouse warming that were observed in tussock and wet sedge tundras (Figure 5.6) are interpreted as resulting from a relatively slow increase in soil nutrient availability in response to modest temperature increases. A similar result was obtained from warming experiments at 20 arctic sites in open-topped chambers as a part of the International Tundra Experiment; initial increases in individual plant growth in the first two years of treatment were frequently not sustained over longer periods (chapter 5).

In the warming and fertilizer experiments in MAT at Toolik (chapter 5), the response of *Betula nana* is striking as this species comes to dominate in every plot. Graminoid plants also respond, but *Betula* and other shrubs have the ability to branch rapidly and grow taller as biomass accumulates. In this community, *Betula* soon shades out other plants except for a few shade-tolerant species like *Rubus chamaemorus*. In the warming experiments in the nonacidic tundra, where *Betula* and other taller shrubs are rare or absent, there is a general increase in abundance of all plant functional types, resulting in an overall increase in biomass and productivity similar in magnitude to that of MAT (Hobbie et al. 2005). In other types of tundra (such as dry heaths), wet sedge, grasses, or sedges are favored by fertilization and warming over at least the first 10–15 years of the experiments (e.g., Gough et al. 2012). But even in these experimental sites where fertilizer treatments have been maintained for over 20 years, the relative abundance of *Betula* appears to be increasing (G. Shaver, personal observation).

Effect of Major Disturbances: Wildfire and Thermokarst Formation

The response of tundra vegetation to wildfire depends strongly on the severity of the burn. The types of response have been well described by Racine et al. (1987) for the North Slope plants after a major fire on the Seward Peninsula (northwestern Alaska) and another fire on the Noatak River on the southern side of the Brooks Range. Lichens and mosses are intolerant and will not survive even the lightest of fires, which typically also consume most or all of the aboveground leaves and shoots. Even after severe burning, which may consume ten or more centimeters of surface organic matter, *Eriophorum* will resprout as the plant rhizomes are protected by the dense, wet tussock. Also, sedges like *Carex* spp. and shrubs such as *Betula*, *Ledum*, and *Vaccinium* will resprout from belowground plant parts if charring is not too deep. A few species, in particular some of the grasses, may survive as seeds. At the other extreme, fires that consume the tussocks and most of the soil organic matter will completely destroy the vegetation communities. However, recovery is rapid: within five to six years after a 1977 fire, the total vascular plant cover reached 50% to 100% of an unburned control. The graminoids often flourish after a fire that removes competing shrubs.

Thermokarst formation will certainly be a major future disturbance of tundra; the effect on streams and lakes of more sediments and nutrients has already been mentioned. A major effect on terrestrial ecosystems will be through rapid colonization of disturbed surfaces by shrubs. An increase in shrubs, discussed in detail in chapter 5, will have impacts on such things as the amount of trapped snow and thus on soil temperatures, on the regional climate, and on animal food and shelter (see the preceding discussion).

Unknown Aspects

As explained in chapter 5, plot-scale and regional-scale models of ecosystem function exist for the Arctic; however, the intermediate scale, which is the scale of hillslopes and small catchments including spatial interactions among neighboring patches of different ecosystem types, is lacking (Rastetter et al. 2004). Temporal modeling is needed to predict ecosystem changes in the 20- to 100-year future. Spatial and temporal scale models are also needed for scaling plot-scale experiments to regions and the pan-Arctic. These longer-term and coarser-scale models will also have to include additional processes that have been relatively little studied at Toolik, such as seed dispersal, germination, and seedling establishment, which are currently rare and slow in undisturbed tundra. Disturbances such as fire and thermokarst may create opportunities for following the time course of disturbance-recovery processes that are poorly understood in the Arctic.

The changes have many implications for the ecological system. For example, in the two types of tussock tundras that have been followed in detail (Vignette 5.5) the increased vascular plant growth reduced the moss and lichen cover in the moister vegetation at Toolik but had no effect on the bryophytes in a mixture of moist and dry tundra at Imnavait Creek. The eventual loss of lichen cover would eliminate an important winter food for caribou in the Toolik region. Studies from other arctic tundra sites suggest that small herbivores are also likely to be impacted by increasing woody shrub dominance as voles, lemmings (Oksanen et al. 2008), ground squirrels (Karels et al. 2000), and marmots all depend very heavily on sedges, grasses, and forbs—plants that are shaded out by increasing dominance of taller woody shrubs. In addition, migratory songbirds that breed on the tundra during the summer months have varied nesting habitat requirements, with some species nesting only in low-stature, graminoid-dominated tussock tundra (i.e., Lapland longspur) and others nesting only at the base of or within tall *Betula nana* or *Salix* spp. shrubs (i.e., Gambel's white-crowned sparrow, redpolls, and American robins) (Boelman, Gough, and Wingfield, unpublished data). Tundra canopies dominated by woody shrubs harbor significantly more arthropod biomass than graminoid dominated communities (Boelman, Gough, and Wingfield, unpublished data), suggesting that increased shrub cover could impact tundra trophic dynamics by altering the base of the food web.

Lakes too would be affected by a change in vegetation to shrubs and trees, which leads to a change in the chemistry of the particulate and dissolved organic carbon entering lakes from the surrounding soils. This is more than speculation because it is known that lakes south of the tree line in Canada contain high concentrations of CDOM (chromophoric dissolved organic matter), which on the one hand absorb light and thus limit the amount of photosynthesis in lakes, but on the other hand protect algae and bacteria from UV radiation (Vincent and Hobbie 2000).

Changes by Year 2100

The slow and subtle changes in the vegetation now in place near Toolik Lake certainly indicate a trend toward a future increase in shrubs. These changes are still small and took decades to develop (Figure 5.V5). The timetable for the change is unknown; therefore, all that can be said right now is that by 2100 the tundra will look a lot more like land on the South Slope of the Brooks Range than at the present time. There will be more and taller shrubs. Trees, most likely the white spruce now found just south of Atigun Pass, will move north at some point. By 2100 the productivity of the tundra will be higher, and the amount of organic matter accumulation in plants and soils will continually increase. Cycling of the nutrients nitrogen and phosphorus will be more rapid.

Stream Ecology Response to Changes in Northern Alaska

Changes in the River Environment

The environmental factors controlling the ecology of the Kuparuk River have remained unchanged over the study period (the mid-1970s to present; see chapter 7). It is an oligotrophic stream with low amounts of algal growth and low amounts of input of organic matter from its watershed. The stream temperature in summer ranges from 7°C to 13°C. Inorganic nutrients are either close to the limits of detectability (soluble reactive phosphate and ammonium are <0.05 μmol L^{-1}) or very low (nitrate plus nitrite 0.1 to 10 μmol L^{-1}). Oxygen is always close to its saturation amount. The one exception to the above lack of change is the McClelland et al. (2007) suggestion that warmer and wetter summers have increased the concentrations and export of inorganic nitrogen in the form of NO_3 from the upper Kuparuk catchment, especially in the last 15 years (Figure 7.37). The reason for this change is unknown, but a good possibility is that there is more microbial nitrification taking place.

The total flow of the Kuparuk River and other streams (chapter 7) has not appreciably changed over the study period (data for the whole Kuparuk River spans 1971 to present). There has been a shift to earlier flow in the spring; most of the spring runoff now occurs in May instead of June, and some flow continues late in the fall. Despite the shift in timing, which spreads the flow over a longer period, the total flow is unchanged. In addition the summer precipitation appears to be more variable in recent years than it was earlier, which increases the probability of drought and of very low stream flows. In times of low flow, the water temperatures are warmer than in times of high flow (Figure 7.7).

The conclusion of chapter 7 is that over the study period, climate or other environmental changes do not appear to have contributed significantly to changes in the stream biota. But there is an effect of temperature on the growth and survival of adult grayling and young-of-the-year (YOY) fish (e.g., Figure 7.24). Each summer differed in the average stream discharge and in the temperature of the water. It was found that adults grew more in cool summers with high discharge, while YOY grew best in warm summers with low flow (Deegan et al. 1999). This makes ecological sense as well because in warm summers the metabolism of adults is raised, and they need more food. The need for more food raises an important question—if stream fish need more food in the future when water temperatures are higher, will insect productivity increase enough to meet this need?

Effects of Major Disturbance

Twice during the 37 years of observations, in 1999 and 2002, the rivers near Toolik flooded during the summer months. In 2002, for example, an August snowfall was followed by rain. The combined precipitation and snow melt caused a tremendous flood in the Kuparuk estimated at >100 $m^3 s^{-1}$, while normal summer flow is 1–5 $m^3 s^{-1}$. The usual flooding at the time of the spring runoff does not do much damage to the streambed because much of the river still contains ice and the streambed is frozen. This summer flood, however, was unique in its intensity. The beds of many streams were completely transformed and eroded, with tremendous effects on the stream insect biota, the basis of the food web for the arctic grayling. The water in Toolik Lake became turbid because of the particulate matter that entered the lake.

A summer drought is another major disruption with important biotic consequences. There have been major drought events in 2005, 2007, 2009, and 2011. During droughts the flow in the Kuparuk River may stop or disappear beneath the stream bed in certain stretches of the river, and grayling are restricted to pools. During the extreme drought in the summer of 2011, the Kuparuk River near Toolik stopped flowing completely from mid-August until mid-September. The ecological consequence was that the arctic grayling could not make their usual end-of-summer migration to the lake in the headwaters of the Kuparuk River where they spend the winter. Luckily, a mid-September rain allowed several thousand grayling to reach the lake.

Response of the Whole Stream System

As noted above, environmental changes occurring in the stream, such as warming, discharge, or chemistry have not changed the stream biota. An increase in the rate of thermokarst formation would add nutrients and sediments to streams; nutrients would increase primary and secondary productivity, but sediments could shade the epilithon and reduce algal productivity. The result of these contrasting effects on the growth of grayling is uncertain. Thermokarst formation would also harm grayling survival if added sediments changed the circulation of oxygen-rich waters in the streambeds, reducing the oxygen necessary for survival of grayling eggs.

As discussed in the section on response of microbial processes to warming, the resultant increase in the availability of nutrients to terrestrial plants appears to be already happening. Eventually some of the increased amounts of nutrients cycling in the soil will move into streams, but there is no clear indication yet of changes in nutrient concentrations in streams (but cf. Figure 7.37). Another exception is described by Hobbie et al. (1999), where a small stream flowing near one of the seven gravel mines in the Toolik Lake catchment supplies 5% of the water entering Toolik Lake but 35% of the phosphate. The underlying permafrost has thawed exposing several meters of previously frozen glacial till to weathering that releases alkalinity (Figure 6.14) and phosphate. A change in the flowpath of groundwater would have the same effect. This phenomenon of increased weathering of previously frozen till is likely happening at a slow rate throughout this region, but the phosphate is being taken up by plants on land and not released into the streams or it is being released and taken up by P-limited stream organisms while the alkalinity is released and because it is more biologically conservative, building up in the lake.

The extensive research based on the nearly 30 years of adding low levels of phosphate to a stretch of the Kuparuk River is summarized in chapter 7. Its aim was to determine the controls on the river biota and river productivity by fertilizing at a low level for the summer months. There was an immediate response by the diatoms, followed by increases in biomass and productivity of all the parts of the food web including the top predator, the arctic grayling. After nearly a decade of treatment, we were surprised to find that a genus of moss (*Hygrohypnum*) was rapidly covering the rocks of the river in the phosphorus-fertilized reach. There is no way of telling whether or not this dramatic shift in the dominant primary producer will happen in the future because of the possibility that all the phosphorus would be removed from the soil water and from small streams by the plants before higher concentrations reached the river.

Changes by Year 2100

The ecology of the upper Kuparuk River is not expected to change dramatically during the remainder of the 21st century. The rivers and adjacent streams will continue to be oligotrophic with low nutrients and low primary productivity. Diatoms on the surface of the rocks and insect larvae will be unchanged by the probable small changes in the water temperature and the indirect effects of higher rates of nutrient cycling in the tundra soils. Even though the precipitation is expected to increase significantly, increases in flow will be within the tolerance of the present organisms.

There are, however, two types of disturbances that may change this scenario and alter the survival of the arctic grayling, the top predator. These, thermokarst formation and drought, have already occurred to a limited extent. A thermokarst, formed on the nearby Toolik River, demonstrated that high amounts of nitrogen and phosphorus as well as eroding soil are added to the river. The soil eroding from a thermokarst would strongly affect the spawning of grayling. If this disturbance came to affect a number of streams, the grayling populations could be threatened. The droughts, which are becoming more frequent, are a larger threat to grayling, however, because if one occurred during the late summer period when the grayling absolutely had to migrate to the overwintering lake, most of the population would be wiped out. Presumably some of the YOY might survive in a few springs or deep holes, but would a population survive the loss of all the adults? Thus, if the late summer droughts became more frequent and lasted longer, it is possible that the grayling, the only species of fish in the river and adjacent streams, could be extinct by 2100. It is also true that populations of grayling are found in lakes near Toolik. Are these genetically different from those found in streams? Would these adapt to stream life?

Lake Ecology Response to Changes in Northern Alaska

Disturbances and Changes in Lakes

The overall conclusion of chapter 8 on lakes was that there were few observed changes or trends in the physics, chemistry, and biology of lakes in the Toolik region during the 37 years of study, from 1975 to the present. Those changes we did observe were driven by changes in disturbance caused by fire, by thermal erosion of permafrost and increased weathering, or by thermokarst formation, rather than by the direct impact of temperature changes.

The ice cover in Toolik Lake typically lasts until mid-June; between 2000 and 2009, the date of the last ice in the lake ranged from June 8 to 1 July. One reason for variation in the date is the snow cover in the late spring. A late-lasting snow cover with a high albedo protects the ice cover; but when the ice is wind-blown and without a snow cover, the melt begins earlier. This variation in ice-out dates, which affects the timing of summer temperatures and stratification in Toolik Lake, and the lack of warmer summer air temperatures (chapter 2) results in no statistically significant warming of lake temperatures. For example, in Figure 8.17 the mean annual epilimnion temperature in Toolik in July from 1975 to 2007 shows no warming.

Direct Response of Lake Organisms to Temperature Change

Algae, the basis of the lake food web, do not show a species shift over the range of temperatures in Toolik and other lakes. The same species of algae are found beneath the ice and when the water temperatures reach 15°C later in the summer (O'Brien et al. 1997). Algae do increase their photosynthesis when the temperature is increased in laboratory experiments. However, as described in chapter 8, algae in natural systems respond much more to changes in light and to changes in nutrient concentrations than to changes in temperature. In fact, primary productivity is strongly limited by low concentrations of both nitrogen and phosphorus (Whalen et al. 2008). In Toolik and other nearby lakes, there is a brief burst of primary production after snowmelt and ice off, but nutrients are soon used up and the productivity falls to a low value. There is no evidence that so far the loading of nutrients into Toolik Lake has changed. The lake chlorophyll content, an indicator of algal productivity, has not changed appreciably from 1985 to the present (Figure 8.18).

Chapter 8 details several temperature effects on animals from lakes near Toolik. A study of effects of increased temperatures on zooplankton from Toolik and other lakes revealed that *Daphnia middendoffiana*, the most common species, is near its thermal maximum in shallow ponds. This species is adapted to thrive in arctic conditions; when warming occurs, it will be replaced by invasive species, probably other *Daphnia*, either from nearby lakes or from lower latitudes.

Natural populations of three species of fish from lakes near Toolik also show an effect from warming. These fish (lake trout, arctic grayling, and arctic char) showed a decrease in their fitness or condition as estimated from the length-weight relationship. In warm summers when epilimnetic temperatures exceeded 15°C for extended periods, a decrease in the slope of the length-weight relationship occurred for all three fish (Figure 8.19). A rise in the epilimnetic temperature would likely result in dramatic declines in fitness. The shallow nature of many of the lakes coupled with the lack of food resources below the epilimnion may narrow the window of suitable habitat and make it difficult for these fish populations to persist.

Effects of Major Disturbances

Both climate warming and fire will increase thermokarst activity in northern Alaska but further data are needed before it is certain that this has happened on the North Slope in the same way as it has already happened on the South Slope. As described by Bowden et al. (2008), these events greatly increase the solute transport to streams and lakes, especially nitrogen and phosphorus (Figure 7.35). In addition, the clarity of recipient lakes is reduced for many years (Figure 7.34). There is no detailed information yet describing the effect on a specific lake except anecdotal information on starving lake trout found in a lake where a thermokarst slump had occurred (M. McDonald, personal communication) and the observation that the clarity of Lake NE14, which was impacted by a thermokarst failure on its shoreline, has been increasing over time since the initial failure made the lake turbid (G. Kling, personal observation).

Response of the Whole Lake System

An initial study on the effects of climate on fish using a bioenergetics simulation model showed that increased lake temperatures would greatly reduce the ability of lake trout to thrive in arctic lakes (McDonald et al. 1996). Lake trout prefer water near 10°C with oxygen concentrations above 6 mg L^{-1} and show stress at temperature above 15°C. The lack of abundant deep-water habitat in most lakes of the region means that usable summer habitat for lake trout would decrease by 30% if epilimnetic temperatures rise by 2°C (Hobbie et al. 1999). This reduction of habitat is exacerbated if nutrient loading to the lakes increases phytoplankton productivity, resulting in decreases in hypolimnetic oxygen concentrations, or if ground water inflows increase, leading to chemical stratification which reduces mixing and also contributes to increased anoxia.

Lakes in Northern Alaska in 2100

If the projections of an increase in air temperature by 0.2°C –0.5°C are correct, then there will be no dramatic change in the ecology of the lakes of the Toolik region by 2100. This is not enough of a change in temperature to increase appreciably the amount of nutrients reaching the lakes or to change the summer temperatures of the lakes to an important degree. However, it is likely that increasing temperatures will further erode the thermal stability of permafrost in the catchments or in the talik beneath lakes, which will lead to continued alterations in lake chemistry (Figure 6.14). Exactly how much the input of nutrients and turbidity from thermokarst formation will change is unknown, but this disturbance, rare at the present time, may increase from climate warming and from fire. It is clear that the most vulnerable part of the lake ecosystem is the fish. Lake trout in particular will be the first to show an effect of climate warming, and this may well happen by 2100.

Summary

Past and Present Changes at Toolik: Ecological Consequences

1. The climate of the whole North Slope has warmed by more than 3°C over the past 60 years. Despite this regional warming, the climate record at the Toolik Field Station shows no statistically significant warming trend in the annual average air temperatures. This could be largely an artifact of the rather brief Toolik record (1989–2010).
2. Significant changes have accumulated over time in the glaciers, soils, streams, lakes, and tundra near Toolik that indicate warming is occurring. These changes include:

    a. Shrinking of mountain glaciers within 30 km of Toolik;
    b. Warming of the permafrost at a depth of 20 m in the soil by 0.8°C over 20 years;
    c. Increasing of vegetation canopy height, leaf area, and "greenness" (satellite view). These are probably related to a change in aboveground biomass and primary production. Based on experimental results, we conclude that there has been an increase in microbial decomposition that releases more nitrogen than previously from the large storehouse in soil organic matter. This change in the release could be caused by an increase in temperatures in the soil, by a longer growing season (in this example the period of time when the microbes are active), or by both.
    d. Changing chemistry in streams and a doubling of the alkalinity in lakes. The most reasonable explanation is that warming has led to increased weathering of previously frozen glacial till either by increasing the thickness of the active layer or by increasing the size of the thaw bulb under streams and lakes, which exposes more mineral soil.

3. In aquatic and terrestrial ecosystems the direct effects of temperature on plants and animals are small, or changes are not sustainable without input of nutrients; direct responses to temperature only will be slow. The associated nutrient responses depend on the temperature-caused changes in nutrient turnover by microbes or on changes related to nutrient inputs from disturbances.
4. Climate-related changes in disturbances including fire but especially those related to permafrost may be more important than the direct effects of climate warming.
5. The response of aquatic ecosystems to warming is linked closely to the response of terrestrial ecosystems through runoff of nutrients, sediments, and organic matter. These factors are very slow to change and are well buffered so that increases in nutrients in runoff are very small; as a result,

the biology of streams and lakes has shown little response to warming, and no drastic change is expected at the century scale. The one exception is the evidence that unusually warm summers have had a negative impact on fish, particularly the arctic grayling and lake trout.

6. The effect of changing seasonality on arctic biology is not discussed in this book; it is under study by several projects and is therefore a work in progress. Topics include the timing of bird arrival and life-history stages, the emergence and peak abundance of their insect prey, the phenology of plant flowering and the timing of emergence of their insect pollinators, the changes in timing of stream flow and fish migration, and the late-season and winter nutrient mineralization and the timing of nutrient uptake by plants.

Future Changes at Toolik: Predictions of Ecological Consequences

Climate models indicate a warmer and wetter environment for northern Alaska by 2100. The warming effects on the terrestrial ecology near Toolik by the end of the century will result in more biomass of grasses and shrubs. The aquatic systems will be somewhat more productive but mostly unchanged. The large, top-of-the-food-web fish are the most vulnerable part of the ecosystem, and some species may disappear. Finally, the warmer and wetter climate will lead to an increase in a major disturbance, thermokarst, caused by the melting of ice in permafrost and resulting in slippage of sections of hillsides, compaction of land, and transport of sediments and nutrients to streams and lakes. The prediction of a wetter environment is under discussion and argument in the scientific community and awaits better understanding and models.

The ecological information collected by the Arctic LTER and associated projects has focused on effects of warming and related changes on communities and species. It is more difficult to predict the ecosystem effects. Yet, it is well known that species-composition changes in terrestrial and aquatic ecosystems have important feedback (or feed-forward) effects that must be considered in making long-term predictions. These effects may cascade through the trophic structure, through changes in carbon and nutrient turnover, habitat structure, food-resource availability and quality, biodiversity, or in the physics of energy balance in ecosystems. As concerns rise that thawing permafrost soils may release greenhouse gases to the atmosphere and accelerate global warming (Serreze et al. 2009), we suggest that it is the ecological interactions in tundra ecosystems which will govern how a changing Arctic system affects the climate at lower latitudes in this century.

References

Adams, H. E., B. C. Crump, and G. W. Kling. 2010. Temperature controls on aquatic bacterial activity and community dynamics. *Environmental Microbiology* 12:1319–1333.

Bowden, W. B., M. N. Gooseff, A. Balser, A. Green, B. J. Peterson, and J. Bradford. 2008. Sediment and nutrient delivery from thermokarst features in the foothills of the

North Slope, Alaska: Potential impacts on headwater stream ecosystems. *Journal of Geophysical Research* 113:1029–1041.

Brown, J., K. M. Hinkel, and F. E. Nelson. 2000. The circumpolar active layer monitoring (CALM) program: Research designs and initial results. *Polar Geography* 24:166–258.

Chapin, F. S. 1983. Direct and indirect effects of temperature on arctic plants. *Polar Biology* 2:47–53.

Chapin, F. S., G. R. Shaver, A. E. Giblin, K. J. Nadelhoffer, and J. A. Laundre. 1995. Responses of arctic tundra to experimental and observed changes in climate. *Ecology* 76:694–711.

Deegan, L. A., H. E. Golden, C. J. Harvey, and B. J. Peterson. 1999. Influence of environmental variability on the growth of age-0 and adult arctic grayling. *Transactions of the American Fisheries Society* 128:1163–1175.

Deslippe, J. R., M. Hartmann, W. W. Mohn, and S. W. Simard. 2011. Long-term experimental manipulation of climate alters the ectomycorrhizal community of *Betula nana* in arctic tundra. *Global Change Biology* 17:1625–1636.

Deslippe, J. R., M. Hartmann, S. W. Simard, and W. W. Mohn. 2012. Long-term warming alters the composition of arctic soil microbial communities. *FEMS Microbiology Ecology* 82:303–315

Gooseff, M. N., A. Balser, W. B. Bowden, and J. B. Jones. 2009. Effects of hillslope thermokarst in northern Alaska. *Eos Transactions: American Geophysical Union* 90(4):29–36.

Gough, L., J. C. Moore, G. R. Shaver, R. T. Simpson, and D. R. Johnson. 2012. Above- and belowground responses of arctic tundra ecosystems to altered soil nutrients and mammalian herbivory. *Ecology* 93:1683–1694.

Hinzman, L. D., N. D. Bettez, W. R. Bolton, F. S. Chapin, M. B. Dyurgerov, C. L. Fastie, B. Griffith, R. D. Hollister, A. Hope, H. P. Huntington, A. M. Jensen, J. J. Gensuo, T. Jorgenson, D. L. Kane, A. H. Lloyd, A. D. McGuire, F. E. Nelson, W. C. Oechel, T. E. Osterkamp, C. H. Racine, V. E. Romanovsky, R. S. Stone, D. A. Stow, M. Sturm, C. E. Tweedie, G. L. Vourlitis, M. D. Walker, D. A. Walker, P. J. Webber, J. M. Welker, K. Winker, and K. Yoshikawa. 2005. Evidence and implications of recent climate change in northern Alaska and other arctic regions. *Climatic Change* 72:252–298.

Hinzman, L. D., C. Deal, A. D. McGuire, I. Polyakov, J. Walsh, and S. H. Mernild. In press. Trajectory of the Arctic as an integrated system. *Ecological Applications*. doi: 10.1890/11-1498.1

Hobbie, J. E., B. J. Peterson, N. Bettez, L. A. Deegan, W. J. O'Brien, G. W. Kling, G. W. Kipphut. 1999. Impact of global change on biogeochemistry and ecosystems of an arctic freshwater system. *Polar Research* 18:207–214.

Hobbie, S. E., L. Gough, and G. R. Shaver. 2005. Species compositional differences on different aged glacial landscapes drive contrasting responses of tundra to nutrient addition. *Journal of Ecology* 93:770–782.

Liu, J. S., N. Hayakawa, M. J. Lu, S. H. Dong, and J. Y. Yuan. 2003. Hydrological and geocryological response of winter streamflow to climate warming in Northeast China. *Cold Regions Science Technology* 37:15–24.

Jia, G. H., H. E. Epstein, and D. A. Walker. 2002. Spatial characteristics of AVHRR-NDVI along latitudinal transects in northern Alaska. *Journal of Vegetation Science* 13:315–326.

Jones, B. M., C. A. Kolden, R. Jandt, J. T. Abatzoglou, F. Urban, and C. C. Arp. 2009. Fire behavior, weather, and burn severity of the 2007 Anaktuvuk River tundra fire, North Slope, Alaska. *Arctic, Antarctic, and Alpine Research* 41:309–316.

Judd, K. E., B. C. Crump, and G. W. Kling. 2006. Environmental drivers control ecosystem function in bacteria through changes in community composition. *Ecology* 87:2068–2079.

Karels, T. J., A. E. Byrom, R. Boonstra, and C. J. Krebs. 2000. The interactive effects of food and predators on reproduction and overwinter survival of arctic ground squirrels. *Journal of Animal Ecology* 69:235–247.

McClelland, J. W., M. Stieglitz, F. Pan, R. M. Holmes, and B. J. Peterson. 2007. Recent changes in nitrate and dissolved organic carbon export from the upper Kuparuk River, North Slope, Alaska. *Journal of Geophysical Research—Biogeosciences* 112. G04S60. doi:10.1029/2006JG000371.

McDonald, M. E., A. E. Hershey, and M. C. Miller. 1996. Global warming impacts on lake trout in arctic lakes. *Limnology and Oceanography* 41:1102–1108.

Mikan, C. J., J. P. Schimel, and A. P. Doyle. 2002. Temperature controls of microbial respiration in arctic tundra soils above and below freezing. *Soil Biology and Biochemistry* 34:1785–1795.

Muskett, R. R., and V. E. Romanovsky. 2011. Alaskan permafrost groundwater storage changes derived from GRACE and ground measurements. *Remote Sensing* 3:378–397.

Nadelhoffer, K. J., A. E. Giblin, G. R. Shaver, and J. A. Laundre. 1991. Effects of temperature and substrate quality on element mineralization in six arctic soils. *Ecology* 72:242–253.

O'Brien, W. J., M. Bahr, A. E. Hershey, J. E. Hobbie, G. W. Kipphut, G. W. Kling, H. Kling, M. McDonald, M. C. Miller, P. A. Rublee, and J. R. Vestal. 1997. The limnology of Toolik Lake. Pages 61–106 in A. M. Milner and M. W. Oswood, editors, *Freshwaters of Alaska*. Ecological Studies, vol. 119. New York: Springer-Verlag.

Oechel, W. C., G. L. Vourlitis, S. J. Hastings, R. C. Zulueta, L. D. Hinzman, and D. L. Kane. 2000. Acclimation of ecosystem CO_2 exchange in the Alaskan Arctic in response to decadal climate warming. *Nature* 406:97–981.

Oksanen, T., L. Oksanen, J. Dahlgren, and J. Olofsson. 2008. Arctic lemmings, *Lemmus* spp., and *Dicrostonyx* spp.: Integrating ecological and evolutionary perspectives. *Evolutionary Ecology Research* 10:415–434.

Pietikäinen, J., M. Pettersson, and E. Bååth. 2005. Comparison of temperature effects on soil respiration and bacterial and fungal growth rates. *FEMS Microbiology Ecology* 52:49–58.

Racine, C. H., L. A. Johnson, and L. A. Viereck. 1987. Patterns of vegetation recovery after tundra fires in northwestern Alaska, U.S.A. *Arctic and Alpine Research* 19:461–469.

Rastetter, E. B., B. L. Kwiatkowski, S. Le Dizès, and J. E. Hobbie. 2004. The role of down-slope water and nutrient fluxes in the response of arctic hill slopes to climate change. *Biogeochemistry* 69:37–62.

Rinnan, R., A. Michelsen, and E. Bååth. 2011. Long-term warming of a subarctic heath decreases soil bacterial community growth but has no effects on its temperature adaptation. *Applied Soil Ecology* 47:217–220.

Rocha, A., and G. R. Shaver. 2011. Post-fire energy exchange in arctic tundra: the importance and climatic implications of burn severity. *Global Change Biology* 17:2831–2841.

Romanovsky, V. E., S. L. Smith, H. H. Christiansen, N. I. Shiklomanov, D. S. Drozdov, N. G. Oberman, A. L. Kholodov, and S. Marchenko. 2011. *Permafrost.* Arctic Report Card, NOAA. www.arctic noaa.gov/reportcard/permafrost.html.

Schimel, J. P., J. Fahnestock, G. Michaelson, C. Mikan, C.-L. Ping, V. E. Romanovsky, and J. Welker. 2006. Cold-season production of CO_2 in arctic soils: Can laboratory and field estimates be reconciled through a simple modeling approach? *Arctic, Antarctic, and Alpine Research* 38:249–256.

Schuur, E. A. G., J. Bockheim, J. G. Canadell, E. Euskirchen, C. B. Field, S. V. Goryachkin, S. Hagemann, P. Kuhry, P. M. Lafleur, H. Lee, G. Mazhitova, F. E. Nelson, A. Rinke, V. E. Romanovsky, N. Shiklomanov, C. Tarnocai, S. Venevsky, J. G. Vogel, and S. A.

Zimov. 2008. Vulnerability of permafrost carbon to climate change: Implications for the global carbon cycle. *BioScience* 58:701–714.

Serreze, M. C., and A. J. Etringer. 2003. Precipitation characteristics of the Eurasian Arctic drainage system. *International Journal of Climatology* 23:1267–1291.

Serreze, M. C., A. P. Barrett, J. C. Stroeve, D. N. Kindig, and M. M. Holland. 2009. The emergence of surface-based Arctic amplification. *The Cryosphere* 3:11–19.

Shiklomanov, N. I., D. A. Streletskiy, F. Nelson, R. D. Hollister, V. E. Romanovsky, C. E. Tweedie, J. G. Bockheim, and J. Brown. 2010. Decadal variations of active-layer thickness in moisture-controlled landscapes, Barrow, Alaska. *Journal of Geophysical Research* 115, G00I04. doi:10.1029/2009JG001248.

Shulski, M., and G. Wendler. 2007. *The Climate of Alaska*. Fairbanks: Alaska. University of Alaska Press.

Smith, L. C., Y. Sheng, G. M. MacDonald, and L. D. Hinzman. 2005. Disappearing arctic lakes. *Science* 308:1429–1429.

Stieglitz, M., S J. Déry, V. E. Romanovsky, and T. E. Osterkamp. 2003. The role of snow cover in the warming of arctic permafrost. *Geophysical Research Letters* 30:1721. doi:10.1029/2003GL017337.

Verbyla, D. 2008. The greening and browning of Alaska based on 1982–2003 satellite data. *Global Ecology and Biogeography* 17:547–555.

Vincent, W. F., and J. E. Hobbie. 2000. Ecology of arctic lakes and rivers. Pages 197–232 in M. Nuttal and T. Callaghan, editors, *The Arctic: Environment, people, policies.* Amsterdam: Harwood Academic Publishers.

Wallenstein, M. D., S. McMahon, and J. Schimel. 2007. Bacterial and fungal community structure in arctic tundra tussock and shrub soils. *FEMS Microbiology Ecology* 59:428–435.

Walter, K. M., F. S. Chapin, III, S. A. Zimov, and J. P. Chanton. 2006. Methane bubbling from Siberian thaw lakes: A positive feedback to climate change. *Nature* 443:71–75.

Wang, M., and J. E. E. Overland. 2012. A sea ice free summer Arctic within 30 years—an update from CMIP5 models. *Geophysical Research Letters* 39, L18501, doi:10.1029/2012GL052868.

Whalen, S. C., B. A. Chalfant, and E. N. Fischer. 2008. Epipelic and pelagic primary production in Alaskan arctic lakes of varying depth. *Hydrobiologia* 614:243–257.

Index

active layer, 91, 100, 279
climate warming effects, 169, 306
relationship with degree days of thawing, 310, 320
summer depth, 167
air temperature
mean annual Toolik, 38
alkalinity, 167, 279
causes of changes, 279
aquatic productivity
middle Holocene, 87
response to fertilization, 269
arctic ecology
glacial geology efffects on, 74–76
arctic ground squirrel, 97
aufeis, 177

bacterial community
changes in soil and water, 153–56
LTER lakes, 257
persistant and transient in lakes, 158
photo-chemical effects, 156
shifts during warming, 311
temperature and production, 160, 153–56
transport and dispersal, 162
benthic animal community
response to fertilization, 272
biomass aboveground
moss, 92
vascular plants, 92
biomass belowground, 92
borehole
locations along Dalton Highway, 307
temperatures. *See* permafrost

C. *See* carbon
carbon cycling
microbial transformations, 150–54
soil, 103–5
soil DOC, DIC production, 148–50
surface water transfer and evasion, 146–48
thermokarst C (in permafrost), 154
caribou, 97
change at the ecosystem level
time and trajectories, 120–26
changes
energy, N balance, disturbance, 126–31
changes by 2100
lakes in northern Alaska, 319
microbial communities and processes, 312
upper Kuparuk River, 317
vegetation, 314

changes, future
  predictions of ecological
    consequences, 321
changes, past and present
  ecological consequences, 320–21
climate, 53
  air temperatures, 304
  climate projection, 53–57
  early Holocene, 85
  general circulation models (GCMs), 53
  late Pleistocene. *See* Pleistocene
  middle-to-late-Holocene, 85
  oscillations, 53
  Pleistocene-to-Holocene transition, 84
  Toolik. *See* Toolik: climate
climate Alaska
  temperature, precipitation, 44
climate change, 304–6
  active layer thickness, 306
  Arctic sea ice, 305
  caveats, 57, 306
  climate projections, 53–57, 306
  fire, 308
  fish, possible effect, 315
  future ecosystem state, 131–32
  glacier shrinkage, 304
  increase in soil weathering, 304
  microbes as key to ecosystem
    change, 310
  microbial activity, 309
  North Slope, 35
  permafrost temperature as indicator, 307
  permafrost temperatures, 304
  precipitation more variable, 315
  water, 304–5
climate norms
  Kuparuk River Basin, 42
climate warming
  microbial activity, 309

Dalton Highway, vii, 5, 14, 123, 186, 307
decomposers
  soil food web, 98–100
decomposition
  acidic vs. nonacidic tundra, 75, 115
  arctic-wide comparisons, 115
  effects of recalcitrant substrates in
    stream, 194
  from lake C vs. from land C, 255
  heterotrophic respiration processes, 103
  litter quality effects, 115
  microbes and N release, 320
  models, 132, 166
  N limitation, 126
  on different age soils, 86
  photochemical, 148, 155, 297
  release of P from lake sediments,
    251, 254
  temperature and moisture effects, 5, 100
DIC, 158, *See* dissolved inorganic carbon
discharge
  effect on algae and insects, 194
  effect on lake stratification, 281
  effect on young fish, 199
  flashy hydrographs from storms, 175
  from snow melt, 46
  Kuparuk River, 186
  long-term trends, 224
  seasonal, in streams, 176
dissolved organic carbon, 144
dissolved organic nitrogen, 144
disturbance
  thermokarst on fish, 316
disturbances
  construction and research camps, 15
  dought effect on fish migration, 316
  effect on microbes, 310
  fire, 14, 308
  floods, 315
  road building, 16
  shape landscapes, 14
  thermokarst formation, 15, 307
DOC, 148–53, 156, 163, 169, 178, 242,
    247–48, 252, 311
DOM, 13, 145, 158–61, 255, 257, 310, *See*
    dissolved organic matter
drought
  effect on fish, 316
  effect on stream flow, 316

ecosystem interactions
  biogeochemistry of hillslopes, 163–66
  biology in aquatic systems, 158
  chemistry in aquatic systems, 156–66
  terrestrial and aquatic systems, 167–69
ecosystem research
  alkalinity and strontium indicate
    weathering, 167–69
  ecosystem interactions, 143–45
  hydrological catchments, 143–45

land-water interactions, 144
tundra, 132–34
element budgets and stoichiometry
soils, 110–11
experiments, 17
algal response to fertilization, 269
fertilization, 109
Kuparuk fertilization, 194
lake trout removal, 276
mesocosms in Toolik, 239, 269
ongoing whole-lake, 275–76
tundra fertilization, 112
warming, 109
whole-lake experiments, 239, 270–75

feedback
shrub increase on climate, 310–11
fertilized plots, 120–26
nitrogen loss over time, 124
fertilized streams
response of algae, insects, fish, moss, 317
fire
as disturbance, 14
effects on C, albedo, thermokarst, 129
fish
grayling, 204–9
response to climate warming, 280, 315
response to fertilization, 273
fish distributions, 268
floods
spring floods vs. summer floods, 315–16
flowering control
Dalton Highway transect, 123
fossil insects, 85
fungi. *See* soil food web
dominant biomass in soil, 100
species shift in warming, 311

geomorphology
glacial geology effects on, 66
glacial geology, 61–65
grayling, 13, 16, 184, 193, 196, 198, 203, 204, 208, 226, 228, 231
gross primary production (GPP)
terrestrial, 103–5
ground ice, 11

herbivores
mammals, birds, insects, other invertebrates, 96–98
herbivory
insect and mammal effects, 96–98
history
foothills, 81–87
International Biological Programme (IBP), 4, 132–34
LTER, 6
Naval Arctic Research Laboratory (NARL), 4
response, resistance, resilience, and recovery (R4D), 6
Toolik Lake research, 5
hyporheic dynamics, 209–13

insects
control by weather, plant growth, 97–98

Kuparuk River Basin (KRB)
climate, 22–24
description, 6–8
location, 22

lake changes
fire, thermokarst effects, 279
lake ecology response to changes, 317–19
lakes
are there responses to change?, 277–79, 311
biological properties, 254–68
bottom-up control, 272
chemistry, budgets, sediments, nutrients, 248–54
description, low arctic, in landscape, 238–40
food web, 254
mercury budget, 296
on glacial surfaces, 239
physical properties, temperature, light, 240–48
predictions about the future, 279–82
responses of whole system to temperature increase, 319
sediment chemistry, fertilization effects, 252–54
stratification, deep, shallow, 246
temperature response fish, 318
temperature responses algae, zooplankton, 318
thermokarst efffects on fish, 319

LTER research project
experiments, 17
observations carried out, 16
process rates, 18
site defined, 1, 9
synthesis, 18

macrozooplankton
crustaceans, 265–67
mercury
atmospheric mercury depletion events (AMDEs), 293–94
biogeochemical cycle, 290–91
budgets of mercury and MMHg, 296–97
loading in Toolik region, 295
mosquitoes as monitors, 292
toxic monomethylmercury (MMHg), 292
methane
losses from lakes, 252
losses from soil, 105
microbial activity
link to plant growth, 309
microbial response
biomass fungi greater than bacterial, 309
microbial transformations, 150–54
microzooplankton
response to fertilization, 274
models
fish bioenergetics, 280
general circulation models (GCMs), 53
intermediate scale models missing, 313
simple arctic model (SAM), 111
spatial interactions ecosystems on hillslope, 132
stream food-web, inverse, 230–32
mosquitoes
as mercury monitors, 292
mycorrhizae
nitrogen uptake, 107–9
mycorrhizal fungi (ECM)
shift in warming experiment, 109

N. *See* nitrogen
NDVI
acidic vs. nonacidic tundra, 75
and summer warming index, 310
as growth indicator, 309
changes in Kuparuk region, 73
definition of, 8
in glacial surfaces, 70
in model, 18
Kuparuk River Basin, 8, 72
of older landscapes, 72
on glacial units, 73
shrubs and greening, 76
net ecosystem exchange (NEE)
soils, 105
net primary production. *See* also primary production
mosses and lichens, 95
roots, 95
stream, 214
stream benthos, 187
stream bryophytes, 181
vascular plants, 94–95
nitrate export
response to change, 315
nitrification in catchment, 315
nitrogen cycling
flux from sediments, 271
lakes, 250–52
soils, 105–9
nitrogen deposition
in rain and snow, 105–9
nitrogen fixation, 105–9, 166
lake, 251
nitrogen mineralization
soils, 107
North Slope of Alaska, 1
N-use efficiency, 119
nutrient response
fertilized experiments, 120–26, 276

oxygen concentrations
climate warmng reduces fish habitat, 280
in fertilized lakes, 271
recovery from fertilization, 274

P. *See* phosphorus
paleoecology
foothills, 81–87
paludification
peat formation, 73–74, 85
permafrost
borehole temperatures, 51
definition of, 11
thawing affects biogeochemistry, 167–69
warming, 307

pH
vascular plants species richness, 96
younger vs. older soils, 100
phosphorus cycling
flux from sediments, 271
lakes, 250–52
release from thawing permafrost, 316
soils, 109–10
phosphorus fertilization
response of algae, insects, fish, moss, 317
photo-oxidation of DOM
respiration, C processing, 153
phytoplankton
and light, 242
demand for nitrogen, 251
response to fertilization, 274, 281
species diversity, 254
Toolik Lake, 258
plant growth
birch response to fertilizer, 313
link to microbial activity, 309
shrub response to warming and fertilization, 313
temperature not a strong direct limitation, 312
plant species diversity
effect of local abiotic conditions, 95
Pleistocene, 77
climate, 81
Pleistocene-to-Holocene transition, 84
climate, 84
pollen record, 81–87
pollutants. *See* mercury
air borne, 15
Arctic and North Slope, 287
mechanism for transport to Arctic, 288
pollution in polar regions
mercury and lead, history, 289–90
poplar
early-Holocene expansion, 85
precipitation
projected, 55
trends at Toolik, 35
predators
jaegers, owls, raptors, foxes, weasels, wolves, 97
primary producers
lake phytoplankton, 258
primary production
effects of enhanced rates, 282
lake, 251, 258–64
stream, 181, 192, 217
surface waters, lakes, 259

Quaternary, 77, 81

relative humidity
Toolik, 39
respiration
effects of compounds, 151
of thermokarst C (from permafrost), 154
photo-oxidation of DOM, 153
soil moisture effects, 152
respiration rates
soil microbes, 103–5

shrub increase, 126–31
shrubs increasing
effect on regional climate, 310
snails
effect on benthic production, 264
snow
annual, 48
measurement problems, 46
snow cover, 46
spring snowmelt, 48
wind redistribution, 50
soil
description, 91
organic matter and nutrients, 100–103
profile, 92
soil food web
effects of nutrients and warming, 99
songbirds
control by climate, vegetation, insects, 98
spatial patterns of meteorological parameters, 35–46
Sphagnum mosses
building peat and organic matter, 95
stream changes
biota, 226–30
discharge, 224
fish, 226–28
nutrient fluxes, 226
past and present, 222–29
stream chemistry
N, P, organic C, 177–80
stream ecology
bottom-up control, 194–99
disturbance as driver, 213–14

stream ecology (*Cont.*)
  disturbance-habitat template, 214–17
  Hershey Creek, 193–94
  Kuparuk River, 186–92
  Oksrukuyik Creek, 192–93
  permafrost and thermokarst effects, 217–22
  recovery, 199–204
  top-down control, 204–7
stream ecology response to change, 315–17
  warmer, wetter summers effect on nitrate, 315
stream food-web, 185
stream microhetertrophs
  bacteria, 182
stream primary producers, 180–82
  bryophytes, 181
stream secondary producers
  insects, fish, birds, 182–84
stream type
  spring, 177
  tundra, mountain, glacier, 174
streams
  floods as major disturbance, 315
strontium isotopes, 279
  link to active layer, 307
structure and function controls
  climatic, other environmental controls, 111–14
  species effects on biogeochemistry, 114–20
succession, 76–77

temperature
  permafrost borehole, 35
  projected, 53
  trends at Toolik, 35
  trends in Kuparuk River Basin, 51
temperature experiments
  International Tundra Experiment (ITEX), 312
thermokarst
  carbon cycling, 154
  description of, 11
  ecosystem response, colonization of shrubs, 313
  effects on nutrients, 222
  features, 217
  fire effects, 129
  fire effects on, 129
  formation, 15, 307
  gully near Toolik, 218
  impact on grayling, 229
  impact on sediment load, 220
  on South Slope, 219
Toolik
  climate, 9, 23–35
  human use of site, 14
  lakes nearby, 13
  location, 1
  soils, age of, 12
  streams, 13
  vegetation, 11
  vegetation history of, 13
Toolik Field Station, 1
  history of, 14
Toolik Lake
  length of ice cover, 318
  limnology of, 13
top-down control
  consumer effects on lake animals, 276–77
  impact of fish predation, 276
top-down controls
  grazing by zooplankton affects phytoplankton, 277
tundra ecosystems
  changes in understanding, 132
tussock tundra
  acidic and nonacidic, 93

ultraviolet light
  and phytoplankton, 242
  effect on zooplankton pigmentation, 267
  increase in lability, 310

vegetation
  response to fire, 313
vegetation changes
  implications for ecological systems, 314
  long-term plots, 130
vegetation communities
  glacial geology effects on, 66–74
vegetation responses to change, 312–14
vegetation succession
  glacial geology effects on, 76–77

Vignette 5.1, 95
Vignette 5.2, 98
Vignette 5.3, 108
Vignette 5.4, 118
Vignette 5.5, 130
Vignette 6.1, 155
Vignette 6.2, 160
Vignette 6.3, 166
Vignette 7.1, 209
Vignette 7.2, 230
Vignette 8.1, 246
Vignette 8.2, 252
Vignette 8.3, 261
Vignette 8.4, 266
Vignette 9.1, 292

water balance
  Kuparuk River Basin, 41
  Toolik Lake, 250
winds
  at Toolik, 41

zooplankton
  predators on, 267
  response to fertilization, 282
  response to warming, 280

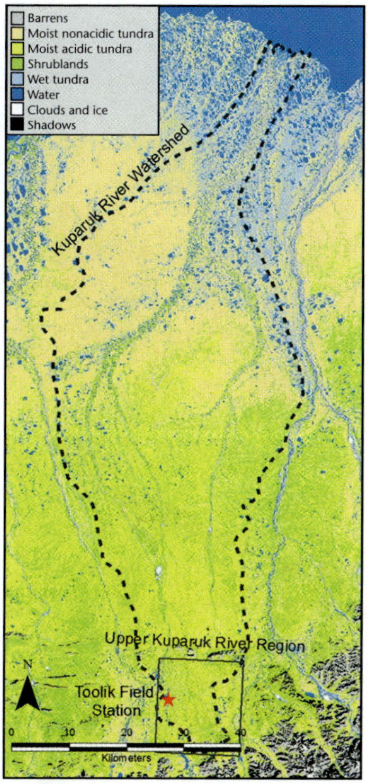

Figure 1.4 The Kuparuk River watershed and the distribution of vegetation. Based on Muller et al. (1998).

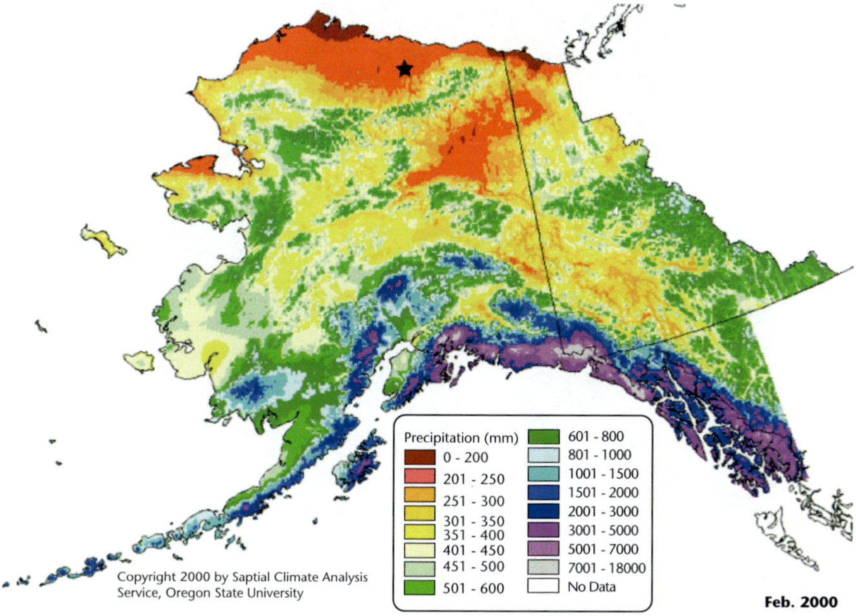

Figure 2.11 Mean Annual Precipitation for 1961 – 2000 (with permission, copyright 2000, PRISM Climate Group, Oregon State University). Approximate location of the Toolik Field Station has been marked with a star.

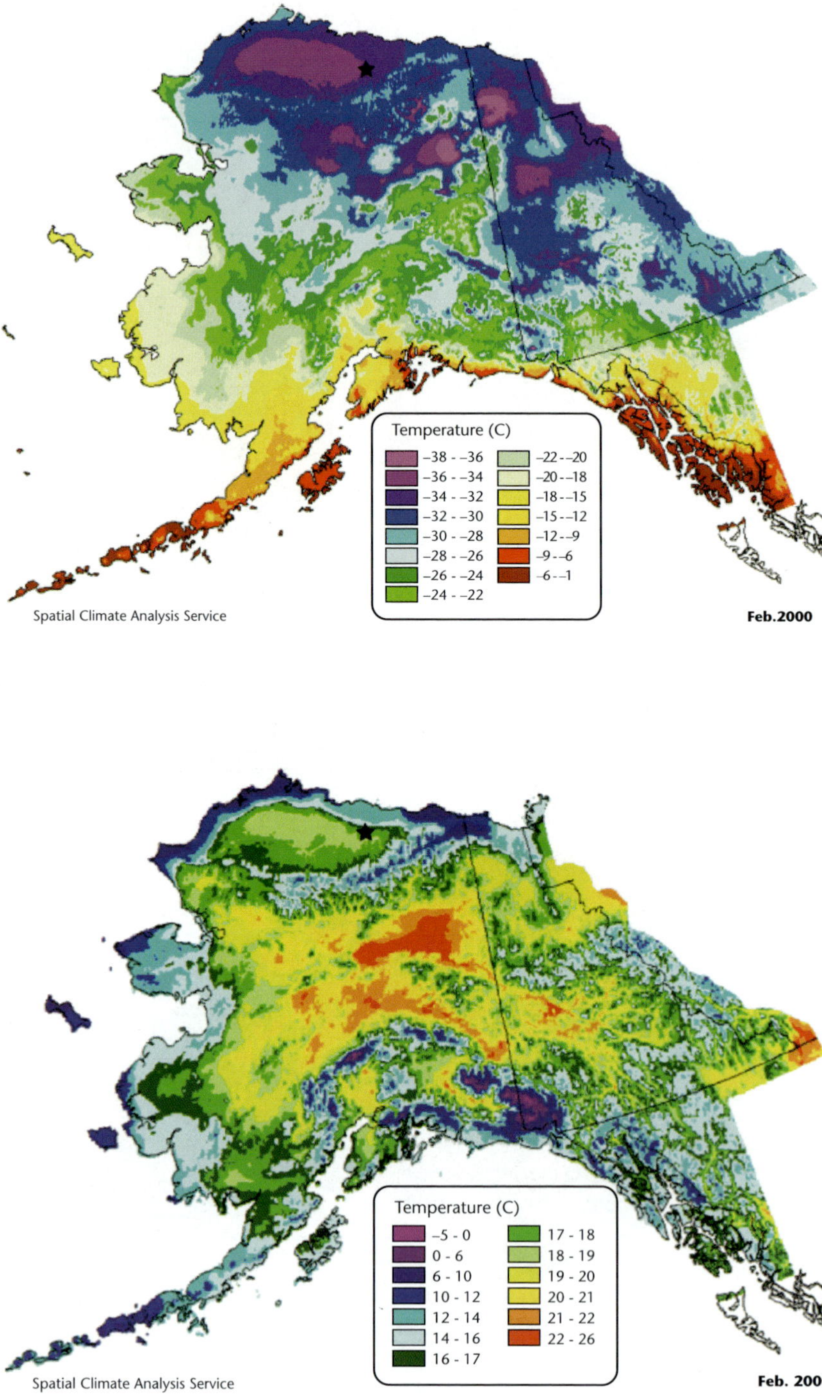

Figure 2.12 (A) January Mean Minimum Temperature for 1961 – 2000 (with permission, copyright 2000, PRISM Climate Group, Oregon State University). Approximate location of the Toolik Field Station has been marked with a star. (B) July Mean Maximum Temperature for 1961-2000 (with permission, copyright 2000, PRISM Climate Group, Oregon State University). Approximate location of the Toolik Field Station has been marked with a star.

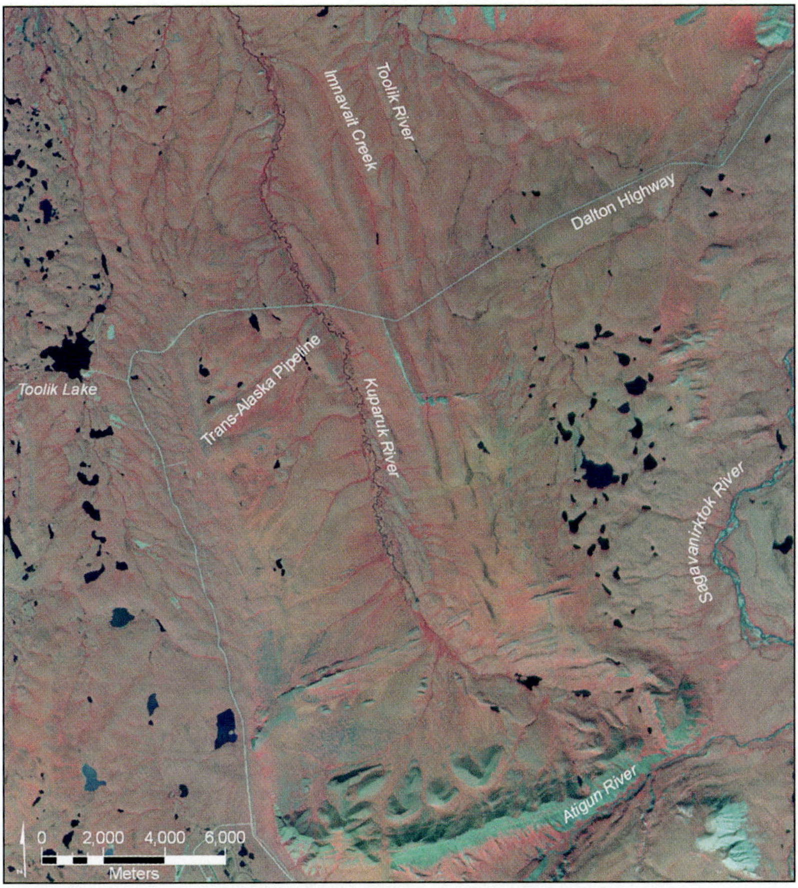

Figure 3.1 False color-infrared SPOT satellite image of the mapped area in the Toolik Lake region. Surfaces glaciated during the Itkillik glaciations have a browner tone and numerous lakes, whereas areas glaciated during the older Sagavanirktok glaciation have a brighter red tone and few lakes (see Fig. 3.2 for glacial boundaries). The brightest red areas, e.g., along streams and water tracks, are densely vegetated with shrubs. More intermediate red tones are generally tussock tundra with varying amounts of shrubs. The more brownish tones are less densely vegetated and include dry tundra, nonacidic tundra, and wet tundra. Rocky areas are either light colored, such as the limestone in the lower right corner or greenish gray as in the sandstones and conglomerates along the north side of the Atigun River. Image taken 28 July 1989.

Figure. 3.3 Glaciated terrain of Sagavanirktok and Itkillik-II age. Top – Sagavanirktok-age moraine, Imnavait Creek vicinity. Note long smooth flanking slopes and continuous vegetation cover broken only by dispersed large, weathered erratic boulders of resistant lithology. The vegetation is a rather homogeneous cover of tussock tundra. The cottongrass, *Eriophorum vaginatum*, is flowering profusely in this photo. Bottom - Drift and outwash surfaces of Itkillik II age, looking northeast from Jade Mountain, west of Toolik Lake. Note narrow moraine crests, steep flanking slopes, and discontinuous vegetation cover on the moraine and kame features. Minor as well as major depositional features are well preserved. (Photos by D. A. Walker.)

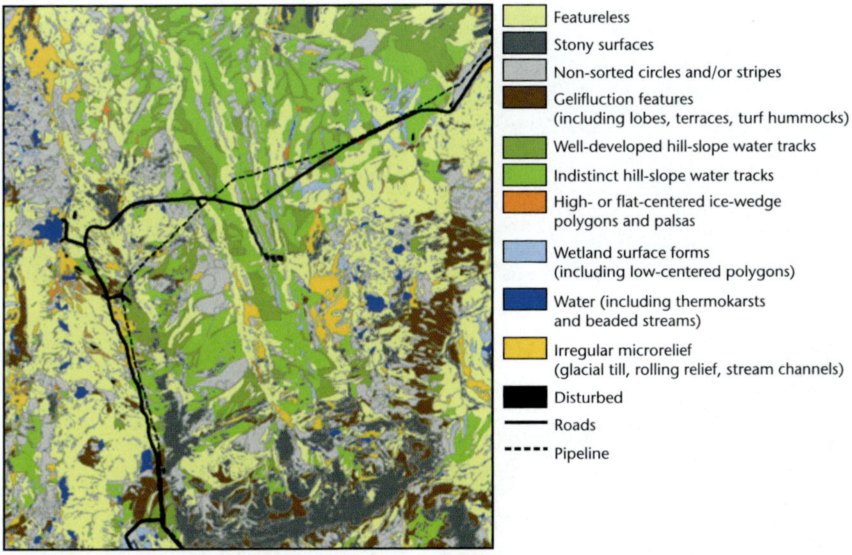

Figure 3.4 Surface geomorphology of the upper Kuparuk River region (based on Walker and Maier 2008)

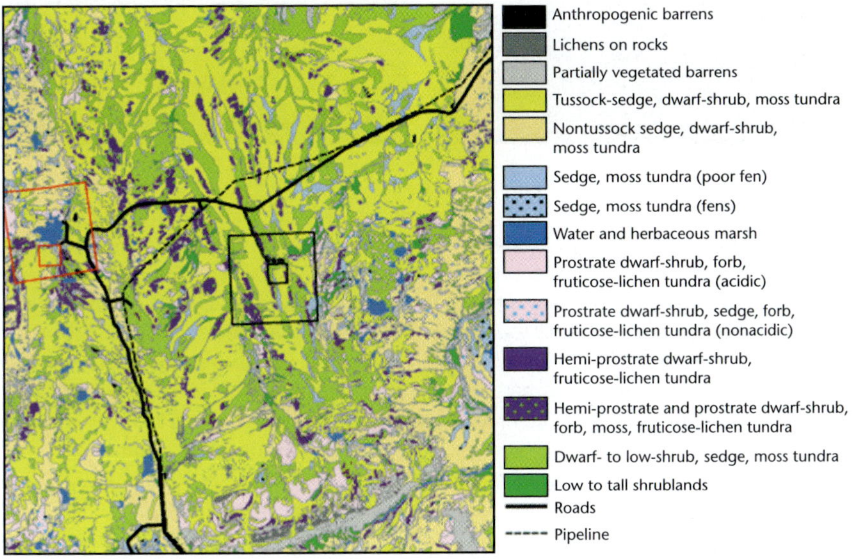

Figure 3.6 Vegetation of the Upper Kuparuk River region. Red rectangles are areas of vegetation maps at Toolik Lake LTER research area (Walker and Maier 2008). Black boxes are areas of maps of research areas at Imnavait Creek (described in Walker and Walker 1996)

Figure 3.7 Common plant communities in the Toolik Lake region: Top left - barrens, *Cetraria nigricans-Rhizocarpon geographicum* comm. on block field of Itkillik-I-age surface; Top right—dry acidic tundra, *Selaginello sibericae-Dryadetum octopetalae* on south-facing Itkillik-II-age kame west side of Toolik Lake; Middle left - snowbed, *Carici microchaetae-Cassiopetum tetragonae* on steep north-facing slope of Itkillik-II-age till deposit; Middle right—moist nonacidic tundra, *Dryado integrifoliae-Caricetum bigelowii* on Itkillik-II-age hill slope west of Toolik Lake; Lower left - shrub tundra, *Sphagno-Eriophoretum vaginati betuletosum nanae* in water track margin; and Lower right—wetland, *Carex aquatilis-Carex chordorrhiza* comm. in rich fen complex on Itkillik II surface. Also see Fig. 3.3 for photo of tussock tundra (*Sphagno-Eriophoretum vaginati* typicum), the most common and zonal vegetation type in the region. Photos by D. A. Walker.

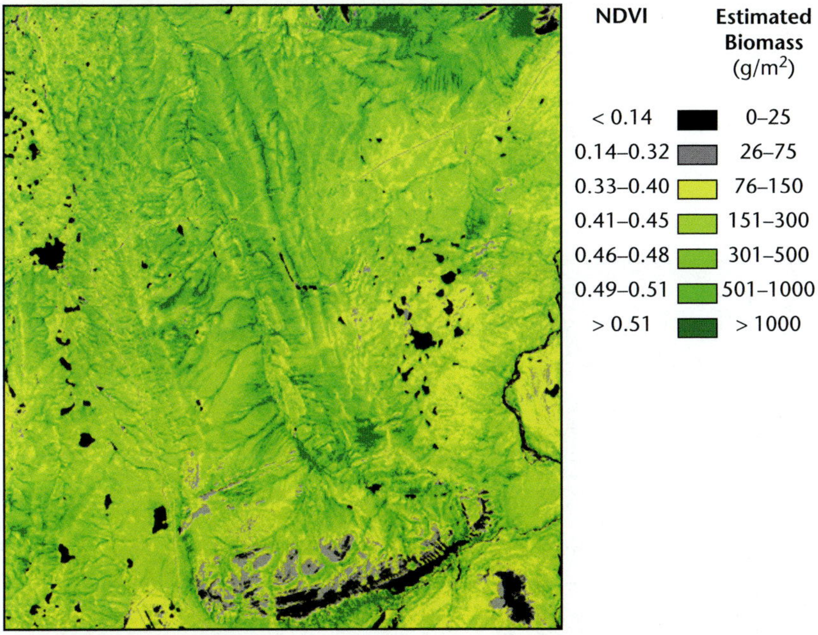

Figure 3.9 NDVI of the Upper Kuparuk River region. The image is from a SPOT-1 multispectral image acquired 29 July 1989. NDVI = (NIR – R)/(NIR + R), where R is the reflectance in the red band (630-690 nm) and the NIR is the reflectance in the near infrared band (760-900 nm). Compare the pattern on the NDVI image with the glacial geology map (Fig. 3.2) and note greater biomass per unit area on the older Sagavanirktok-age surfaces. Based on Shippert et al. (1995).

Figure 7.30 Examples of hillslope failures induced by thermokarst terrain. (A) glacial thermokarst, note the exposed massive ice and people for scale. (B) retrogressive thaw slump, widest dimension is about 100 m. (C) an active layer detachments slide, length is about 800 m. (D) gully thermokarst, length is about 150 m. Photo credits W. B. Bowden except (C) by M. Gooseff (reproduced with permission from the American Geophysical Union).

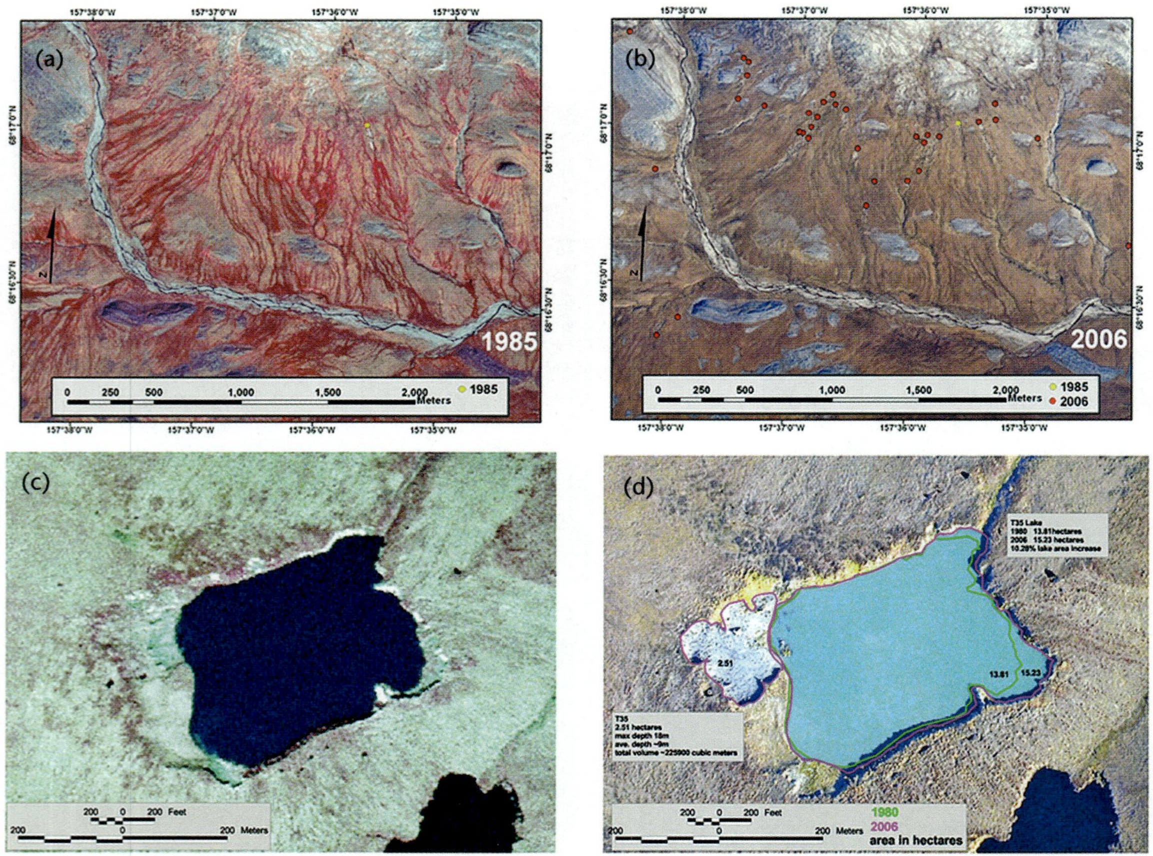

Figure 7.31 Thermokarst distribution near Feniak Lake: Changes in the past 25 years (A) 1985 Alaska High-Altitude Photography (AHAP) photo (color-infrared) showing 1 active thermokarst feature in 1985 (green dot) and 27 active thermokarst features in 2006 (yellow dots, see panel B). Note the dark red strips of riparian vegetation (mainly *Salix* spp.) along water tracks in this image. (B) 2006 aerial photo (real-color) showing the same location and features as in panel A. (C) 1980 AHAP photo of lake in process of expansion. (D) 2006 airphoto with graphics showing 26 years of substantial change (1980-2006). From Gooseff et al. (2009) with permission from the American Geophysical Union